# Deindustrialization
and Plant Closure

# Deindustrialization and Plant Closure

Paul D. Staudohar
California State University, Hayward

Holly E. Brown
University of California, Berkeley

**Lexington Books**
*D.C. Heath and Company/Lexington, Massachusetts/Toronto*

*Library of Congress Cataloging-in-Publication Data*
Deindustrialization and plant closure.

   Bibliography: p.
   Includes index.
   1. Plant shutdowns—United States.   2. Plant
shutdowns.   3. Plant shutdowns—Law and legislation—
United States.   I. Staudohar, Paul D.   II. Brown,
Holly E.
HD5708.55.U6D45   1986   338.6'042   86-45615
ISBN 0-669-14037-6 (alk. paper)
ISBN 0-669-14038-4 (pbk. : alk. paper)

Published simultaneously in Canada
Printed in the United States of America
Casebound International Standard Book Number: 0-669-14037-6
Paperbound International Standard Book Number: 0-669-14038-4
Library of Congress Catalog Card Number: 86-45615

The paper used in this publication meets the minimum requirements of
American National Standard for Information Sciences—Permanence of
Paper for Printed Library Materials, ANSI Z39.48-1984.
∞ ™

86  87  88  89  90  8  7  6  5  4  3  2  1

*Dedicated with affection and respect to Clark Kerr and Dale Yoder*

# Contents

# Figures and Tables

## Figures

## Tables

# Preface

Plant closure is not a new problem in the United States. It is an inevitable consequence of continual industrial restructuring. Corporate disinvestment in manufacturing facilities caused extensive dislocation during the Great Depression years of the 1930s. Since that time, especially during recurring bouts with economic recession, plant closure has continued to plague workers and their communities. In the past, however, deindustrialization has proceeded on a selective basis. That is, particular industries and geographic areas have borne the brunt of decline—such as shoes and textiles in New England, lumber in the Northwest, meatpacking in the Midwest, coal in Appalachia, and iron ore in northern Minnesota. Today, deindustrialization and its attendant problem of plant closure has a new face. A broad cross section of manufacturing industries is shifting production activities to locations in the Sunbelt and outside the United States. Some industries confronted with declining domestic markets are shrinking toward complete shutdown. Because the incidence of plant closure has spread to more industries over a wider geographic area in the United States, the problem merits greater public attention today.

What are the causes of deindustrialization? How can it be avoided? Why do plant closures occur? What impacts do they have on workers and their communities? Are the existing protections adequate? If not, what new policies are needed? These are some of the questions to which the twenty-six chapters in this book provide answers. Underlying the development of policies dealing with deindustrialization and plant closure are two major opposing views, those of protectionists, who perceive a need for additional legislation to protect displaced workers and their communities, and those of free-market proponents, who believe that existing protections are adequate and recommend avoidance of government interference.

We think that this book is unique in presenting a balanced perspective indicating both the pro-legislation and free-market approaches. Included in the book are chapters covering: (1) an overview of deindustrialization and plant closure; (2) impact of plant closure on firms, workers, and communities; (3) policies of management, unions, and government for dealing with the problems; (4) perspectives on plant closure from foreign countries that shed light on American solutions; and (5) assessment of state laws and proposed federal legislation. An extensive bibliography is included at the end of the book for additional reading on the subject.

The methodologies in the readings are essentially descriptive, with both quantitative and qualitative analyses presented. The approach generally is from a social and behavioral science perspective and is interdisciplinary. Among the academic disciplines represented in the readings are business, economics, law, political science, public administration, sociology, and psychology. Because of the interdisciplinary nature of the subject matter, it is expected that a diverse group of scholars will find the book attractive. Plant closure is a highly visible topic in the print and broadcast media and is the subject of lively debate among policymakers. Therefore, management, union, and public officials should find the book valuable.

Several persons provided helpful inputs into the textual material and framework for the readings. Anne T. Lawrence, a postdoctoral fellow in the Department of Sociology at Stanford University, wrote an original essay for the readings. She also consulted with us on the structure and content of the rest of the book. Joan Lichterman, Senior Editor at the Institute of Governmental Studies, University of California, Berkeley, edited the manuscript text material. We had the assistance of three outstanding librarians—Nanette Sand, Clara Stern, and Allison Shock—all from the Institute of Industrial Relations, University of California, Berkeley. Philip Shapira made several suggestions during the writing, while he was a Research Specialist in the Department of City and Regional Planning at the University of California, Berkeley. For the 1986–87 term, he was appointed as a Fellow in the Office of Technology Assessment of the United States Congress. We are also grateful for the guidance and insights of Bruce Katz, Editor at Lexington Books, and for the superior production editing of Marsha M. Finley.

# Introduction

T his book presents a collection of readings and text that focus on the debate over deindustrialization and plant closure. It raises crucial questions that frame this debate and seeks to provide answers that can be translated into policy development. To enable the reader to direct his or her attention more fully to the contents of the readings, this introduction outlines the fundamental nature of the debate. The introduction is divided into four areas: (1) defining the problem, (2) debate over the extent of the problem, (3) divergent views on solutions, and (4) other policy alternatives.

## Defining the Problem

Major structural changes are sweeping through industrial societies around the world. The less developed nations are emerging as centers of production for a broad range of manufactured goods. Multinational firms, many of them American, are relocating portions of their operations in nations with low labor costs. In advanced countries, particularly the United States, the economies are shifting away from basic industries and toward services, information systems, and specialized high-tech production. The consequences of these changes in the United States are unemployment, plant shutdowns, community disintegration, and a host of related ills that tear at the nation's social fabric.

The problem comes generally under the heading of structural unemployment. This type of unemployment is caused primarily by technological change and the relocation of capital. The workers unemployed by these changes often cannot match up with job vacancies because of lack of skill or geographical constraints. Duration of structural unemployment for a particular worker depends on the availability of jobs in the local labor market, his or her ability to develop new skills, opportunities for retraining, and willingness to relocate geographically.

In the United States and other democratic societies, capital is able to move freely. That is, capital shifts readily among industries and firms and even across national boundaries. Most economists agree that restraints on capital mobility are bad for economic growth. Hardly any would advocate putting a crimp on technological change. Robots and computer-aided production facilities are here to stay. But what about the effects of technological change and capital mobility on the workers and their communities? This is the other side of the coin, and it points up the basic problem. That problem is reconciling the freedom of the market with social welfare objectives—in other words, allowing capital mobility and technological change, toward investment, disinvestment, and industrial restructuring, while at the same time minimizing the social costs of dislocation and structural unemployment.

Just as plant closure is subsumed under the broader heading of structural unemployment, the solution to the problem is part of a broader topic called industrial policy. This is government policy designed to promote economic growth by stimulating the industrial base through monetary grants, loans, and tax relief. Japan is often cited as an example of a country with a successful industrial policy. Through its Ministry of International Trade and Industry (MITI), the Japanese government targets industrial sectors that enhance the country's competitive advantage. MITI channels government seed funds into firms that have good potential for success in international markets. Japan's rapid economic growth and low unemployment have sparked a debate in the United States over what kind of industrial policy would be appropriate for this country.

Some advocates would adopt a policy of "reindustrialization," providing government aid to heavy industry such as automobiles, steel, and rubber, to revitalize these industries. Others would target only industries that have good possibilities for sustained growth, such as computers and silicon chips. Still others contend that what is needed is greater investment in infrastructure, such as power supplies and transportation resources, as well as new government-sponsored financial institutions that would create an environment favorable to industrial development. During the debate over government action, the Reagan administration has pursued an industrial policy of withdrawing government from regulating the economy as a means of stimulating economic growth. Although this policy has accompanied economic growth, it is not aimed directly at deindustrialization and plant closure. Only a few states have developed industrial policies on these issues, either with laws that regulate plant closure or policies that seek to attract new investment.

## Debate over the Extent of the Problem

Some analysts think deindustrialization and plant closure are transitory phenomena that reflect natural adjustments in the economy and labor force. They argue that foreign inroads into domestic markets, like business cycles, come and go. Firms can and do adjust to downturns, and even workers who are permanently displaced move on to new jobs in growing segments of the labor market. Eventually the economy adjusts and employment opportunities grow. This viewpoint has some validity, as readings in the book indicate.

Others argue that plant closure and dislocated workers are chronic problems that stem from fundamental change in the operation of international markets. Foreign manufacturers of steel, automobiles, farm implements, shoes, textiles, apparel, and a wide range of electronic devices now have levels of productivity that are closer to and in some cases exceed those of American workers, with much lower wage scales. This gives foreign products a significant advantage in competing with American-made goods. Will the advantage dwindle over time as the wages of foreign workers catch up? Many observers think not, because the gap is so large. As figure I-1 indicates, despite wage moderation in the United States the gap remains large, and compared to many countries it has gotten larger.

What if the future for many U.S. manufacturing industries is a continued erosion of markets and number of workers? In this event many plants are going to be shut down even during periods of sustained economic growth. Other manufacturers may expand, especially those that are high tech with capital-intensive cost structures, but the geographic locations of these enterprises are often not the same as the delining ones. Nor are the skills required necessarily the same. The American economy has for many years experienced these kinds of dislocation problems, but the magnitude, extent, and nature of the change that is occurring overshadows anything we have seen in the past. This change is structural and could have serious long-term consequences. A relatively affluent middle class of Americans is based largely on heavy industry that is in decline. It is questionable whether the nation can afford a further loss of this industrial capacity.

What is the extent of plant closure? The precise magnitude is unknown, because there is no government agency that records plant closures nationwide. However, there are reliable data on some individual states, enabling researchers to provide reasonably good estimates of the number of closings and workers affected. As indicated in the readings, these data show that plant closure is extensive in numerous industries throughout the nation.

## Divergent Views on Solutions

What makes the topic of this book both interesting and controversial is that there are good arguments, based on extensive research, that suggest widely divergent approaches to resolving the problems of deindustrialization and plant closure. The propriety of solutions depends on the question of whether deindustrialization is reality or myth. Those who believe no new policies are needed contend that the current trend reflects merely an inevitable recasting of the

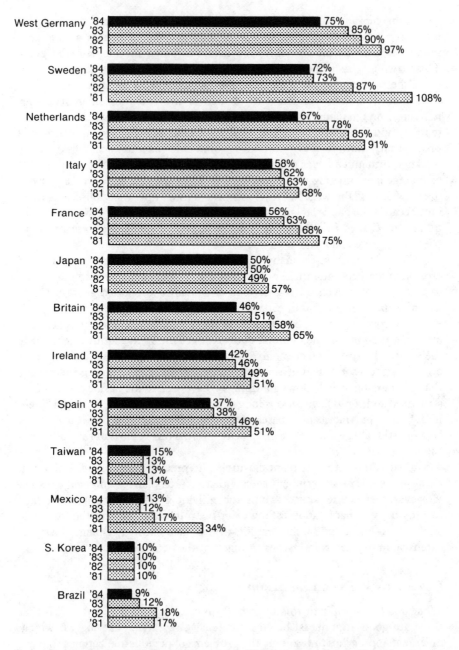

*Source:* Bureau of Labor Statistics. Data reported in *Wall Street Journal*, July 17, 1985, p. 6.

**Figure I-1. Hourly Pay Levels Abroad in Manufacturing Industries as Percentages of U.S. Average**

American economy toward new growth industries. Capital investment, they argue, is channeled into industries and firms that promise the highest rates of return. Disinvestment occurs when it is no longer economically feasible to continue operation. This is a policy already, contend the free-market advocates, and it requires no "solution."

Proponents of the free-market approach thus argue that the best policy is passive, not active. They believe what is needed is less government influence, not more. Let the market take its course and work things out, and economic growth will provide jobs to absorb the structurally unemployed. These advocates are disciples of the great Harvard economist, Joseph A. Schumpeter. What he called "creative destruction" occurs when new areas of economic development cause the old ones to die from removal of the capital necessary for their survival.

The free-market advocates further contend that change in policy is not required because adequate help for displaced workers already exists. For example, all states provide unemployment insurance and job placement services. Moreover, the federal Trade Adjustment Act of 1974 gives extra financial support to workers who lose their jobs as a result of foreign competition. In addition to government protection, some companies give adequate notice of plant closure to workers and seek to help them get retrained and place them in jobs. Unions too seek to protect workers in the event of a shutdown.

The protectionists argue, however, that existing measures are woefully inadequate. Unemployment insurance benefits will not long help workers who are permanently displaced. Similarly, the amount of funds provided under the Trade Adjustment Act has been cut drastically during the Reagan administration, so relatively few deserving persons are helped from this source. Additionally, although some firms have an enlightened management policy to cushion the blow of a plant closure, too many others simply want to cut their losses without assuming any responsibility for the workers or their communities.

For their part, unions can do little because plant closure is a management prerogative over which there is no legal requirement to bargain. Unless there are specific labor contract provisions that preclude closure—and such provisions are rare—companies are relatively free to close all or part of their operations. The common-law criteria that courts have applied to limit closure are not usually troublesome to employers. So long as the company's motivation is based on reasonable economic grounds, the closure is undertaken in good faith, and the company is not obviously trying to subvert a union, firms are legally allowed to close or relocate operations at will.

Thus, the protectionists argue, victims of plant closure are left with insufficient protection in too many cases, and new laws are required, not to restrict closure but to minimize its impact. Adequate notice should be provided to workers, who should get severance pay and continue to receive health and welfare benefits for a certain period after closure. These advocates also urge guarantees of greater opportunities for retraining, so workers can re-enter the labor force in a good job. As to the added cost of these protections to government and employers, proponents point to the high social cost of closures to workers and their communities despite existing protections.

## Other Policy Alternatives

The debate over plant closure and proposed legislation has sparked a large number of creative ideas for change. These suggestions come not from the free-market advocates, who are satisfied with the status quo, but from observers who are outraged by the dislocations from plant closure. Many of these ideas are found in the readings in this book. Although the quality of these ideas is good, in some cases they run aground on political shoals. The conservative shift in American political philosophy makes it impractical to expect adoption of solutions that involve heavy government expenditure or involvement. Although plant closure laws are being enacted in more states, their provisions are limited. The nature, content, and impact of these laws are discussed in the readings, along with extensive discussion of foreign experiences with laws that regulate plant closure.

Voluntary solutions are also discussed in the readings. The extent of capital flight can be reduced by joint ventures involving management, unions, and communities that restructure operations. Some of the measures that have worked are wage reductions, elimination of restrictive work rules, and work sharing (spreading work among employees to reduce unemployment). In some cases, employee buyouts have been used to good effect either for total takeover of an ailing plant or through an employee stock ownership plan that allows for increased worker participation in the business over time. The key to all these programs is cooperation. Instead of looking at potential disinvestment as a confrontation, the parties to successful joint ventures view them as cooperative attempts to save the plant and revise operations for higher rates of return on investment. Ideally, this cooperation would not begin when the company announces its intention to shut down a plant, but would be an ongoing process that regularly brings the parties together to discuss ways to continue productive operation.

If a company determines that plant closure or a permanent reduction in the workforce is necessary, it is vital that groups affected join together to help alleviate the economic crisis. It may not be easy to promote harmony among these diverse parties—management, workers, unions, government, and communities—but it can be done. Effective programs require a spirit of teamwork, mutual respect, and trust. Joint ventures based on these attitudes may result in a uniquely American solution, in which the effects of plant closure are gradually reduced through a combination of law and voluntarism that mitigates the impacts of technological change and capital mobility by increasing worker protections.

# Part I
# Overview of Deindustrialization and Plant Closure

The readings in this part provide an orientation to themes that are explored in greater detail in later sections. During the past decade, as the pace of changing technology and international competition has quickened, the incidence of plant closure has grown. Plant closures present numerous difficulties for the displaced workers, their families, and the community. The extent of deindustrialization and its impact is examined in the opening reading by Barry Bluestone. A professor of economics at Boston College, Bluestone is one of the leading advocates of greater government attention to plant closure. After an analysis of the dimensions of the problem and its historical roots, Bluestone presents his views on how government can seek to relieve the hardship of plant closure. His selection is valuable because it presents an insightful overview of key issues.

In the next excerpt, H. Craig Leroy presents a view that contrasts with Bluestone's. Based on his analysis of research studies, Leroy finds that the data on plant relocation are inconclusive rather than pointing clearly to the need for increased government involvement. Leroy describes the free-market position, which has been adopted by many economists, and he argues against government involvement in the market to ward off or alleviate plant closure. This position is explored more fully in later chapters, but Leroy's excerpt is usefully presented here because it summarizes the essence of the free-market approach. H. Craig Leroy is Staff Economist at the Connecticut Business and Industry Association.

Although Bluestone's and Leroy's selections provide an interesting summary of the government involvement vs. free-market approaches, they do not provide the kind of quantitative analysis that allows deeper penetration into the causes and consequences of deindustrialization. Such analysis is found in the next excerpt, from Robert Z. Lawrence, a Senior Fellow at the Brookings Institution. Based on data, Lawrence mounts a powerful argument against the notion that the main cause of depressed industries in the United States is foreign competition. Instead, he points to slack demand and an overvalued dollar as the principal causes of deindustrialization. Thus the unemployment is perceived as cyclical rather than permanent. Lawrence's analysis is important because industrial policy prescriptions flow from the nature of causes. If industrial policies are aimed at causes that do not exist or are relatively unimportant, their usefulness is reduced.

It is not widely agreed among economists, however, that the causes of deindustrialization are as transitory as Lawrence describes them. Economist Bluestone, in the next selection, presents evidence on individual industries and regions that supports the deindustrialization thesis. He finds that despite the stability in aggregate manufacturing employment during the past decade, important industries and geographical areas have been confronted with significant job losses. The key problem, as Bluestone sees it, is how effectively workers in these declining industries and areas can be absorbed into growth areas of the labor market. In contrast to Lawrence, who concludes that deindustrialization is a myth, Bluestone finds that it is a major social and economic problem.

In the section's last selection, Lee Iacocca presents his views on the need for a strong industrial base in America. Iacocca is uniquely qualified to comment on deindustrialization because of his heroic efforts to save Chrysler Corporation from bankruptcy. As chief executive of the company, he led a cooperative venture with the federal government and the United Automobile Workers, demonstrating that a major automobile manufacturer could not only be preserved but revitalized. Iacocca identifies the implications to the nation of a long-term decline in major industries such as autos, steel, and textiles. His ideas shed light on the debate over what to do about plant closure and how to reindustrialize America.

# 1
# Deindustrialization and Unemployment in America

*Barry Bluestone*

T he economic issues we face today (e.g., unemployment, productivity
decline, slow growth, the abandonment of community) are among the
most difficult issues we have ever had to face. For somebody like me,
who attended college in the 1960s and was a charter member of Students for a
Democratic Society, the issues we fought over then were much easier than the
ones that now must be faced. The issues in the 1960s concerned clear-cut
fundamental human values. There was a definite right and wrong to the issues,
and one hardly needed a sophisticated mathematical model to tell one from the
other. There was, for example, no question at all about the sanctity of voting
rights. In 1965, when friends and I drove down to Montgomery, Alabama, to
march with Robert Moses of the Student Non-Violent Coordinating Committee and Martin Luther King, Jr., there was no question about what was right
and what was wrong. There was no need to debate fine philosophical points or
consult a computer before one could decide what needed to be done. For many
of us, Vietnam was similar. There was a moral right and a moral wrong to that
war and one did not need a Ph.D. to figure it out.

In the 1980s, the issues are extraordinarily more complex. I wish there
were some simple analyses and simple solutions to unemployment, economic
growth, income distribution, and international economic equity, but there are
none. The Kennedy tax cut, the development of the Great Society programs of
Lyndon Johnson, and the War on Poverty all seemed to work during the 1960s.
Unemployment plummeted to less than 4 percent, the incidence of poverty was
cut nearly in half, and our standard of living increased dramatically. In the
1970s, however, these same programs failed to bring any additional improvement. Inflation soared, and along with high unemployment we learned about a
new phenomenon: stagflation.

In 1980 the country turned to Reaganomics, a sharp departure from earlier
policy. After two years of supply-side economics, it seems fair to call this

Barry Bluestone, "Deindustrialization and Unemployment in America," in *New Perspectives
on Unemployment*, ed. by Barbara A.P. Jones (New Brunswick, N.J.: Transaction Books, 1984),
pp. 27–42.

experiment a substantial failure. Reaganomics never did make much sense theoretically: cutting taxes and boosting defense spending do not mix very well; they create massive deficits. Unfortunately, at this point the Democrats have no alternative. The problems of unemployment, inflation, and falling incomes are indeed complex, and simple solutions simply will not do.

To begin to think about solutions, we need first to analyze what has been going on in the economy. My colleague, Bennett Harrison of MIT, and I have been trying to do this for a number of years. In 1980, the Progressive Alliance, a group of trade union, civil rights, environmental, women's rights, and antinuclear groups led by Douglas Fraser of the United Automobile Workers and Coretta Scott King, asked us to produce a pamphlet on the problem of plant closings. They requested that we assess the extent of the problem and the degree to which workers and communities suffered as the result of runaway shops and the permanent closing of factories, stores, and offices.

Ben and I took up the task. But because academics will be academics, our pamphlet took six months to complete, and when it was done it ran a total of 336 pages! It was hardly the kind of thing you could hand out at the factory gate. What prompted us to dig so deeply into the subject was our shock at finding how extensive was the problem and how devastating were its consequences.

To estimate the number of jobs affected by plant closings, we used data from the Dun & Bradstreet Corporation. Because the federal government collects practically no data on corporate investment and disinvestment, we were forced to rely on this private source. We had data only for 1969 through 1976, but using our computers we were able to answer a very important question: Of all the jobs that existed in the private sector in 1969, how many had disappeared as a result of plant (and store and office) closings by 1976? The number that came out of the computer was staggering. Between 1969 and 1976, 22.3 million jobs had disappeared as a result of plant closings and the interstate and overseas movement of business establishments. This was equivalent to nearly 39 percent of all the jobs that had existed in 1969.

When we extrapolated these numbers to the entire decade of the 1970s, we concluded that somewhere between 32 and 38 million jobs had disappeared in this ten-year period. So staggering was the number that we ran the data through the computer several times before we were convinced we had not made a computational error.

There were other rather remarkable surprises in the data. Massive job loss in the Frost Belt—from the New England states through to the industrial Midwest—was expected, but we were hardly prepared for the results we found for the Sun Belt. It turns out that despite all of the hoopla about the booming South and Southwest, the number of jobs lost to plant closings during that 1969–1976 period was nearly the same as those that disappeared in the deindustrialized North. The Sun Belt lost 11 million jobs, whereas the Frost Belt lost

11.3. During that period, over 1 million new jobs were created in Georgia by openings of new business establishments, but the state lost 587,000 jobs when plants closed down or moved away.

These extraordinary numbers reflect an amazing phenomenon. Among all business establishments existing in 1969, including the small "mom and pop" corner drugstore, the probability of being out of business by 1976 was greater than 50 percent. If we restrict the sample to include only manufacturing firms with 100 or more employees, the bankruptcy rate was still 30 percent. Only seven out of ten establishments in business in 1969 were still in business 7 years later. Moreover, all of this occurred before, not after, the wave of plant closings later in the decade in basic industries such as auto, steel, and tires. What may be even more important is the fact that most of the large-scale closings occurred not because parent companies literally went out of business; in many cases, profitable establishments were closed down only because they were not profitable enough. Corporate managers decided they could increase earnings by closing down one set of operations and opening up another set somewhere else, often in a totally different business.

## The Impact of Plant Closings on Workers and Their Communities

What is the impact of these closings? One way to answer this question is to analyze the earnings of those who lost their jobs. To do this we used the Social Security Administration's Longitudinal Employer-Employee Data file (LEED), which contains information on 1 percent of all workers covered by Social Security for the years 1957–1975. The data show quite convincingly that workers who lose their jobs permanently in such industries as auto and steel continue to be at an earnings disadvantage for years to come. After 2 years, the ex-autoworker makes 43 percent less than autoworkers who kept their jobs. Even after 6 years, the earnings loss is nearly 16 percent. Similar results are found for steelworkers, those in meat packing and aerospace, and even those who worked in lower-wage industries such as men's clothing. The process of "creative destruction" that the famous Harvard economic historian, Joseph Schumpeter, wrote about in the 1940s is not bearing fruit. Workers are not being freed from lower-productivity, lower-wage jobs for work in higher-productivity, higher-wage jobs. The opposite is occurring. Workers are skidding downward in the occupational spectrum, not moving up to better jobs and a better standard of living.

The high-tech revolution is a case in point. High-technology jobs in the computer, medical instrument, and business systems industries are supposed to replace the jobs the country is losing in the old mill-based and smokestack industries. But do they? Using the LEED file again, we looked at New England's blossoming high-tech sector and asked, "What happened to all the workers who lost their jobs in the old mill-based industries in the region?" Did these workers find their way into high technology? The answer was a resounding no. Between 1957 and 1975, 833,000 workers were employed in the old mill-based industries (apparel, textiles, shoes, rubber goods, and the like) sometime during the period. By 1975, 674,000 no longer worked in those industries, largely because the firms had closed down. Some of the companies relocated in the South; some went to Singapore, South Korea, or Brazil. What happened to the workers? Of the 674,000 no longer in these industries, fewer than 3 percent were able to make the transition to the new high-tech sector. Five times as many (16 percent) skidded downward into retail trade and low-wage service jobs like those at K-Mart or McDonald's. Many left the labor force altogether, unable to find suitable employment. Others were forced to leave the region to search for jobs (often unsuccessfully) in other parts of the country. In short, the high-tech revolution, despite its great promise, has held out little hope for the victims of deindustrialization. As one displaced worker told a Boston TV newscaster, "My dream used to be to live better at age 50 than I am now at age 35. Now my dream is simply to survive."

## The Other Costs of Unemployment

The costs of unemployment go far beyond the loss of income. Professor Paula Rayman of Brandeis University and I recently completed a study of unemployed aircraft workers in Hartford, Connecticut. We interviewed over 200 workers, many of whom had worked at Pratt & Whitney, the nation's leading jet engine manufacturer. Workers told us that during their unemployment they suffered from chronic insomnia, headaches, and stomach ailments—all as a result of the personal and family tension associated with income insecurity. To cope with the financial drain of job loss, 36 percent of our sample depleted their entire savings. The first unexpected expense, caused by an illness or accident, placed some of these workers on the brink of bankruptcy. Three of the eighty workers who answered a detailed personal questionnaire reported they ultimately lost their homes to eviction or foreclosure. One worker I interviewed during the study told me that after 18 years in the plant he was given only a day's notice of layoff. Stunned, he picked up his tools and went home heartsick. When his wife asked him what he had done to lose his job, he went berserk and beat her up. It was the first time in twenty years of marriage that he had ever laid a hand on her.

Other researchers have taken a broader look at the social consequences of unemployment. Dr. Harvey Brenner of Johns Hopkins University has statistically correlated a 1 percent increase in the aggregate unemployment rate sustained over a period of six years with:

37,000 total deaths

920 suicides

650 homicides

500 deaths from cirrhosis of the liver

4,000 admissions to state mental hospitals

3,300 admissions to state prisons.

This is what unemployment does to workers and their families.

The impact on the community is also serious. In smaller towns where a single company dominates the local economy, a plant closing shrinks the community tax base. As a result, public services from police and fire protection to education and recreation suffer. In larger cities like Detroit, Buffalo, and St. Louis, the loss of an entire industry such as auto or steel often means the same thing. As a result, during the 1970s these three cities lost more than 20 percent of their population, with the richer fleeing to the suburbs where better public services were maintained and the poorer to God-knows-where to seek any kind of job at all.

The impact on the national economy is also not to be underestimated. The Bureau of Economic Analysis of the U.S. Department of Commerce recently estimated that for every 1-point increase in the unemployment rate sustained over a year, the nation loses $68 billion in output (gross national product) and $20 billion in tax revenues and must spend an additional $3.3 billion on unemployment benefits, public assistance, food stamps, and other programs to aid the jobless. If plant closings have been responsible for boosting the unemployment rate by just three points (out of the current 10.1 percent), then closings have accounted for nearly $200 billion—a fifth of a trillion!—in foregone output and contributed nearly $70 billion to the federal deficit.

Of course, the cost of plant closings and unemployment is not borne evenly across society. Greg Squires, who works for the United States Civil Rights Commission in Chicago, points out that black workers and their families are hurt disproportionately. Blacks are more concentrated in urban areas, particularly in the industrial Midwest, where the number of closings has been especially high, and blacks are concentrated in those industries undergoing rapid deindustrialization—auto, steel, and tires. Moreover, plants are moving from central cities to suburban and rural areas where few blacks currently live. In the years before Chrysler Corporation nearly went bankrupt, it was the largest employer of nonwhites in the country. It was responsible for nearly 30 percent of all manufacturing jobs in Detroit. When the corporation shrank in size after reorganization, many blacks were the victims. When a large chain of industrial laundries closed down in Detroit and moved to a rural community in Ohio, again the big losers were black. Forty percent of the Detroit work force were black; after relocation to rural Ohio, the company's black employment fell to 2 percent. This is surely part of the reason why black unemployment rates remain double the national average.

## The Historical Roots of Deindustrialization

Deindustrialization—a systematic decline in the industrial base—is happening to large chunks of America. It has had a more severe impact on the North than on the South, on blacks more than on whites, but the entire nation is really the victim. Why have we all of a sudden lost our ability to thrive?

To understand deindustrialization requires that we trace American economic history from World War II. At the end of the war, the United States was the only major country left with its economy intact. Tremendous pent-up savings as a result of rationing during the war created the aggregate demand needed for a postwar spending spree at home. The signing of the Bretton Woods agreement in 1944 provided the basis for a new international economic order that gave the U.S. dollar reserve currency status worldwide. This helped to make the United States the largest exporter of goods and the largest investor of multinational capital. The combination of a domestic consumer-led boom at home and an open world market for our exports and investments was responsible for nearly two decades of unprecedented growth in the American economy, marred only temporarily by recessions in 1954, 1958, and 1961.

With the boom in economic growth came a boom in profits. American corporations were more profitable after World War II than at any time in history. By the early 1960s, it is estimated, the average rate of return on assets in the entire economy approached 16 percent. Stockholders were happy, and corporate managers were pleased with their own apparent performance.

Inevitably, in this context of buoyant profits, labor and the rest of the community came to demand their share of the rapidly expanding economic pie. What followed the war, then, was a heightened struggle over the terms of the "social contract" between labor and capital and over the extension of the government-provided "social wage" or social safety net. Labor demanded higher wages, more fringe benefits, and greater input in the decisions on the speed and control of production. As one remarkable indication of the growth in the social contract, we can simply consider the physical size of negotiated labor-management agreements. The original collectively bargained agreement between the United Autoworkers Union (UAW) and General Motors (GM) was one and one-half pages in length, and then only because it was typed triple spaced in large boldface type. Today's UAW-GM contract, including the master agreement plus local agreements, covers something like 14,000 pages and includes clauses on payment for every job, the pace of every machine, fringe benefits from holiday time to dental insurance, and conditions governing the subcontracting of work outside the GM system. Although they struggled against such an encroachment upon what they considered to be their prerogatives, businesses found that the new social contract was affordable and in some sense even productive. It provided a general level of labor peace that permitted continued economic growth and profitability.

This period also made affordable a veritable explosion in the social wage. Income transfers to the disadvantaged through unemployment benefits, workers' compensation, public assistance, social security, disability insurance, food stamps, and other programs mushroomed at both the state and federal level. Medicaid and Medicare were added to the federal government's responsibility to the poor and the elderly. On top of these transfers came new government regulation: occupational health and safety rules, environmental legislation, equal opportunity laws, pension protection, and so on. The corporate sector fought these provisions, but learned (for the most part) to live with them and still make more than a tidy profit.

All of this changed rather dramatically near the end of the 1960s. International competition, which had hardly made a dent in American production since World War II, blossomed in the 1970s. The rebuilding of the European and Japanese economies was finally accomplished, in part through U.S. government support under the Marshall Plan and the generous granting of U.S. technology to foreign companies through licenses and joint ventures provided by American corporations. By the middle of the last decade, the rest of the developed world could easily compete with the United States in steel, autos, tires, petrochemicals, electronics, and even high technology.

Japan, of course, provides the best example. In 1960, the Japanese exported worldwide 38,809 motor vehicles. That is equivalent to the output from one single American auto assembly plant operating for no more than eight weeks with the usual two shifts. By 1980, Japan exported 2.3 million cars to the United States alone and 6 million vehicles worldwide. In building 11 million cars that year, they exceeded the total U.S. output by almost 20 percent. Moreover, the Japanese excelled in international trade across a broad range of goods. In 1979 Japan exported $26 billion worth of goods to the United States. The United States, at best, could export $17.5 billion to them. Since that time the trade deficit has grown even larger.

It is more than the total amount of imports and exports that is alarming, however. What Japan exports to us and what we export in return is the real shocker. Here is a list of the top six exports from Japan to America:

Motor vehicles

Iron and steel plates

Truck and tractor chassis

Radios

Motorcycles and motorbikes

Audio and video tape recorders.

Here is the list of the leading exports from the United States to Japan:

Soybeans

Corn

Fir logs

Hemlock logs

Coal

Wheat

Cotton.

It is only when we get to the eighth most important export, aircraft and aircraft engines, that one can find an American-manufactured product going to Japan. It is almost as though an underdeveloped country were trading its agricultural products for the high-tech products of a developed nation.

Japan became the leading world exporter during the 1970s, but Western Europe was not far behind. Moreover, the "Newly Industrialized Countries" (NICs), including Taiwan, South Korea, Mexico, and Brazil, are now adding to the world's capacity to produce manufactured goods. The result is overcapacity in one key sector after another—steel, auto, electronics, and now perhaps even computers. The competition to sell one's products became fierce during the 1970s, putting tremendous pressure on prices. America's hegemony in the world market ended.

The impact on American corporate profits was severe. It is estimated that the average rate of return on all assets during the period 1963–1966 was 15.5 percent. By 1967–1970 the average profit rate had declined to 12.7 percent, and it continued to fall. In the following three years it was down to 10.1 percent, and finally from 1975 to 1978 it fell to 9.7 percent. The rise in international competition meant a rapid decline in profitability. In particular industries, the profit squeeze was even more dramatic. Profits in the auto industry fell by 65 percent between the early 1960s and the middle 1970s. In radios and TV, the decline was nearly 70 percent; in farm machinery, 51 percent; in electrical equipment, 49 percent; and in steel, 39 percent.

Now here is the crucial point. If you are a corporate manager and you see your profits eroding by one-third or more, and your very job description reads "maximize profits" very much in the same way that the job of a major league batter is to maximize RBIs, what do you do? You have to find either a way to increase total revenues or a way to cut total costs.

Some firms attempted to boost revenues by introducing new products or new production processes or by reorganizing their work plans. Many more, it turns out, shifted their attention almost exclusively to finding ways of cutting costs—in particular, labor costs and their tax burdens. Ultimately, it was precisely in the search for a cost-cutting strategy that management hit upon the tactic that resulted in the deindustrialization of America. The strategy they discovered was capital mobility. The way to cut labor costs, add flexibility to the production process, and reduce tax liability was simple: Move (or merely threaten to). Shifting capital from one corporate division to another was one tactic; disinvesting in one industry to invest in another was a second; moving from the North to the South was a third; moving from urban to rural areas was a fourth; and, of course, there was a rash of multinational activity with American corporations relocating domestic operations abroad.

Innovations in transportation and communications provided the "permissive technological environment" that made the capital mobility strategy viable. The development of wide-body jet transports such as the Boeing 747, the McDonnell-Douglas DC-10, and the Lockheed L-1011 made it possible to move commodities, components, and executives at nearly the speed of sound. Pratt & Whitney, for example, the world's leader in aircraft engines, produces its F-100 military model in its East Hartford, Connecticut, plant. One-half of the turbine rings for this engine are supplied by a small firm in South Glastonbury, Connecticut, only 6 miles away. The other one-half is imported from an equally small shop in Tel Aviv, Israel, nearly 6,000 miles away. During the infamous February blizzard of 1978, Pratt & Whitney found it easier to fly the Tel Aviv parts into its East Hartford plant (where it has an airport runway capable of landing the largest of the air transports) than to get parts by truck from its local supplier.

World sourcing of this type is more than a futuristic fairy tale; it is a present-day reality. The same is true in the auto industry, the computer industry, and in machine tools. My U.S.-assembled Volkswagen Rabbit has an engine from Brazil, wheels from Portugal, a radiator from Canada, a fuel injection system from West Germany, and an alternator and windshield from the United States. The "world" car, the "world" airplane, and the "world" computer terminal have been made possible by the permissive technological environment.

Satellite-linked computerized communications may be even more important to the capital mobility strategy. To take advantage of global production and global sourcing requires closely knit coordination of factory and sales activity. The new communications technology permits managers to be in immediate contact with each other and allows engineers to manage design and production tasks half a world away. It seemed like science fiction to me, but recently I was interviewed in Boston by an Australian correspondent in Sidney, Australia, and an economist in Melbourne. The communications link involved two satellites including Westar IV, four dish antennae, and Lord knows how much electronics. Except for the one-second delay due to the enormous distances between us, the interviewers sounded as if they were in my living room. Normal telephone conversation is not as clear and static-free.

The ability to move things and people at nearly the speed of sound and information at nearly the speed of light makes the corporate environment markedly new, but even the opening of the Erie Canal in 1830 led to capital mobility. The canal ran 353 miles from Albany to Buffalo, New York, and at best one could move cargo along the route at 3.8 mph; however, this was fast enough to lead to "runaway shops." Factories closed down in lower New York and opened up in new towns like Rochester, Syracuse, and Utica. Today, General Electric can close down a sixty-year-old steam iron plant in Ontario, California, and expand operations at their plants in Singapore and Brazil. So it goes. If GE can make 16 percent rates of return on its South American investments and only 14 percent in the United States, one can bet that sooner or later the U.S. operation will be closed down, despite the fact that it is making a respectable profit in the old location.

## The Object of the Strategy

What does the frenetic mobility of capital buy for American firms? The answer is the weakening or destruction of the old social contract and the old social wage. By moving or threatening to move, corporations are in perfect position to force one group of workers to compete directly with another. In such a climate of economic insecurity, the company can demand and will often receive major contract concessions. The so-called give-backs that have recently been granted by the auto and steel unions to their respective industries reflect not so much the current recession, but the longer-run threat of capital mobility. If the unions thought the jobs would come back as soon as the recession ended and production picked up, they would be extremely unlikely to grant such concessions. The capital mobility option is therefore the key.

Pratt & Whitney has used this strategy to destabilize the Machinist's Union in Connecticut. Instead of expanding one of their five plants in that state, they instead moved part of their production to a new facility located in the rural area around North Berwick, Maine, 200 miles away from their main assembly plant. The new nonunion plant produces the same parts as its unionized Southington, Connecticut, factory. Now when the union workers complain about working conditions, Pratt & Whitney simply threatens to move more production to Maine. In the process, the workers at both plants have lost almost all of their bargaining power. This "parallel production" strategy is analogous to the "multiple sourcing" ploy that the company forces on its parts suppliers, and it has the same effect: it empowers the powerful corporation and destabilizes the work force and the smaller companies that deal with it. What better way to lower one's costs and increase one's profit? The company obtains lower wages, fewer work rules, generally less uppity workers and, from its suppliers, cheaper prices.

In the same way that a corporation can use capital mobility to play one group of workers against another, it can force entire communities to compete for survival. By threatening to move, a corporation can "persuade" a community to offer it tax abatements and sometimes outright subsidies. Indeed, companies have become rather bold in their demands for a "good business climate." In 1976, the business community of Massachusetts was able to persuade the citizens of the state to vote down statewide referenda on a progressive income tax, a bottle bill, municipal-owned electric power, and utility rate regulation simply by running ads in the newspaper suggesting that such legislation would signal a "bad business climate" and therefore lead to job-destroying corporate disinvestment.

In the pursuit of the elusive "good business climate," states and local communities have fallen over each other offering ever greater sacrifices to the business community. The General Motors "Poletown" plant is a case in point. GM announced in 1980 that it was preparing to close down its last two production facilities permanently in the City of Detroit, but that it would be willing to build a brand-new Cadillac assembly plant in the city if Detroit were willing to make some concessions to the corporation. Otherwise the new plant would go to a southern location. What did GM demand of Coleman Young, the mayor of Detroit? First, the corporation wanted two-thirds of a square mile of land in the middle of the city. The area they wanted cleared was one of the most integrated in the city (51 percent white [Polish], 49 percent black) where over 3,000 people lived. Mind you this was not a blighted area. Second, they wanted the city at taxpayer expense to relocate these 3,000 people, tear down and compensate 160 small businesses in the area, knock down a 170-bed hospital, remove three nursing homes, redirect two expressway on-off ramps, move a railroad right of way, and do something about the 2-acre Jewish cemetery in the middle of the plot. The corporation wanted the city then to clear the land to a depth of 10 feet below grade so they would not have to worry about underground water, sewer, telephone, and gas lines. Finally, if the city agreed to all of this and then gave GM a 12-year 50 percent local tax abatement, the corporation would agree to build in Detroit. A conservative estimate of the cost to the city (including state and federal contributions) was $450 million dollars. What is interesting is that the City Council of Detroit, including its progressive and socialist members, voted unanimously to give in to GM's outrageous demands. The people were removed by eminent domain; the houses, churches, businesses, nursing homes, and hospital were bulldozed; and the plant has been built. Ironically, because of the depression in the auto industry, the plant is not occupied. Moreover, even if the plant were to go into full operation, only 6,200 jobs would be created, fully 600 less than the number eliminated by the closing of the other two GM facilities. Why did Detroit agree to such a lopsided deal? With unemployment approaching 20 percent, it really had no alternative.

Detroit is obviously not alone in its attempt to attract or retain industry. International Harvester notified Ft. Wayne, Indiana, and Springfield, Ohio, that it was going to close a plant in one of these two cities. By bidding to keep the jobs in their city, each community could determine which plant would be shut down. After much negotiation, Ft. Wayne offered the corporation $30 million in tax abatements, loan guarantees, and various subsidies to be paid by taxpayers. This should have won the day except for the fact that Springfield, Ohio, offered $31 million. International Harvester will close the Ft. Wayne facility in April. Such poker games are now determining who becomes unemployed.

## What's to be Done?

The basic question we need to ask is the following: If the capital mobility option is being used to "discipline" labor and force communities to make concessions to business, how do we save communities from deindustrialization? What should be part of a progressive economic program?

Unfortunately, there is no blueprint for economic survival. At best there is a loose bunch of economic ideas that need to be discussed and debated. First of all, we need to reconstruct the social wage. It is impossible to build a growing economy simply on the basis of fear. The Japanese have been able to develop an economy that is based on economic security, at least for a substantial portion of the labor force, and we should also be able to do so. This means that we must reconstruct the unemployment benefit program, expand and improve—not gut—public assistance programs, and extend social legislation to cover plant closings. In particular, we need to insist on prenotification of impending closings and on the providing of severance pay to workers according to seniority. Some twenty states have such bills pending in their legislatures, and the Congress has considered, but not passed, the Ford-Riegel bill, which would accomplish this at the national level.

Legislation on plant closing will not stop capital from moving or solve the problem caused by the capital mobility option, but it will provide workers and communities with some time and resources to plan once a plant closing has been announced. With adequate notification, it may be possible to save a plant or at least find alternative employment for those whose jobs are threatened. Worker buyouts are another possibility that deserves attention.

We must caution, however, that rebuilding the social wage and extending it to plant closings is important medicine but no panacea. Such legislation is defensive in nature, defending workers against the vagaries of the economic system and the sometimes dire certainties of management prerogatives. To truly rebuild the economy will take new progressive policies well beyond those provided by fiscal and monetary instruments.

A full employment federal tax and expenditure budget and an expansionary Federal Reserve Board strategy are sine qua nons for an economic renaissance, but these aggregate demand policies cannot deal effectively with the sectoral and regional specific dislocation occurring in the auto, steel, textile, and apparel industries or in cities like Detroit, Youngstown, and Buffalo and in parts of the South. To smooth the transition from mill-based and smokestack industries to high-tech and services will require specific industrial policies that assist workers to acquire new skills and help communities acquire new jobs or retain the ones they have.

With the success of Chrysler, we know it is possible to devise industrial policy instruments that can save part of the old manufacturing base and at least partially cushion the loss of jobs in particular communities. What we need to do is develop a national planning mechanism and a national development bank that can provide assistance to workers, industries, and communities not merely on an emergency ad hoc basis, but according to a carefully developed and reviewed plan. A new partnership between the private sector and government is needed.

The first step in generating such a plan must be taken at the local level, not at the top. Resources need to be made available to every community facing dislocation so that each can prepare redevelopment strategies. Such a strategy, including an assessment of community resources and community needs, has already been carried out for southeastern Michigan. The sophisticated methodology for creating this plan, including the use of occupation and industry data, demographic surveys, and "input-output" analysis, should be exported to other communities.

But this is only the start. We need to coordinate these plans at the federal level, assess what types of public infrastructure (from roads and bridges to health care facilities and new vocational education institutions) need to be constructed at all levels of government, and make sure that everyone affected by these plans has the opportunity to express an opinion. It is an ambitious agenda, but a necessary one.

The most important thing about the 1980s is that they have to be a decade of immense experimentation. Somewhat as in the 1930s, we must be willing to try out new ideas and pass new legislation. Knowing in advance that some of our experiments will fail, we must nonetheless not fear to try. In the 1980s we must bring together our concern for freedom and justice with our expertise for making the economy work better and more equitably. This is the struggle before us. It is extremely hard work, but I for one think we are up to it.

# 2

# The Free-Market Approach

*H. Craig Leroy*

A re plants actually closing their doors in the North to relocate in the South? One important study suggests that they are not. David Birch of the MIT Program on Neighborhood and Regional Change used data supplied by the Dun and Bradstreet Corporation to trace what happened to employment, corporate affiliation, age, and location of 5.6 million business establishments between 1969 and 1976. According to Birch, "Virtually none of an area's employment change is due to firms moving out, in the sense of hiring a moving van and relocating."[1] Birch found that only 1.5 percent of job losses in the Northeast and Midwest were caused by runaway plants, while only 1.2 percent of the South's employment gain was produced by companies that moved into the region from other parts of the country. In other words, the effect of firm migration appeared negligible relative to the job base for most localities.

In another study, two Harvard professors corroborated Birch's findings on firm relocation. Robert A. Leone and John R. Meyer found virtually no evidence that manufacturers are leaving the Frost Belt for the Sun Belt. Analyzing 400,000 manufacturing firms during the 1971-1976 period, Leone and Meyer reported only 2,381 that moved far enough to change their telephone area codes. Futhermore, 86 percent of the plants in the Northeast that moved remained in the region. Although manufacturing employment has indeed declined in the Northeast/Midwest, the decline has been less than 1 percent per year. Richard B. McKenzie concluded that "the 'destruction' of Northern manufacturing jobs has, generally speaking, been more than offset by expansions in job opportunities in other sectors of the Northern economy." As he pointed out, "employment in every Northern and Midwestern state has grown over the last ten or fifteen years. Even employment in relatively depressed states like Ohio and depressed cities like Gary, Indiana, grew during the 1970s, quite slowly and irregularly at 1 to 2 percent per year."[2]

H. Craig Leroy, "The Effects of Plant-Closing Legislation," *Journal of Contemporary Studies* 6, no. 3 (summer 1983), pp. 78–85.

Birch's second significant finding concerned the constancy of job losses. Every area in the U.S. appears to lose jobs at about the same rate, regardless of the strength of the local economy. This average loss rate—about 8 percent per year—means that one-half the jobs in any area must be replaced every five years if the number of jobs is to remain stable. Birch concluded that slow growth results not from a higher rate of job loss, but rather from a failure to compete effectively for new jobs.

Barry Bluestone and Bennett Harrison have offered a different view. These authors attempted to show that closings and layoffs have had a severe effect on the Northeast and Northcentral regions of the country. They believe that capital (as distinct from manufacturing establishments per se) is indeed moving from the Frost Belt to the Sun Belt and argued that Birch's findings are invalid because of certain limitations in the Dun & Bradstreet data he used. They pointed, for example, to problems with the definition of business deaths and company migration. Deaths are in reality disappearances, not confirmed closings. Company migrations can be identified only for those firms for which Dun & Bradstreet provides verification. Bluestone and Harrison argued that some reported "deaths" are really relocations. Furthermore, the closing of a firm at one location may result in the expansion of the company at another location owned by the same conglomerate parent. Consequently, some plant closings and plant contractions of firms represent capital shifts from one region to another.

Another pair of investigators, John S. Hekman and John S. Strong, have attempted to reconcile these conflicting views. Looking at the historical regional imbalance in U.S. manufacturing, they accounted for the higher rate of growth in the Sun Belt by the fact that the Northeast and Northcentral regions have historically had a disproportionate share of manufacturing employment. This movement from the traditional manufacturing centers can be termed "outward" as much as "sunward." That is, decentralization or outward movement should be understood as a dispersion from urban to nonurban areas. Hekman and Strong pointed to the fact that smaller labor market areas of New England gained in manufacturing employment relative to the large manufacturing areas during the 1968–1978 period. They concluded that production is indeed "moving" to the South and West (and to other rural areas). Yet this shift is occurring *not* through plant relocations, but, much as Birch found, "through a greater rate of plant openings in those areas."

In fact, Hekman and Strong found that the Sun Belt has actually had a slightly higher *rate* of plant closings than the North. The North has indeed experienced many more plant closings than the South, but that is because the North simply has more factories.

In sum, data on plant movement are far from clear. The phenomenon of the classic runaway shop appears to be rare in relation to total job loss; nonetheless, to a degree, capital does appear to be dispersing from its traditional regions of highly concentrated manufacturing.

## The Job Creation Process

But there is a broader issue. A dynamic, growing economy must be able to react to changes in technology. Innovations and inventions must be fostered and incorporated into a production system. Technological change is usually available only in the form of newly produced capital equipment; in order to capture the benefits of technological progress, a business must actively introduce the new machinery, equipment, or plant into its operations.

In a market system such as ours, the incentive to invest in new capital equipment comes from the desire to increase profits by the introduction of new, low-cost production techniques. New purchases of capital equipment boost productivity. What is important, therefore, is not only the absolute size of a state's capital stock, but also the age of its capital. This is why so much is heard about the significance of capital formation. If a local economy is to continue expanding while remaining competitive in the marketplace, capital investment must constantly reoccur. This is especially true for businesses involved in advanced technologies.

Many state officials hope that the emergence of these high-tech companies, committed to heavy expenditures for research and development and employing a skilled labor force, will compensate for the loss of heavy manufacturers. Because of rapid changes in the composition of demand, many states are at present left with a mix of industries out of step with changes occurring in the marketplace, and many are going through a somewhat painful restructuring of their local economies. This process has put severe strains on state and local officials as they attempt to cope with a changing tax base and periodic spells of high unemployment while at the same time trying to understand what exactly is happening to their local economy.

Much of this can be explained, as Hekman and Strong noted, by looking at the long-standing regional imbalances of manufacturing in the United States. During the era of industrialization, manufacturing activity tended to be concentrated because of the economic advantages inherent in the clustering of production facilities, mainly because of interindustry linkages and the need to keep transportation costs as low as possible.

However, technological advances are now allowing a higher degree of decentralization. The greater efficiency in trucking and air freight services, as well as the near-completion of the interstate highway system, have contributed to dispersal. Similarly, new developments in miniaturization and lightweight materials have cut transportation costs. Innovations in communication systems through advances in information storage, retrieval, and transmission also have reduced the economic advantages of concentrating business activity. Finally, as production processes have become more standardized and demand less specialized labor, firms have been able to locate in less industrialized regions where wages, as well as other costs such as energy, are lower.

As a result, heavy manufacturing activity has been spreading out from its traditional centers since World War II. In the long run, there is little that state governments can do to stem the tide of movement toward less costly regions. Inevitably, many of the old-line manufacturing states will cease to compete successfully for industries in which production processes have become standardized and for which lower wages or energy costs or transportational advantage are dominant concerns. The attempt to retain, through governmental assistance, industries no longer competitive in the marketplace is a questionable use of scarce public resources.

The overall health of our economy depends on the ability of capital to flow to the most productive enterprises and away from the inefficient ones. As McKenzie notes, competition is destructive, but it is "creative destruction, a process whereby consumers get more of what they want at more favorable prices. Plant closings may just as well signal a growing dynamic economy as a dying one."[3] New enterprises begin, old jobs die, new products are conceived and developed, and new jobs are created. For labor this means that some jobs and skills become obsolete and new jobs are created that require new skills. Moses Abramowitz remarks:

> The pace of growth in a country depends not only on its access to new technology, but on its ability to make and absorb the social adjustments required to exploit new products and processes. Simply to recall the familiar, the process includes the displacement and redistribution of population among regions, and from farm to city. It depends on the abandonment of old industries and occupations, and the qualifications of workers for new, more skilled occupations.[4]

Encouragement should be given to this adjustment process. States need not be insensitive to problems caused by displacement. Indeed, positive actions by a state government to retrain displaced workers for new, emerging occupations send a clear signal to the private sector.

It is important to understand that unlike most other industrialized nations, the United States has never pursued anything approaching a comprehensive manpower policy that consciously views the quality and productivity of the labor forces as integral to its economic objectives. It is true that the manpower policies of the 1960s began with a focus on retraining skilled workers threatened by industrial automation, but this focus was quickly redirected to the structurally unemployed as a push for equal opportunity began to dominate domestic policy. This trend, narrowing the target population to only the disadvantaged, continued through most of the 1970s. Only recently has the federal government begun to shift its emphasis in manpower programs to training for industry's needs and addressing the problems of the "displaced worker." However, state governments themselves can develop manpower programs targeted to displaced workers. In sum, what a state government must do is decide how it can best encourage new start-ups of business and expansions of existing industry, creating new job opportunities for its citizens in the process.

## Importance of Business Climate

It is here that one must stress the impact of plant-closing legislation on a state's overall economic climate. Prosperity requires confidence in the state of the economy and in the action of government. Businesses need to predict with some degree of certainty what attitudes and trends are likely to prevail over the long term. A healthy economy is more likely to arise from policies directed toward stimulation of business growth and job creation than from negative restrictions.

To understand why economic climate is important, it is necessary to consider what factors individual firms examine in deciding whether or not to commit to new capital expenditures. Investment proposals are evaluated not only on the basis of the expected cash benefits that will accrue to the firm from the investment, but also in light of estimated risk of ever receiving the projected cash flows. A number of factors, some more important than others, shape this estimation of risk: the expected inflation rate, the level of interest rates, federal tax policies, and the like. Each contributes to the calculation of the overall risk of an investment project. A firm examines the expected return on an investment in the light of the risk in order to evaluate the viability of a potential project.

Ultimately, the private sector's assessment of the stability and predictability of a given state's policies will affect the risk factor assigned to potential capital expenditures within the state. In many cases, the effect of an uncertain or unfavorable view of the state government by the business community might be small in comparison to other factors, such as a possible change in the price of raw materials, but in some cases it could be the final determinant in the business decision. In other words, a marginally attractive investment proposal might be held off or cancelled due to the perceived governmental climate. This is why a state's economic climate receives such attention and discussion within the business community.

Areas growing economically are doing so because of their perceived vitality and opportunity. Although all areas of the country appear to lose jobs at a roughly constant rate, many localities are growing rapidly because of a greater rate of business openings and expansions. The secret is to create an environment imbued with a sense of economic boundlessness, encouraging new enterprises and greater capital investment by established companies. This can be accomplished only by fostering a favorable attitude toward government in the business community and by providing assurance about the attitudes and trends likely to prevail over the long term.

Of course, public officials may routinely reiterate their long-term policy of stimulating economic development, but it is important to remember that the long run is only a sequence of short runs. Pursuing short-run policies that deviate from long-run goals ultimately means abandoning the long-run objective. Frequent changes in policy only make businesses uncertain about the probable duration of current policies, increasing the risk of committing new investments in a particular state.

There is an additional consideration. According to Birch's analysis, 80 percent of all new jobs were created by firms with 100 or fewer employees. Moreover, approximately 80 percent of all jobs were created by establishments four years old or younger. As Birch commented, "Job creators are the relatively few younger companies that start up and expand rapidly in their youth, outgrowing the single 'small' designation in the process."[5] It follows that to promote the creation of new jobs, state governments need to nurture and assist the thousands of small and medium-sized companies that provide the ongoing employment base in a locality. Plant-closing legislation sends a discouraging message to these employers as they face the decision of whether and where to expand operations. A recent survey by the Council of State Planning Agencies confirmed that how a state government is perceived does influence expansion and new plant location decisions. "Political climate" was rated quite high by a number of companies in deciding where to locate new production facilities.

Plant-closing legislation will also severely hamper government's efforts to attract out-of-state and foreign companies. In a Conference Board survey on locational decisions of companies, most executives stated that the state's business climate was an important determinant to them, but more so in the avoidance of a location with a problem than as an attraction to a community or state with a good business climate. Passing plant-closing legislation in effect raises a red flag of warning to the private sector. Once a state government begins interfering with basic business decisions, a clear message is sent to existing and potential companies: what harmful legislation is next?

## Alternative Measures

Certain economists believe that the American economy is going through a major transition as the baton of growth has been passed from heavy industry to high technology. Some regions of the country are further along than others. Innovation appears to occur in bursts or waves. The history of capitalism is marked by violent bursts of activity contrasted by times of stagnation in which the inefficient and outmoded enterprises are weeded out and resources are freed for new endeavors.

This is not to say that pain is not involved. Unemployment is indeed a serious problem. There are positive actions a legislature can take to help, such as strengthening job-training programs to assist the transition of workers into new emerging occupations. By contrast, plant-closing laws, as Hekman and Strong have pointed out, "are a kind of unwieldy instrument relative to the desired tasks. They influence business decisions that they are not meant to influence by discouraging some plant openings."[6]

It should be remembered that whereas large corporations tend to make use of complex quantitative financial analyses to evaluate proposed investment projects, small companies deciding to buy a new machine or to expand do not usually rely on mathematical equations. The proprietor's intuitive feeling about how good a place is to do business has a strong influence.

Yet it is a difficult task for state officials to consider the long-run implications of their actions. The time horizon of government officials may reach only as far as the next election; and when economic conditions are difficult, pressure to adopt counterproductive legislation grows. In every statehouse there are a number of individuals who advocate actions to protect state industries and jobs regardless of the long-range consequences. The challenge is to resist the temptations of such short-run but ultimately spurious solutions.

## Notes

1. David L. Birch, *The Job Generation Process* (Cambridge, Mass.: MIT Program in Neighborhood and Regional Change, 1979).

2. Richard B. McKenzie, "Myths of Sunbelt and Frostbelt," *Policy Review* (spring 1982).

3. Ibid.

4. Moses Abramowitz, "Welfare Quandaries and Productivity Concerns," *The American Economic Review* (March 1981).

5. David L. Birch, "Who Creates Jobs?" *The Public Interest*, no. 65 (fall 1981), p. 14.

6. John S. Hekman and John S. Strong, "Is There a Case for Plant Closing Laws?" *New England Economic Review* (July-Aug. 1980), p. 46.

# 3
# Is Deindustrialization a Myth?

*Robert Z. Lawrence*

For the first time since World War II, employment in U.S. manufacturing has fallen for three consecutive years. The 10.4 percent decline from 1979 to 1982 is the largest since the wartime economy was demobilized. The current slump is also unusual because international trade has made an important contribution: normally, the volume of manufactured goods imported falls steeply in a recession, yet from 1980 to 1982, it rose by 8.3 percent. Normally, U.S. manufactured exports reflect growth in export markets abroad, yet despite a 5.3 percent rise in these markets from 1980 to 1982, the volume of U.S. manufactured exports dropped 17.5 percent.

Are these developments the predictable consequences of three years of demand restraint and a strong dollar, or do they result from deep-rooted structural changes?

There are widely held views that the recession has simply dramatized a secular decline in the U.S. industrial base. One of these views blames U.S. producers for the trend: Americans fail to produce quality goods because managers are myopic and care only about short-term profits, workers lack discipline and are shackled by work rules, and labor and management look on one another as adversaries. Others blame the U.S. government. On the one hand are those who fault it for excessive interference—for restrictive regulatory practices that have raised production costs; for faulty tax rules that have discouraged investment, savings, and innovation; and for trade protection that has slowed adjustment to international competition. On the other hand are those who blame government neglect. The United States, they contend, has failed to plan and coordinate its industrial evolution; it ought to have policies to promote industries with potential and to assist those in decline. Finally, there is also the more fatalistic view of the decline in U.S. manufacturing as the inevitable result of the rapid international diffusion of U.S. technology.

Robert Z. Lawrence, "The Myth of U.S. Deindustrialization," *Challenge* (Nov.-Dec. 1983), pp. 12–21.

While some argue that particular deficiencies have become worse over time, others point to changes in the environment that have made U.S. structural flaws increasingly costly. As long as competition was primarily domestic, it is argued, weaknesses were obscured. As global trade expanded, however, U.S. firms were forced to meet foreign competitors staffed with superior work forces and managers and backed by superior government policies.

## Four Unfounded Assumptions

The perceived effect of international competition has grown to the point that it is frequently cited as the major source of structural change in the U.S. economy and the primary reason for the declining share of manufacturing in U.S. employment. This change is viewed with some alarm, both because manufacturing activity is considered intrinsically desirable and because of the adjustment costs associated with the shift. In addition, some argue that this decline in U.S. comparative advantage does not result from an inevitable process of technological diffusion or from changes in factors of production, but from the industrial and trade policies adopted by other nations. Without similar policies, some contend that the United States will eventually become an economy specialized in farm products and services—"a nation of hamburger stands."

Although the role of the deficiencies in U.S. policies and practices in retarding productivity growth over the past decade remains unresolved, the links between these deficiencies, trade performance, and shifts in economic structure have not been convincingly demonstrated. There are several implicit assumptions in the current discussion about U.S. industrial performance that I will show to be inappropriate. First, the policy discussion often presumes that rapid productivity growth will increase the share of resources devoted to an activity, that "higher productivity will create jobs." It implicitly assumes the existence of elastic demand. As the experience of U.S. agriculture has demonstrated, however, rapid productivity growth in the face of limited markets may have the opposite effect. Indeed, the declining employment in Japanese manufacturing in the 1970s and the contrasting rise in U.S. employment suggest that manufacturing productivity may be *negatively* associated with employment.

Second, the discussion presumes that a decline in the international technological lead in a particular area will reduce the resources devoted to that activity. It assumes implicitly that an erosion in absolute advantage will lead to an erosion in *comparative* advantage. Yet, even though foreign productive capacities are converging to those of the United States, the U.S. *comparative* advantage in high-technology products has actually increased.

Third, the discussion implicitly presumes that the trade balance can decline indefinitely. It ignores the automatic adjustment mechanisms that tend to keep the trade balance in goods and services within fairly narrow bounds. An increase in imports eventually leads to an increase in exports. When global demand shifts away from U.S. products, it creates an excess supply of American goods and an excess demand for foreign goods. Because the relative price of U.S. goods may have to fall to restore the trade balance, this will increase the resources devoted to export production, for a decline in the terms of trade entails providing more exports for any given volume of imports. Indeed, the decline in U.S. terms of trade associated with the real devaluations of the dollar between 1973 and 1980 contributed to the rise in U.S. employment due to trade over that period.

Fourth, international trade is neither the only nor the most important source of structural change. In many cases, trade has simply reinforced the effects of demand and technological change. At least five factors have had important effects on the U.S. industrial base:

1. The share of manufactured products in consumer spending has declined secularly because of the pattern of demand associated with rising U.S. income levels.

2. Some of the long-run decline in the share of manufacturing in total employment reflects the relatively more rapid productivity growth in this sector.

3. Because the demand for manufactured goods is highly sensitive to the overall growth rate of GNP, manufacturing production has been slowed disproportionately by the sluggish overall growth in the global economy since 1973.

4. Shifts in the pattern of U.S. international specialization have arisen from changes in comparative advantage; these, in turn, result from changes in relative factor endowments and production capabilities associated with foreign economic growth and policies.

5. Short-run changes in U.S. international competitiveness have come from changes in exchange rates and cyclical conditions both at home and abroad.

The appropriate choice of policy hinges on the relative impacts of these various factors on current U.S. industrial performance. Given the radical changes in the world economy after 1973, the period from 1973 to 1980 is the most relevant sample for current policy discussions. The data for this period show stagnation, volatile exchange rates, and increasing government intervention in trade; it is during these years, it is alleged, that foreign industrial policies damaged the U.S. manufacturing base. The data also allow comparison of U.S. industrial performance with those of other major industrial countries in a period when comparative performance is less heavily influenced by relative stages of development.

Observations for 1973 to 1980, however, may be unduly influenced by the different cyclical positions prevailing in the endpoint years. Because capacity utilization in manufacturing was similar in 1970 and 1980, U.S. data for the entire decade are used to provide a second, cyclically neutral measure of structural changes. Observations for 1970–1980 are still influenced by changes in the real exchange rate of the dollar in these years. As measured by the International Monetary Fund relative U.S. export prices for manufactured goods were 13.5 percent lower in 1980 than in 1970. In evaluating the results, therefore, it should be kept in mind that the U.S. trade performance during the 1970s depended in part upon this price-adjustment process.

## Defining Deindustrialization

The contention that declining U.S. international competitiveness has induced the country's deindustrialization is wrong on two counts. First, in the most relevant sense, the United States has not been undergoing deindustrialization. Second, over the period 1973 to 1980, the net impact of international competition on the overall size of the U.S. manufacturing sector has been small and positive.

The term "deindustrialization" needs further elaboration. What is industry? Does it, for example, include the construction and mining sectors or refer more narrowly, as we will interpret it here (partly for reasons of data availability), to the manufacturing sector alone? Does "deindustrialization" refer to a drop in the output of industry or to the inputs (e.g., capital and/or labor) devoted to industry? Does it refer to an absolute decline in the volume of output from (or inputs to) manufacturing or simply to a relative decline in the growth of manufacturing outputs (inputs) as compared to output (inputs) in the rest of the economy?

Because industrial policy is generally concerned with easing adjustment, absolute deindustrialization with respect to factors of production would probably be the definition that fits current policy concerns about the manufacturing sector. While a declining share of output or employment could change the relative power of industrial workers or the character of a society, an absolute decline in industrial employment means much greater problems of adjustment.

This distinction is relevant: measured by the size of its manufacturing labor force, capital stock, and output growth, the United States has not experienced absolute deindustrialization over either the periods 1950–1973 or 1973–1980. Employment in U.S. manufacturing rose from 15.24 million in 1950 to 16.8 million in 1960, 19.4 million in 1970, 20.1 million in 1973, and 20.3 million in 1980. The capital stock in manufacturing grew at an annual rate of 3.3 percent from 1960 to 1973 and 4.5 percent between 1973 and 1980. Output in manufacturing increased at a 3.9 percent annual rate between 1960 and 1973 and a 1.1 percent annual rate from 1973 to 1980.

Judged by the output share of goods, the United States was no more a service economy in 1980 than it was in 1960. In 1960, 1973, and 1980, the ratio of goods output to GNP (measured in 1972 dollars) was 45.6, 45.6, and 45.3

percent, respectively. Similarly, the ratio of value added in manufacturing to GNP (in 1972 dollars) was actually somewhat higher in 1973 than it was in 1950. Nonetheless, from 1950 to 1973, the shares of expenditure, employment, capital stock, and research and development (R&D) devoted to the manufacturing sector declined. Factors on both the demand and supply sides account for this. As U.S. incomes have risen, Americans have allocated increasing shares of their budgets to items in the service sector such as government services, education, medical care, finance, and real estate services. At the same time, productivity in manufacturing has increased more rapidly than elsewhere in the economy. Although the more rapid growth in manufacturing productivity has led to slower increases in manufacturing prices, the demand stimulated by the relative price decline for manufactured goods has not offset the fall in the share of resources devoted to value added in manufacturing. As a result, overall real industrial output has risen about as rapidly as GNP, but the shares of employment and capital in manufactured goods have declined.

From 1973 to 1982 there has been a marked acceleration in the rate at which manufacturing's share of output and employment has fallen. This should have been expected, given the slow overall growth in GNP and the fact that growth in labor productivity (output per man-hour) fell less in manufacturing than in the rest of the economy. The demand for manufacturing output is particularly sensitive to fluctuations in income. The demand for goods, particularly durables, is inherently more sensitive to short-run income fluctuations than the demand for services, because many such purchases can be easily postponed. Thus the generally slow growth in U.S. GNP from 1973 to 1980 was reflected in disproportionately slow growth in the manufacturing sector.

The relationship between the growth of manufacturing and the overall growth of the economy can be summarized quite well by using standard econometric analysis. The results show that industrial performance is a magnification of performance in the overall economy: if GNP grows at 1.7 percent per year, there will be no increase in manufacturing production, but for each percentage point increase (decrease) of GNP growth above 1.7 percent, manufacturing output will rise (fall) by 2.2 percentage points. The analysis shows that, based on the estimates from 1960 to 1973, one can forecast industrial production for the period 1973 to 1982, taking actual GNP as given, with remarkable accuracy. Thus, there is no puzzle in explaining aggregate manufacturing production. It is almost exactly what one should have expected given the performance of the total economy.

## Employment, Plant, and R&D

Although the overall level of manufacturing output has matched its historic relationship with GNP, the relationship between the growth of outputs and inputs has changed. As a result of the decline in productivity growth in manufacturing since 1973, given rates of output growth are now associated with somewhat higher rates of employment and capital growth. Analysis indicates that, taking manufacturing output as given, manufacturing employment growth has been about 1.36 percent per year higher than it would have been without the decline in manufacturing productivity growth. Thus employment has actually held up better than might have been anticipated from past relationships.

Probably the most commonly provided reason for poor U.S. manufacturing performance is the failure of businesses to invest in new plant and equipment. Nonetheless, although there has been a marked decline in the growth of the capital-labor ratio in the economy overall since 1973, the measured growth of the net capital stock in manufacturing has been remarkably rapid. Although the ratio of the net capital stock to full-time equivalent employees in manufacturing grew at about 2.03 percent per year from 1950 to 1973, it grew at 3.8 percent per year from 1974 to 1980. This supports the view that automation has actually accelerated. And, while historically the ratio of the net capital stock in U.S. manufacturing to the net stock in the rest of the economy declined (from 0.30 in the 1950s to 0.26 in the 1960s to 0.237 in 1973), since 1973 the capital stock in manufacturing has actually grown more rapidly than in the rest of the private economy.

The 1970s saw a much-publicized decline in the growth of real R&D expenditures. While real R&D spending increased 3.1 percent per year from 1960 to 1973, it fell to a 2.5 percent annual growth rate from 1973 to 1980. However, this does not reflect a similar drop in real R&D spending in U.S. industry. Between 1960 and 1972, such spending in manufacturing grew 1.9 percent per year. From 1972 to 1979 (the latest data available), it accelerated to 2.4 percent. A similar pattern is evident in industry hiring. While the number of scientists and engineers employed in industry R&D grew at 1.6 percent between 1960 and 1973, growth from 1973 to 1980 averaged 3.2 percent per year.

The increased commitment of plant, equipment, and R&D expenditures makes the decline in productivity growth in U.S. manufacturing since 1973 particularly puzzling. One question is whether the capital stock is accurately measured. Mismeasurement could be due to an increase in capital and R&D devoted to meeting regulatory requirements such as safety and pollution control, which do not show up as output. Subtracting Commerce Department estimates of the net capital stock devoted to reducing air and water pollution from the net capital stock in manufacturing lowers the growth in manufacturing capital from 4.5 to 4.2 percent per year. A second reason might be the premature retirement of capital that has become economically obsolete in changed economic conditions.

Nonetheless, as these data make clear, there has been no erosion in the U.S. industrial base. The decline in employment shares has been the predictable result of slow demand and relatively more rapid labor productivity growth in manufacturing because of an acceleration in capital formation. Paradoxically, the slow (absolute) growth in productivity has required unpredictably large increases in employment, plant and equipment, and R&D.

## The Global Perspective

Proponents of a radical change in U.S. industrial policies contrast the ad hoc and laissez-faire policies of the United States with the systematic and interventionist practices abroad. Although conceding there are marked differences in the degree to which foreign practices have succeeded, they argue that the conscious policy of managing the decline of older industries and the rise of new industries has been superior to the U.S. approach, which is marked by malign neglect. Similarly, the broader provision of social services in European economies, the more extensive rights to their jobs enjoyed by workers, and the greater restrictions on plant closings have all been held up as worthy of emulation. Opponents of such policies argue that they will delay adjustment, for the government is most likely to be captured by forces seeking to preserve the status quo, and strictures on mobility are likely to retard adaptation.

It is particularly important that international comparisons be made on the basis of performance since 1973, for policies that succeeded amid strong global growth and economic expansion might not be appropriate for the current era of stagnation.

The 1972–1974 commodity boom and the inflation that accompanied it brought a new era. All developed countries have been plagued by low rates of investment, slow growth, and inflation. The problems associated with high inflation and energy shocks have destroyed investors' confidence. They have learned from their experiences in 1974 (and again in 1979) that a political disruption in the Middle East or a sudden increase in domestic inflation may at any time force their government to adopt policies that bring on a recession, leaving them with excess capacity. The rate of investment has slumped, the growth of the heavy manufacturing industries has been cut, and consumption expenditures have risen as a share of GDP. Industries with long gestation periods for investment, such as steel and shipbuilding, have been particularly hard hit by the post-1973 slump. There is insufficient demand for the products of plants that were built on the basis of overoptimistic projections of market growth in the late 1960s.

By a wide variety of indicators, the relative performance of U.S. manufacturing since 1973 has improved. The declines in the growth of manufacturing production, productivity growth, employment, and investment in manufacturing were all smaller here than in other industrial nations. (See table 3–1 for rates of growth for GDP and industrial production in the major industrial economies.) Although U.S. growth was among the slowest prior to 1973, since that time it

**Table 3-1**
**Output Growth in Major Industrial Economies, 1960–80**
*(average annual rates of change in percent)*

| | USA | | W. Germany | | France | | Japan | | United Kingdom | | OECD | |
|---|---|---|---|---|---|---|---|---|---|---|---|---|
| | GDP | Manufac-turing Production | GDP | Manufac-turing Production | GDP | Manufac-turing Production | GDP | Manufac-turing Production | GDP | Manufac-turing Production | GDP | Manufac-turing Production |
| 1960–1973 | 4.0 | 5.4 | 4.5 | 5.2 | 5.6 | 5.0 | 10.5 | 12.5 | 3.1 | 3.0 | 5.0 | 6.0 |
| 1973–1980 | 2.3 | 1.8 | 2.3 | 1.1 | 2.7 | 1.3 | 3.6 | 2.9 | 0.9 | -2.2 | 2.5 | 1.7 |

*Source:* Organization for Economic Cooperation and Development

has been quite typical for a developed country. Although trailing Japan, U.S. industrial production grew more rapidly than in Germany, France, or the United Kingdom.

It is in Europe rather than in the United States that employment is undergoing absolute deindustrialization. Compared with historical trends, industrial production in Japan was abnormally strong, whereas industrial production in Europe is unusually weak. Econometric analysis relating industrial production to GNP in European countries from 1960 to 1973 demonstrates a substantial overprediction of the level of industrial production in 1980. In the case of Japan, the analysis underpredicts industrial production (by 12 percent in 1980).

According to disaggregated statistics for industry (*OECD Industrial Production*, various issues) U.S. output growth from 1973 to 1980 for food, textiles, apparel, chemicals, glass, and fabricated metal products was more rapid than that of either Germany or Japan. Primary metals were an exception. Although U.S. growth lagged behind Japan in the various engineering categories, it trailed German growth only in basic metals production and transportation equipment.

A closer look at several important measures of industrial vitality gives a similar picture of the supposed American deindustrialization.

## Employment

The employment record of the U.S. manufacturing sector may come as a surprise to those concerned about deindustrialization: from 1973 to 1980, the United States increased its employment in manufacturing more rapidly than any other major industrial country, including Japan. Moreover, because the average workweek declined more rapidly abroad, the relatively larger growth in U.S. manufacturing employment is even more conspicuous. A comparison between the United States and Japan indicates that Japanese employment in sectors such as transportation, electrical machinery, nonelectrical machinery, chemicals, and nonferrous metals grew less rapidly or declined more than that in the United States (see table 3–2).

As the case of Japan makes clear, in the current global environment of relatively slow growth in demand, rapid increases in productivity do not necessarily increase employment. Indeed, compared with the United States, the faster increases in Japanese productivity have entailed the more rapid process of labor-force deindustrialization. In the case of Europe, employment opportunities in manufacturing have decreased because faster productivity growth has been combined with relatively slower growth in output.

## Capital Formation

If we compare gross fixed investment in manufacturing in the United States with that of industrialized European countries, the sluggish growth of such

**Table 3–2**
**Employment by Three-Digit ISIC:**
**United States and Japan**
(*annual average percent change*)

| ISIC | | 1960–73 | 1973–80 |
|---|---|---|---|
| 321 Textile products | USA | 2.1 | -2.5 |
| | Japan | -1.1 | -5.7 |
| 341 Paper | USA | 0.9 | 0.0 |
| | Japan | 0.7 | -0.5 |
| 342 Printing | USA | 1.3 | 2.2 |
| | Japan | 3.7 | -0.1 |
| 351–52 Chemical | USA | -0.7 | 0.3 |
| products | Japan | -3.0 | -1.9 |
| 371 Iron, steel | USA | -0.4 | -2.0 |
| | Japan | 1.7 | -2.9 |
| 372 Nonferrous metals | USA | n.a. | 0.3 |
| | Japan | n.a. | -2.0 |
| 381 Fabricated metal | USA | 2.1 | 0.9 |
| products | Japan | 5.1 | -1.5 |
| 382 Nonelectrical | USA | 3.0 | 2.7 |
| machinery | Japan | 4.5 | -1.4 |
| 383 Electrical machinery | USA | 2.2 | 1.2 |
| | Japan | 5.2 | -0.7 |
| 384 Transportation | USA | 1.0 | -0.9 |
| | Japan | 4.5 | -1.2 |
| 3 All manufacturing | USA | 1.8 | 0.3 |
| | Japan | 2.6 | -1.2 |

Source: *United Nations Yearbook of Industrial Statistics*
n.a.: not available

**Table 3–3**
**Gross Fixed Investment in Selected OECD Countries**
(*1973 = 100*)

| | 1963 | 1970 | 1973 | 1978 | 1979 |
|---|---|---|---|---|---|
| United States total | 64 | 83 | 100 | 105 | 107 |
| manufacturing | 63 | 93 | 100 | 133 | 144 |
| W. Germany total | 64 | 91 | 100 | 99 | 107 |
| manufacturing | 71 | 118 | 100 | 88 | n.a. |
| France total | 46 | 82 | 100 | 102 | 106 |
| manufacturing[a] | n.a. | 89 | 100 | 101 | 101 |
| Belgium total | 62 | 92 | 100 | 110 | 110 |
| manufacturing | 72 | 103 | 100 | 74 | 72 |
| Netherlands total | 54 | 95 | 100 | 102 | 103 |
| manufacturing[b] | 60 | 109 | 100 | 102 | 105 |
| United Kingdom total | 62 | 92 | 100 | 98 | 97 |
| manufacturing | 90 | 122 | 100 | 112 | 113 |

Source: *OECD National Accounts* (1951–1980).

[a]Mining, manufacturing, and utilities.

[b]Mining, manufacturing, and utilities plus construction.

## Table 3–4
## Manufacturing Growth Rates of Capital Stocks in Selected Industrial Countries
*(average annual rates of change in percent)*

|  | 1960–73 | 1973–79 |
|---|---|---|
| United States | 3.1 | 3.8 |
| Austria | 5.0 | 4.3 |
| W. Germany | 6.9 | 2.6[a] |
| Sweden | 4.3[b] | 3.4[a] |
| United Kingdom | 3.5 | 2.4 |

*Source:* Economic Survey of Europe in 1981, United Nations.
[a]1973–1978.
[b]1963–1973.

investment in Europe is apparent (table 3–3); only in France was it above its 1970 levels in 1979. Compare the ratios of European investment in manufacturing with overall gross fixed investment in those countries: in contrast to the United States, most of the European economies are allocating proportionately less of their new capital formation to industrial production than they did in 1970.

Just as an automobile may be decelerating and yet going faster than another, so one country may have a declining growth rate for investment with a capital stock growing at a relatively faster rate. Thus capital stock measures are required. In table 3–4 I report such estimates, gathered by the United Nations. They indicate that in contrast to its previous performance, the U.S. capital stock in manufacturing grew as rapidly as those in Europe.

*Research and Development*

Since 1972, the United States has maintained its share in R&D spending among industrial countries, thereby reversing the relative decline of the late 1960s and early 1970s, when U.S. government-funded R&D was cut back while R&D spending in other major countries advanced rapidly. From 1972 to 1980, the growth in business-funded R&D in the United States has been similar to the growth in France, Germany, and Japan; although government-funded R&D in the United States has not grown at the Japanese pace, it has exceeded the rise in support provided by the governments of France, Germany, and the United Kingdom.

According to OECD estimates, by a wide variety of indicators the United States continues to dominate other industrial countries in its commitment to R&D. In 1977, for example, spending on R&D in U.S. manufacturing was equal to about 6.5 percent of the domestic U.S. industrial output. By contrast, spending on manufacturing R&D in Japan, the United Kingdom, and Germany amounted to 3.7, 5, and 4.0 percent, respectively, of industrial output. Indeed, privately funded American R&D alone was equal to 4.4 percent of manufacturing product. In absolute terms in 1979, measured at purchasing power parity levels, the United States spent about 1.5 times as much as Japan, Germany, France, and the United Kingdom combined and employed about 1.3 times as many scientists and engineers. By contrast, manufacturing employment in these countries in 1979 was 1.5 times that in the United States. The OECD has also ranked industrial countries according to the percentage of manufacturing output spent on R&D in a variety of industry groups during the 1970s. The United States ranked first in manufacturing overall and in the electrical, aerospace, machinery, and transportation categories.

*Productivity*

Measured both in terms of the ratio of total output to all inputs and in output per man-hour, U.S. productivity growth in manufacturing, as in the economy as a whole, has slowed down in the period since 1973. Over the same period, however, there has been an even larger slowdown in foreign productivity growth, both in manufacturing and in the whole economy. Careful studies have been unable to provide convincing explanations for these slowdowns, and I will not attempt an investigation here. It should, however, be noted that despite some convergence since 1973, the U.S. productivity growth rate in manufacturing remains the slowest of any major industrial country.

Measured by output per man-hour, however, the United States continues to be the world's most productive manufacturing nation. According to A.D. Roy, for example, output per employed worker-year in U.S. manufacturing in 1980 was about 16 percent higher than in Japan, 21.7 percent higher than in Germany, and 31.3 percent higher than in France. To be sure, America no longer leads in all industries. According to the 1981 White Paper on International Trade issued by the government of Japan, Japanese productivity levels in 1979 were above those of the United States in steel (108 percent above U.S. levels), general machinery (11 percent higher), electrical machinery (19 percent), transportation equipment (24 percent), and precision machinery and equipment (34 percent).

## Achieving Structural Change

Some economists have unfavorably contrasted the U.S. failure to promote industrial adjustment with the explicit adjustment policies followed in Europe and Japan. It is therefore of some interest to compare the shifts in the U.S.

industrial structure with those in other major economies to determine whether in fact U.S. industrial adaptation has been lagging. To explore this question I have used the matched set of data collected by the United Nations. These provide fairly disaggregated information on industries at the three-digit ISIC level. First, I selected the group of industries that is generally considered to have high-growth potential. They are characterized by relative intensity in R&D and by rapid rates of technological innovation. The sample includes chemicals, plastic products, machinery, and professional instruments and typically provides up to about 35 percent of manufacturing employment in major industrial nations. Next, I calculated the share of total manufacturing employment these industries accounted for in the United States, Germany, and Japan and compared growth in these shares between 1973 and 1979 (see table 3–5).

Although employment shares in all three countries increased, the 8.9 percent rise in the U.S. share far exceeded those of both Japan (up 0.6 percent) and Germany (up 3.0 percent). A similar analysis was performed for a group of slow growers—labor-intensive industries such as textiles, apparel, leather, footwear, and furniture, and capital-intensive industries such as metals, metal products, and shipbuilding. This group also typically accounted for between 30 and 35 percent of total employment. In this case, Germany had the most rapid decline in the share of employment (9.2 percent), while Japan and the U.S. had shifts

**Table 3–5**
**Employment Shares in Manufacturing of High- and Low-Growth Sectors: United States, Germany, and Japan**

| Shares | United States | W. Germany | Japan |
|---|---|---|---|
| High-growth industries[a] | | | |
| 1973 | 30.4 | 39.7 | 31.0 |
| 1979 | 33.1 | 40.9 | 31.2 |
| Percent change in share[b] | 8.9 | 3.0 | 0.6 |
| Low-growth industries[c, d] | | | |
| 1973 | 34.0 | 32.8 | 37.5 |
| 1979 | 32.0 | 29.8 | 35.1 |
| Percent change in share[b] | −5.9 | −9.2 | −6.4 |
| Of which: Labor-<br>intensive[c] | | | |
| 1973 | 19.2 | 15.1 | 21.6 |
| 1979 | 17.3 | 13.1 | 20.4 |
| Percent change in share[b] | −9.9 | −13.2 | −5.5 |
| Capital-intensive[b] | | | |
| 1973 | 14.8 | 17.6 | 15.9 |
| 1979 | 14.7 | 16.7 | 14.7 |
| Percent change in share[b] | −0.7 | −5.1 | −7.5 |

*Source: United Nations Yearbook of Industrial Statistics,* 1977 and 1980 editions.

[a]Industrial chemicals, plastic products, machinery, electrical machinery, and professional goods.

[b]$(1 - 2)/1 \times (100)$.

[c]Textiles, apparel, leather, footwear, wood products, and furniture.

[d]Iron and steel, nonferrous metals, metals products, and shipbuilding.

quite similar in magnitude. Although the United States moved out of labor-intensive industries faster than Japan, the drop in the Japanese share of the capital-intensive group exceeded that of the United States.

These results should, of course, be treated with some caution because of the relatively aggregative nature of the industry divisions and the possible discrepancies in national classification schemes. Nonetheless, they contradict assertions of America's relative failure to shift resources toward high-growth sectors. They indicate that the United States has been about as successful as Japan in reducing the role of the low-growth group.

## The European Dilemma

I have pointed to the marked contrast in European economic performance before and after 1973, a contrast that is particularly evident in data on industrial performance. European manufacturing production has declined by more than might have been expected, given GNP. Employment has fallen and productivity growth slowed. While Germany has been relatively successful in shifting out of slow-growing industries, it has been less successful in moving into new ones.

European governments have assumed much greater responsibility than those in Japan or the United States for providing steady increases in standards of living, and a much greater degree of job tenure is provided in Europe than is common in the United States. In the 1950s and 1960s these guarantees were relatively costless, for rapid growth in demand facilitated job retention, while rising productivity growth made higher wages affordable. With the shocks and slow growth in the 1970s, however, governments were forced to make good on the guarantees. Partly because they were backstopped by generous social payments by schemes such as indexation, growth of European real wages exceeded the pace warranted by changes in productivity and the terms of trade. This squeezed profits, discouraged investment, and slowed growth. While manufacturing employment declined, the service sectors in Europe were unable to provide employment for new labor force entrants and those displaced from manufacturing. By contrast, the slowdown in the growth of U.S. manufacturing employment opportunities was more than matched by the expansion in services.

Whereas European unemployment rates have been considerably lower than those in the United States for most of the period since World War II, by 1982 the average unemployment rates in the United States and the European community (EC9) were 9.7 and 9.5 percent, respectively. Structural unemployment, however, seems much higher in Europe. According to the OECD, in the United States in 1982 about 16.6 percent of the unemployed had been without a job for more than six months. By contrast, in Germany, France, and the United Kingdom, the long-term unemployed were 38.1, 55.8, and 45.7 percent, respectively, of the total unemployed. In 1979, males over the age of 45 constituted 36 percent of all unemployed German males, whereas in the United States, older males were 17 percent of the unemployed. Similarly, older women were 29 percent of the unemployed in Germany, and 15 percent of those in the United States.

## The Importance of Relative Prices

Compared with its postwar track record, since 1973 the U.S. manufacturing sector has fared relatively well in comparison with other industrial countries. This might have been expected, given the relative exhaustion of catch-up gains that others could enjoy by adopting U.S. techniques. The U.S. performance may also be ascribed to its greater flexibility in a period marked by external shocks. In particular, U.S. real wage growth has been more adaptable and U.S. labor more mobile. The American share of manufacturing employment in high-growth industries has increased more rapidly than that of Germany or Japan. There are, therefore, strengths as well as weaknesses in the U.S. industrial system.

Flexible exchange rates have been important to U.S. trade performance. From 1973 to 1980, partly because of the real devaluation of the dollar, foreign trade provided a net addition to output and jobs in U.S. manufacturing. From 1980 to 1982, the erosion in relative price competitiveness has been the source of the declines in employment due to manufactured goods trade. Changes in the real exchange rate are effective in moving the current account toward equilibrium determined by expenditure patterns. In 1970 and 1980, the current account was a similar percentage of GNP. This stability was achieved in part by growth in the manufactured goods trade balance because of real devaluation. In the 1980s, the shift toward large full-employment government deficits unmatched by lower private absorption entails a current account deficit as foreign savings help finance the government deficit. This is accomplished in part by a manufactured goods trade deficit achieved through real appreciation. If these trade deficits are viewed as undesirable, policies to lower full-employment government deficits should be considered.

The decline in the manufactured goods trade balance over the past two years is not the result of a sudden erosion in U.S. international competitiveness brought about by foreign industrial and trade policies. It is predictable, given previous trends and current levels of economic activity and relative prices. Given a continuation of trends in U.S. and foreign trade policies and growth patterns, in the absence of relative price changes, the U.S. trade balance in manufactured goods would register small annual declines. If required for overall external equilibrium, these declines could be offset by minor improvements in relative U.S. prices.

There has not been increased turbulence in the demand for industrial workers across manufacturing industries. The recent rise in dislocation is principally related to the slow overall growth in employment rather than an increase in structural change at any given growth rate.

The perceptions of an absolute decline in the U.S. industrial base and the belief that foreign competition has made a major contribution to that decline stem from the reinforcing effects of U.S. trade and domestic growth and the nature of adjustment difficulties associated with declines in industries adversely affected. The troubled industries are large and highly unionized, and the average plant size is large. Workers displaced from several of these industries face the prospect of considerably lower wages.

The U.S. comparative advantage in unskilled-labor and standardized capital-intensive products has been declining secularly. Additionally, because of slow domestic economic growth, the home market for those products has not expanded rapidly, but the U.S. comparative advantage in high-technology products has strengthened while the demand for high-technology products has grown relatively more rapidly in a climate of stagnation. In general, however, structural changes in the U.S. economy during this period arose mainly from domestic factors.

Let me offer an overall conclusion based upon this evidence. If changes in industrial policy are adopted, they should be made on the grounds that they improve productivity and stimulate economic growth. They should not be undertaken because of fears, based largely on confusion about the sources of economic change, that policies that appear inadvisable on domestic grounds are required in order to compete internationally.

# 4

# In Support of the Deindustrialization Thesis

*Barry Bluestone*

I n spite of the current economic recovery, there is widespread agreement
among economists that high levels of unemployment are becoming a fix-
ture of modern industrial society. Unemployment in the United States
averaged 4.5 percent during the 1950s and only slightly higher, 4.8 percent,
during the following decade. In the 1970s it rose to 6.2 percent, but this
appeared to be merely prologue to rates that now may remain in the 7.0 to 8.0
percent range. Even the most optimistic forecasts indicate that jobless rates will
not fall below 6 to 7 percent for the remainder of the current decade.

A significant, but not precisely known, share of unemployment has been
attributed to a phenomenon now popularly termed *deindustrialization*—a widespread, systematic disinvestment in the nation's basic industrial capacity. Using
data from Dun & Bradstreet on the manufacturing sector, an establishment with
100 employees or more had only a 70 percent probability of surviving to the
year 1976, conditional on its operation in 1969. Using an updated version of the
same Dun & Bradstreet file, Candee Harris and her colleagues at the Brookings
Institution have found comparable results for 1978–82. Schmenner has demonstrated a similar phenomenon among establishments owned by 410 of the
largest manufacturing corporations in the United States. Between 1970–72 and
1978, these large corporations relocated, shut down, or divested over 21 percent
of the 12,000 or more establishments they owned and operated at the beginning
of the period. Of this number, more than half were simply shut down (8.4
percent) or relocated (3.7 percent) rather than sold to new owners. These same
corporations opened 1600 new plants and acquired nearly 3400 during this
period, but for the most part they were in new industries and different regions,
providing little employment opportunity for those immediately affected by the
closings.

Barry Bluestone, "Is Deindustrialization a Myth? Capital Mobility versus Absorptive Capacity in
the U.S. Economy," *The Annals*, American Academy of Political and Social Science 475 (Sept.
1984), pp. 39–51.

Deindustrialization has been said to be particularly acute in metropolitan areas where, according to Varaiya and Wiseman, only a fraction of the investment needed to stabilize existing production-worker employment took place in the 1970s. Controlling for the vintage of the capital stock in the manufacturing sectors of 77 standard metropolitan statistical areas (SMSAs), they found that in 1976 it would have taken an additional $16 billion in investment to maintain production employment in these urban areas. Nationwide the capital spending rate would have had to triple to maintain urban manufacturing employment at existing gross investment/employment ratios. Because of the higher capital retirement rates in the Northeast and North Central regions, investment fell short of employment stabilizing levels by as much as 85 percent in the older SMSAs. Overall, the investment rate was below the level needed to maintain manufacturing employment in 60 of the 77 urban areas.

One can presume that similar evidence prompted the editors of *Business Week* in 1980 to declare with uncharacteristic alarm,

> The U.S. economy must undergo a fundamental change if it is to retain a measure of economic viability let alone leadership in the remaining 20 years of this century. The goal must be nothing less than the reindustrialization of America. A conscious effort to rebuild America's productive capacity is the only real alternative to the precipitous loss of competitiveness of the last 15 years, of which this year's wave of plant closings across the continent is only the most vivid manifestation.[1]

## Is Deindustrialization a Myth?

Any cursory analysis of the steel, automobile, textile, apparel, or footwear industry, or simply a field visit to Detroit, Buffalo, Youngstown, or Akron, would seem to leave little doubt that capital investment has been insufficient to maintain basic industry or mitigate the apparent abandonment of entire communities. Yet in the past year the claim of deindustrialization has come under intense scrutiny with the result that some researchers now suggest that deindustrialization is, in fact, not occurring and therefore does not pose a serious problem. One of the most careful studies in this regard, undertaken by Robert Z. Lawrence of the Brookings Institution, concludes that U.S. deindustrialization is simply "a myth." A more journalistic account by a *Forbes* magazine correspondent, James Cook, suggests that deindustrialization is, for the most part, irrelevant because the United States has been a service economy, not an industrial manufacturing society, for at least 40 years. "Most of the breast-beating about our industrial decline," Cook contends "is a kind of masochism—much like the hysteria in 18th-century Britain that building roads fit for stagecoaches would weaken the national fiber and lead to a decline in equestrian skills."[2]

Lawrence's argument is based primarily on aggregate output, investment, and employment trends and on comparisons with manufacturing activity in other developed countries. Essentially, for Lawrence, whether deindustrialization has or has not taken place depends on whether the absolute number of manufacturing jobs has declined and whether the United States is deindustrializing faster or slower than the Europeans and the Japanese.

If we rely on this definition, the data clearly contradict the deindustrialization hypothesis. Indeed, between 1973 and 1980, U.S. manufacturing output increased in real terms by 12.3 percent. In Germany the increase was only 8.9 percent; in France, 9.9 percent; and in all of the Organization for Economic Cooperation and Development (OECD) countries, 11.8 percent. Only Japan's output growth of 22 percent exceeded that of the United States, prompting the conclusion that it is Europe rather than the United States that is undergoing a process of deindustrialization. U.S. output growth in food, textiles, apparel, chemicals, glass, and fabricated metal products was more rapid than that of either Germany or Japan while U.S. growth in nonelectrical machinery, electrical equipment, and professional and scientific equipment outstripped that of Germany.

Data on capital expenditures in the manufacturing sector also seem to question the deindustrialization phenomenon. Between 1973 and 1979, purchases of plant and equipment increased at an average annual rate of 3.8 percent from $26.9 billion to nearly $35 billion, in 1972 dollars. After accounting for depreciation, new net investment increased even more rapidly, averaging 6.9 percent per year.

Even more striking is the alleged evidence on employment. Lawrence expresses it bluntly: the United States was the only OECD country to have a positive growth rate in aggregate manufacturing work hours between 1973 and 1980—seven-tenths of 1 percent per year. France, Germany, England, Denmark, the Netherlands, and Sweden all experienced annual percentage declines in manufacturing hours of 2 percent or more. Even Japan reduced its manufacturing time by nearly three-fourths of a percentage point. More rapid growth in productivity in Japan and Europe, combined with the relatively slower growth in European output, is responsible for what appears to be worker deindustrialization everywhere but the United States.

The Brookings research also questions the widely held belief that the growth in foreign trade has promoted a decline in U.S. employment. Between 1970 and 1980, Lawrence notes, the volume of U.S. manufactured goods increased by 101.5 percent while the growth in manufactured imports rose by only 72 percent. Thus it is entirely possible that U.S. export trade more than offset any job loss associated with imports. Moreover, using a 1972 input-output model, Lawrence attributes most of the decline in employment in such industries as automobile and steel to reduced domestic absorption rather than imports. Between 1973 and 1980, according to his calculations, 20.4 percent of the decline in automobile output can be traced to a decline in aggregate sales while only 3.5 percent is due to changes in the net trade balance. Hence whatever deterioration has occurred in employment levels in basic manufacturing presumably can be attributed to changes in domestic activity. This, in turn, is a result of depressed aggregate demand rather than any long-term, secular decline in these industries. Thus it is assumed that continued macroeconomic growth in the economy will be sufficient to solve the unemployment problem.

## Deindustrialization Once Again

Neither the data nor the aggregate trends reported by Lawrence can be disputed on face value. Although the proportion of the total work force in manufacturing declined from 26.2 to 22.1 percent between 1973 and 1980, total manufacturing employment in the latter year—20.2 million—remained almost identical to its 1973 level. Relative to services and trade, manufacturing employment declined precipitously, but in absolute aggregate terms, the claim of deindustrialization does not seem to be sustained.

There is, however, a serious error in this formulation of the problem. From a social efficiency or social cost perspective, the aggregate trend in employment is inadequate to prove or disprove deindustrialization if interindustry and interregional worker mobility is insufficient to clear labor markets. What counts in an economy where mobility is imperfect are the trends in specific industries and regions. There is, for example, no disputing the fact that worldwide employment in manufacturing is expanding rapidly, but if it is declining sharply in the United Kingdom, the growth in other countries would not in any serious way offset the costs imposed on Britain.

Likewise private and social costs are imposed on workers and communities within the United States to the extent that those dislocated from declining industries in particular regions cannot find employment in equally productive jobs in other sectors. The level of imposed costs is a positive function of the rate at which employment is declining in particular sectors and regions and a negative function of the economic system's capacity to absorb dislocated workers into other areas of the economy. For this reason the velocity of sectoral and regional specific deindustrialization and the overall absorptive capacity of the economy are the proper phenomena to study. It is in this regard that deindustrialization is no myth.

## Sectoral and Regional Deindustrialization

The actual employment performance of key sectors of the economy is disclosed in table 4-1. Although the flat trend, as opposed to a declining trend, in total

**Table 4-1**
**Change in Total Employment and Number of Production Workers in the United States, by Industry, 1960-80**

| | Industry | Total Employment (percentage) | | Production Workers (percentage) | | Production Worker Average Wage (dollars) |
|---|---|---|---|---|---|---|
| | | 1960-73 | 1973-80 | 1960-73 | 1973-80 | 1980 |
| 20000 | Total manufacturing | 16.7 | .13 | 17.9 | -4.7 | 7.27 |
| 30000 | Durable goods | 25.7 | 2.0 | 23.2 | -3.8 | 7.75 |
| 40000 | Nondurable goods | 12.6 | -2.4 | 9.9 | -5.9 | 6.55 |
| SIC 33 | Primary metals | 6.3 | -9.7 | 5.6 | -13.5 | 9.77 |
| SIC 35 | Machinery (except electrical) | 41.2 | 19.9 | 35.3 | 14.9 | 8.00 |
| SIC 361 | Electrical distribution equipment | 14.8 | -11.2 | 29.2 | -16.0 | 6.96 |
| SIC 362 | Electrical industrial apparatus | 32.3 | -.5 | 40.8 | -3.8 | 6.91 |
| SIC 363 | Household appliances | 27.3 | -17.9 | 31.3 | -18.1 | 6.95 |
| SIC 365 | Radio/TV receivers | 42.6 | -27.2 | 46.5 | -30.9 | 6.42 |
| SIC 367 | Electronic components and accessories | 75.9 | 25.6 | 60.4 | 17.7 | 6.05 |
| SIC 371 | Motor vehicles | 34.9 | -20.3 | 57.6 | -25.3 | 9.85 |
| SIC 372 | Aircrafts and parts | -15.8 | 24.6 | -23.1 | 24.6 | 9.28 |
| SIC 38 | Instruments and related products | 33.1 | 27.6 | 27.9 | 22.4 | 6.80 |
| SIC 22 | Textile mill products | 9.2 | -15.4 | 6.1 | -16.2 | 5.07 |
| SIC 23 | Apparel and other products | 16.6 | -12.8 | 13.8 | -10.5 | 4.56 |
| SIC 28 | Chemicals and allied products | 25.3 | 6.3 | 19.7 | 2.1 | 8.30 |
| SIC 301 | Tires and tubes | 25.1 | -12.6 | 25.1 | -16.4 | 9.74 |
| SIC 314 | Footwear | -24.6 | -22.0 | -26.8 | -23.2 | 4.42 |
| SIC 531 | Department stores | 87.1 | 5.8 | 86.6 | 7.6 | 4.95 |
| SIC 58 | Eating and drinking establishments | 84.6 | 51.2 | — | 48.7 | 3.69 |
| SIC 6000-6999 | Finance, insurance, and real estate | 53.9 | 26.6 | 45.5 | 24.3 | 5.79 |
| SIC 7000-8999 | Services | 74.3 | 37.8 | — | 35.8 | 5.85 |
| | Total Employment | 41.7 | 16.9 | | | |

*Source:* U.S., Department of Labor, Bureau of Labor Statistics, *Employment and Earnings Statistics for the United States, 1909-1980* (Washington, D.C.: Government Printing Office, 1981).

manufacturing employment is confirmed by the small (+0.13 percent) change in the number of jobs between 1973 and 1980, production workers did not fare anywhere near as well, and employment in certain key sectors fell sharply.

Using regression equations that controlled for deviations from potential gross national product, estimates of potential employment levels for the year 1980 were projected under the assumption of prior (1960–72) trends in employment growth after accounting for the recession conditions in 1980. These are reported as ratios of actual to projected employment in table 4–2 along with the actual 1973–80 change in total and production worker employment.

For total manufacturing, the ratio appears to confirm the Brookings conclusion. In fact, manufacturing employment after 1973 was slightly above its secular trend line once the business cycle effect is statistically included. Total employment was 6.6 percent higher than what might have been expected given 1980 business conditions.

**Table 4–2**
**Ratio of Actual Employment to Projected Trend Employment in 1980**

| | | Actual Percentage Change, 1973–80, in | |
| | | | Production Worker |
| Ratio | Industry | Employment | Employment |
|---|---|---|---|
| 1.853 | Aircraft and parts | +24.6 | +24.6 |
| 1.461 | Electronic components | +25.6 | +17.7 |
| 1.318 | Instruments | +27.6 | +22.4 |
| 1.266 | Machinery (except electrical) | +19.9 | +14.9 |
| 1.168 | Eating and drinking places | +51.2 | +48.7 |
| 1.134 | Durable manufacturing | +2.0 | −3.8 |
| 1.046 | Primary metals | −9.7 | −13.5 |
| 1.066 | All manufacturing | +.1 | −4.7 |
| 1.028 | Services | +37.8 | +35.8 |
| 1.023 | Electrical industrial equipment | −.5 | −3.8 |
| .990 | Finance, insurance, and real estate | +26.6 | +24.3 |
| .980 | Chemicals | +6.3 | +2.1 |
| .975 | Nondurable manufacturing | −2.4 | −5.9 |
| .974 | Radio/TV receivers | −27.2 | −30.9 |
| .939 | Electrical distribution equipment | −11.2 | −16.0 |
| .936 | Footwear | −22.0 | −23.2 |
| .934 | Apparel | −12.8 | −10.5 |
| .920 | Textile mill products | −15.4 | −16.2 |
| .898 | Motor vehicles | −20.8 | −25.3 |
| .810 | Household appliances | −17.9 | −18.1 |
| .792 | Tires and inner tubes | −12.6 | −16.4 |
| .784 | Department stores | +5.8 | +7.6 |

*Source:* The data on employment change are calculated from U.S., Department of Labor, Bureau of Labor Statistics, *Employment and Earnings in the United States, 1947–1980* (Washington, D.C.: Government Printing Office, 1982).

*Note:* The ratio is calculated from regression equations that control for cyclical components in the employment series. A ratio value of 1.000 indicates an industry in which the long-term trend in employment between 1973 and 1980 is identical to the trend that existed between 1960 and 1972. A ratio value greater than 1.000 indicates an industry where the post-1972 employment growth was higher than the 1960–72 trend. A ratio value less than 1.000 indicates an industry where employment in the later period was below the trend line for the earlier period.

Within eleven of nineteen individual industries, however, employment was below trend, and in four of these—motor vehicles, household appliances, tires and inner tubes, and department stores—actual employment was no more than 90 percent of its projected value. Virtually all of the industries with an actual/ projected ratio less than .98 experienced declines in measured total employment and in the number of production workers. The manufacturing sectors that appear to have become deindustrialized during the 1970s were:

Tires and inner tubes

Household appliances

Motor vehicles and parts

Textile mill products

Apparel

Footwear

Electrical distribution equipment

Radio/TV receivers and

Chemicals.

There were some surprises. Despite the widespread concern with the recent collapse of the steel industry, the actual loss in jobs during the 1970s was 4.6 percent less than the long-run trend would have suggested. This is explained by the fact that even during the 1960s, the trend in primary metals employment was flat and all of the variance was due to cyclical demand variables. Deindustrialization in the steel industry simply began earlier than in other industries. On the other hand, department store employment was nearly 22 percent below trend. After experiencing the most rapid growth rate of all industries between 1960 and 1972, the boom in department store jobs came to an abrupt halt in the 1970s with the saturation of most retail markets.

Up to this point, the analysis has been presented in terms of national aggregates. Yet it is precisely within particular regions that much of the dramatic employment activity is taking place. This can be illustrated by tracing employment trends in four key frostbelt states and four large sunbelt states: respectively, Massachusetts, New York, Michigan, and Ohio, and North Carolina, Georgia, California, and Texas.

Table 4–3 presents data on the percentage change in total employment between 1973 and 1980 in major industries in these states. A sharp decline in basic manufacturing is clearly evident in Michigan, Ohio, and New York, where the total manufacturing job loss ranges from 10 to 17 percent. On net, Michigan lost over 200,000 manufacturing jobs in this eight-year period, nine-tenths of them in durables. Ohio and New York each experienced a net loss of

**Table 4–3**
**Change in Total Manufacturing Employment in Selected Industries in Selected States: 1973–80**
(*percentage*)

| Industry | United States | Massachusetts | New York | Michigan | Ohio | Georgia | North Carolina | Texas | California |
|---|---|---|---|---|---|---|---|---|---|
| Total Manufacturing | 0.13 | 6.4 | -10.3 | -17.3 | -11.0 | 3.3 | 2.7 | 31.5 | 20.6 |
| Durable goods | 2.0 | 20.0 | -4.8 | -19.0 | -13.1 | 7.8 | 17.2 | 43.0 | 23.3 |
| Nondurable goods | -2.4 | -9.6 | -15.4 | -10.0 | -5.9 | 0.9 | -4.1 | 17.8 | 15.1 |
| Selected Industries | | | | | | | | | |
| Primary metals | -9.7 | — | -24.4 | -27.7 | -20.0 | — | — | 27.4 | -2.2 |
| Fabricated metals | -1.1 | — | -10.0 | -22.9 | 10.4 | 4.6 | 26.3 | 29.1 | 16.4 |
| Nonelectrical machines | 19.9 | 42.8 | 6.3 | -7.3 | -2.0 | 36.6 | 36.8 | 77.2 | 43.9 |
| Electrical equipment | 13.6 | 22.6 | -1.1 | -14.9 | -19.2 | 30.9 | 17.1 | 88.2 | 45.0 |
| Transportation equipment | 11.0 | 7.8 | -12.7 | -22.8 | -18.6 | 3.8 | 101.2 | 23.4 | 5.8 |
| Instruments | 27.6 | 24.0 | 4.4 | 46.1 | -3.7 | — | — | 43.8 | 60.3 |
| Textiles | -15.4 | -16.7 | -34.1 | — | — | -8.0 | -15.0 | -26.6 | — |
| Apparel | -12.8 | -10.0 | -22.3 | — | -22.3 | -3.7 | 1.8 | 8.2 | 17.5 |
| Chemicals | 6.3 | -10.8 | -6.5 | 5.5 | 9.8 | 13.4 | 8.5 | 25.6 | 16.3 |

*Sources:* U.S., Department of Labor, Bureau of Labor Statistics, *Employment and Earnings for States and Areas, 1939–78*, Bulletin 1370–13, 1979 (Washington, D.C.: Government Printing Office, 1980); and idem, *Supplement, 1977–81*, Bulletin 1370–16, 1982 (Washington, D.C.: Government Printing Office, 1980).

over 150,000. In contrast, California increased its manufacturing base by over one-fifth during this sluggish economic period, while Texas increased its base by nearly one-third and its durables sector by 43 percent. Recall that nationwide, net manufacturing employment increased by a mere 0.13 of 1 percent.

Regional shifts in the location of particular industries are notable. Michigan lost nearly 28 percent of its primary metals industry and 23 percent of its jobs in fabricated metal operations; Texas, on the other hand, enjoyed 27 and 29 percent growth in these two sectors. Similar shifts, often of even greater magnitude, are found in nonelectrical machinery, electrical and electronic apparatus, and transportation equipment. The move to the sunbelt of high-value-added/high-wage durable manufacturing jobs is hardly a myth.

## Capital Velocity, Absorptive Capacity, and Structural Unemployment

Properly analyzed, then, the statistics reveal not an absolute decline in manufacturing output, investment, and employment, but a substantial amount of capital mobility from one sector to another and from region to region. The problem is economic turbulence. In the spirit of Joseph Schumpeter's notion that economic progress proceeds through a process of "creative destruction," there are many who claim that such capital restructuring is fundamental to the health of the economy. It presumably frees up scarce resources from lower-value production for use in higher-value activities. In fact, some argue that America's malaise is the result of too little capital mobility, not too much.

There is a serious flaw in this argument. More capital mobility is likely to be optimal only if the resources freed up are reemployed in activities of equal or greater productivity. If productive resources are left idle as a consequence of capital restructuring, there are potential efficiency losses and serious questions of equity. In essence, an increase in the velocity of capital mobility requires a sufficient improvement in the absorptive capacity of the economy to assure that the process of creative destruction is, on net, beneficial to society. To a great extent, the entire deindustrialization issue turns on this point, for deindustrialization, like poverty, is relative. It takes on meaning only once it is considered in relation to the reabsorption process.

Whether the absorptive capacity of U.S. labor markets is sufficient is an empirical question. Here the data on the reemployment outcomes of displaced manufacturing workers suggest that for many workers reabsorption is highly problematic. Those who are displaced from such high-value-added/high-wage industries as aircraft, steel, and motor vehicle production run an extremely high risk of suffering permanent losses in their earnings streams.

Each worker's loss in earnings after displacement is a function of what new employment opportunity is available. This is well illustrated by an analysis of displaced New England aircraft industry workers carried out with the LEED file at the Social Welfare Research Institute at Boston College. Between 1967 and 1972, 31 percent of the workers in this industry were displaced as a result of the sharp downturn in this sector and a substantial increase in subcontracting to other regions. Of the 18,300 displaced, 600 were able to locate new jobs in aircraft, but only by migrating out of New England. Sixty-five percent (11,900) located jobs in other primary-sector industries, 11.5 percent (2,100) found jobs in secondary-sector industries, and 20.2 percent (3,700) either found no job at all or worked outside the Social Security system.

The results of this analysis are reported in table 4–4. Those who stayed in the aircraft industry by migrating to other regions had only 78 percent as much nominal earnings growth as those who were able to keep their New England aircraft jobs. Those forced into other primary-sector industries, including most durable manufacturing, wholesale trade, and public utility industries, experienced only 33 percent as much earnings growth. Finally, the more than 1 in 9 relegated to secondary-sector industries of nondurable manufacturing, retail trade, and personal services experienced an absolute 26 percent earnings loss. For them annual earnings in nominal terms fell from an average of $6054 to $4468. After controlling for inflation, these workers earned in 1972 only 59 percent of their 1967 aircraft wages.

Further analysis of the LEED file suggests that downward mobility into the secondary sector is not at all uncommon. This can be seen by following the job mobility patterns of the 833,200 workers in New England whose principal activity in 1958 was to work in traditional mill-based industries, for example,

## Table 4–4
## Earnings Trajectories of Those Displaced from the New England Aircraft Industry (1967–72)
*(in current dollars)*

| New Employment | Number | Percentage | Earnings Growth (1967–72) as a Percentage of Earnings Growth of Continuously Employed New England Aircraft Workers |
|---|---|---|---|
| In aircraft, outside region | 600 | 3.3 | 78 |
| Other primary-sector industries[a] | 11,900 | 65.0 | 33 |
| Secondary sector industries[b] | 2100 | 11.5 | absolute 26% earnings loss |
| Not covered by Social Security[c] | 3700 | 20.2 | — |
| Total | 18,300 | 100.0 | — |

*Source:* Special tabulations of Social Security LEED file prepared by Alan Matthews and Barry Bluestone, Social Welfare Research Institute, Boston College, Sept. 1979.

[a]Primary industries include most durable manufacturing, wholesale trade, public utilities, and some services.

[b]Secondary industries include most nondurable manufacturing, retail trade, and lower-skill requirement, higher-turnover personal services.

[c]Not covered by Social Security includes those who were no longer in the labor force in 1972 or who worked in jobs not covered by Social Security.

apparel, textiles, and shoes. In the period after 1958, 674,000 left the mills. Of this number, only 18,000, or less than 3 percent, were able to locate jobs in the growing high-technology sector in the region by 1975. Another 2000 had migrated to high-tech jobs in other states. But more than five times as many—106,000—ended up in service and retail trade industries, almost all of which paid significantly lower wages.

The general decline in earnings following dislocation is to a great extent a function of the relative earnings levels in what are the currently growing and declining industries. Although there has been substantial employment growth in some higher-wage sectors (for example, nonelectrical machinery and aircraft and parts), many of the most rapidly growing industries are in the lower-paying manufacturing and nonmanufacturing sectors. Employment in the electronic components industry rose by 75.9 percent between 1960 and 1972 and then by another 25.6 percent between 1973 and 1980. But the average production worker at $6.05 per hour in 1980 earned a weekly salary only 61 percent as high as that of an average employee in the primary metals industry. Essentially, 163 electronic components jobs were needed to compensate for the wage-bill loss of 100 steelworkers. Similarly, it takes two department store jobs or three restaurant jobs to make up for the earnings loss of just one average manufacturing position.

The overall decline in real U.S. gross average weekly earnings by nearly 13 percent between 1973 and 1980 reflects the changing composition of jobs in the economy much more than inflation. In manufacturing, real earnings declined by only 8 percent, and in industries like auto and steel, where there is substantial negotiated cost-of-living protection, the loss was minimal. Sector-specific deindustrialization therefore can seriously erode the size of the real wage bill even when aggregate employment in manufacturing remains constant.

Over the next decade, the national unemployment rate may fall as more jobs in the service and trade sectors are created, but the decline in unemployment may not do very much for standards of living, because many of the new jobs pay significantly less than those that are disappearing.

## Conclusion

The significance of deindustrialization can therefore be evaluated only in terms of how rapidly and how successfully workers dislocated from so-called sunset industries are reemployed in growing, sunrise industries. The absolute magnitude of output, investment, and employment in manufacturing—or the extent to which employment is ahead of or behind that in other developed nations—is by itself not a very useful measure of deindustrialization.

The proper measure is a function of how rapidly employment is shrinking in certain sectors of the economy and how rapidly workers are being reabsorbed into equivalent jobs. The data on sector- and region-specific dislocation combined with the information on the downward mobility of job losers strongly suggest that the velocity of capital mobility and the absorptive capacity of the economy are not well synchronized. To rectify this problem requires either a massive effort at improving absorptive capacity through retraining and migration, a slowing of dislocation through various sectoral and regional industrial policies, or both. In this regard, deindustrialization is neither a myth nor a trivial social and economic problem.

## Notes

1. "The Reindustrialization of America," *Business Week*, special issue (June 30, 1980), p. 58.

2. James Cook, "You Mean We've Been Speaking Prose All These Years?" *Forbes* (April 11, 1983), p. 143.

# 5
# Industrial Policy

*Lee Iacocca*

Right now, our biggest industrial employers are in autos, steel, electronics, aircraft, and textiles. If we want to save millions of jobs, we've got to preserve these industries. They're the ones that create markets for the service sector as well as for high technology. They're also critical to our national interest. Can we really maintain the backbone of our defense system without strong steel, machine tool, and auto industries?

Without a strong industrial base, we can kiss our national security goodbye. We can also bid farewell to the majority of our high value-added jobs. Take away America's $10 to $15-an-hour industrial jobs and you undercut our whole economy. Ooops—there goes the middle class!

So we've got to make some basic decisions. Unless we act soon, we're going to lose both steel and autos to Japan by the year 2000. And worst of all, we will have given them up without a fight.

Some people seem to think that this defeat is inevitable. They believe we should even hasten along the process by abandoning our industrial base and concentrating instead on high technology.

Now, I don't for a moment dispute the importance of high technology in America's industrial future. But high tech alone won't save us. It's important to our economy precisely because so many other segments of American industry are its customers.

Especially the auto industry. We're the ones who use all the robots. We've got more computer-aided design and manufacturing facilities than anyone. We're using computers to get better fuel economy, to clean up emissions, and to get precision and quality in the way we build our cars.

Not many people know that the computer industry's three biggest customers (excluding defense) are GM, Ford, and Chrysler. There can't be a Silicon Valley without a Detroit. If somebody is producing silicon chips, somebody else has to use them. And we do. There's now at least one computer on board every car we build. Some of our more exotic models have as many as eight!

Lee Iacocca, with William Novak, *Iacocca: An Autobiography* (New York: Bantam Books, 1984), pp. 327–330, 332–334.

You can't sell your silicon chips in a brown paper bag down at the hardware store. They've got to have a use. And America's basic industries are the users. Close us down and you close down your market. Close down autos and you close down steel and rubber—and then you've lost about one of every seven jobs in this country.

Where would that leave us? We'd have a country of people who serve hamburgers to each other and silicon chips to the rest of the world.

Don't get me wrong: high technology is critical to our economic future. But as important as it is, high tech will never employ the number of people that our basic industries do today. That's a lesson we should have learned from the demise of the textile industry. Between 1957 and 1975, 674,000 textile workers were laid off in New England. But despite that region's booming high-tech industries, only 18,000 of those workers—about 3 percent—found work in the computer industry.

Nearly five times as many ended up in lower-wage retail trade and service jobs. In other words, if you lost your job in a textile mill in Massachusetts, you were five times as likely to end up working at K-Mart or McDonald's than at Digital Equipment or Wang. You just can't take a forty-year-old pipefitter from Detroit or Pittsburgh or Newark, put a white coat on him, and expect him to program computers in Silicon Valley.

So the answer isn't to promote high technology at the expense of our basic industries. The answer is to promote *both* of them together. There's room for all of us in the cornucopia, but we need a concerted national effort to make it happen.

In other words, our country needs a rational industrial policy.

These days, "industrial policy" is a loaded term. It's like yelling "fire!" in a crowded theater. A lot of people panic whenever they hear the phrase.

Don't they want America to be strong and healthy? Sure they do. But they want it to happen without any planning. They want America to be great by *accident*.

The ideologues argue that industrial policy would mark the end of the free enterprise system as we know it. Well, our wonderful free enterprise system now includes a $200 billion deficit, a spending program that's out of control, and a trade deficit of $100 billion. The plain truth is that the marketplace isn't always efficient. We live in a complex world. Every now and then the pump has to be primed.

Unlike some people who talk about industrial policy, I don't mean that the government should be picking winners and losers. The government has proved again and again that it's not smart enough to do that.

And I don't want the government interfering in the operations of my company—or any other company, for that matter. Believe me, the existing regulations are bad enough.

As I see it, industrial policy means restructuring and revitalizing our so-called sunset industries—the older industries that are in trouble. Government must become more active in helping American industry meet the challenge of foreign competition and a changing world.

Almost everyone admires the Japanese, with their clear vision of the future; the cooperation among their government, banks, and labor; and the way they lead from their strengths. But whenever somebody suggests that we ought to follow their lead, the image suddenly shifts to the Soviets and their five-year plans.

But government planning doesn't have to mean socialism. All it means is having a game plan, an objective. It means coordinating all the pieces of economic policy instead of setting it piecemeal, in dark rooms, by people who have only their own vested interests at heart.

Is planning un-American? We do a great deal of planning at Chrysler. So does every other successful corporation. Football teams plan. Universities plan. Unions plan. Banks plan. Governments all over the world plan—except for ours.

We're not going to make progress until we give up the ridiculous idea that any planning on a national level represents an attack on the capitalist system. Because of this fear, we're the only advanced country in the world without an industrial policy.

Here's my six-point program that could form the basis for a new industrial policy.

First, we should provide for energy independence by 1990 by taxing foreign energy, both at the port and at the pump, in order to restore the conservation ethic and rekindle investments in alternate sources of energy. We must not be lulled by the current depressed demand. OPEC will always act in its own interest, and that interest will always be served best by high prices and tight supplies. The American people are willing to pay a price for energy independence. They know it can't be achieved without a sacrifice.

Second, we should provide for specific limits to Japan's market share for certain critical industries. We should declare a state of economic emergency for those industries and unilaterally set aside the restrictive General Agreement on Tariffs and Trade (GATT) provisions during this period. We don't have to apologize for taking this commonsense approach to trade with Japan. At this point in our history, we can't afford a trading partner who insists on the right to sell but who refuses to buy.

Third, as a nation, we've got to face reality on the costs and funding mechanisms for federal entitlement programs. They're studying this to death in Washington because it's a political hot potato. But the answer has always been right in front of our noses: we can't continue to pay out more than we take in, and that will mean some very painful adjustments.

Fourth, America needs more engineers, scientists, and technicians. On a per capita basis, Japan graduates about four times as many engineers as we do (but we graduate fifteen times as many lawyers!). Special education grants and loans should be provided for high-technology fields of study. The Soviets and the Japanese are both dedicated to building up their technological competence—and we are not keeping up.

Fifth, we need new incentives to increase research and development efforts in the private sector and to accelerate factory modernization and productivity in critical industries. One approach is to offer investment tax credits for R&D and twelve-month depreciation write-offs for productivity-related investments.

Finally, we need to establish a long-term program for rebuilding America's arteries of commerce—our roadways, bridges, railroads, and water systems. Our infrastructure, which is vital to any strengthening and expansion of our industrial power, is deteriorating at an alarming rate. Something must be done. Such a program could be partially funded by the OPEC energy tax. It would also provide a major buffer from the future employment dislocation that will inevitably result from productivity gains and industrial automation.

To put all these programs into practice, we should set up a Critical Industries Commission—a forum where government, labor, and management could get together to find a way out of the mess we're in. We have to learn how to talk to each other before we can take joint action.

This tripartite coalition would recommend specific measures to strengthen our vital industries and to restore and enhance their competitiveness in international markets.

Let me make clear that I am not proposing a welfare system for every company that gets into trouble. *We need a program that kicks in only when troubled American companies have agreed to equality of sacrifice among management, labor, suppliers, and financial backers.* It worked for Chrysler, and it can work for the rest of America.

When an industry or a company comes looking for help, as I did five years ago in Washington, the commission should ask on behalf of the taxpayers, who are going to take the risk: "What's in it for us?" What's in it for the people? In other words, "What are management and labor bringing to the party?"

I've lived through this, and it's simple. It's management agreeing to do something *before* the government does *anything*—such as loan guarantees, or import restraints, or investment tax credits, or R&D help. Management might have to agree to plow back its earnings into job-creating investments—in *this* country. It might have to agree to profit sharing with its employees. It might even have to agree to keeping a lid on prices.

As for the unions, they would have to come out of the dark ages. They'd have to agree to changes in the many work rules that hamper productivity, such as having 114 job classifications in assembly plants where about 6 would do fine. They might even have to agree to restraints on the runaway medical costs that are now built into our system.

If neither management nor labor is going to make sacrifices, then the meeting's over. You can't expect to get government help if you're not willing to get your own house in order. In other words, there's no free lunch. Whoever applies for assistance will have to understand that there are strings attached.

If all of this sounds a little like a Marshall Plan for America, that's exactly what it is. If America could rebuild Western Europe after World War II, if we could create the International Monetary Fund and a dozen international development banks to help rebuild the world, we ought to be able to rebuild our own country today. If the World Bank, which is a profit-making institution, can successfully help out underdeveloped countries, why couldn't a new national development bank do as well in helping out troubled American industries?

Maybe what we need is an *American* Monetary Fund. What's so terrible about a $5 billion national development bank to get our basic industries competitive again?

Early in 1984, the Kissinger Commission requested $8 billion for the economic development of Central America. Now, I always thought that Central America meant places like Michigan, Ohio, and Indiana. (Shows you how simpleminded I am!) What about *our* Central America? How can we spend $8 billion to strengthen the economies of other countries while neglecting ailing industries in our own backyard?

Some people say that an industrial policy is nothing more than lemon socialism. If it is, I'll take a crateful—because unless we act fast, our industrial heartland is going to turn into an industrial wasteland.

# Part II
# Impact of Plant Closure

A s indicated in Part I, the question of whether deindustrialization is actually occurring in the United States has sparked controversy. Also controversial is what to do about the problem, that is, the government involvement versus the free-market approach. There is little if any controversy, however, over the impact that plant closure has on workers and their communities. It is clear that jobs are lost and new ones, especially high-paying jobs, are difficult to find. Families suffer along with unemployed breadwinners, with increased stress a distinct byproduct of plant closure. Communities suffer a reduction in tax base and may experience a social malaise that diminishes local incentive. Thus plant closure has numerous consequences, none of them good for workers or the community.

In the first excerpt, from Barry Bluestone and Bennett Harrison's well-known book, *The Deindustrialization of America*, the impact of plant closure is portrayed graphically. They cite numerous studies and examples of the devastating effects that plant closure has on jobs, income, and health. Barry Bluestone is a professor of economics at Boston College. Bennett Harrison is a professor of political economy and planning at Massachusetts Institute of Technology.

The next selection summarizes five case studies involving the effect of plant closure on large-scale layoffs. Author Herbert Hammerman, from the Bureau of Labor Statistics, examines shutdowns that took place in the early 1960s. From 1961 to 1969, the United States had the longest sustained peacetime boom in its history. Although this overall economic prosperity may have helped to ease the burden of permanent layoff for the employees studied, many workers were still unable to find jobs. Several points are made in the reading that one will find noted in virtually all research on the impact of plant closure: (1) victims of closures are, on the average, unemployed for longer periods than others who are laid off; (2) new jobs pay significantly less than the old ones; and (3) laid-off workers are relatively immobile in the labor market, choosing instead to remain where they are for a variety of reasons such as home ownership, relocation costs, and family ties.

Candee S. Harris, Research Analyst at the Brookings Institution, next provides a quantitative examination of the extent of job loss from plant closures. This is a valuable study because it gives a good idea of the current scope of the problem nationally as well as how many new employment opportunities are being created. The industries and geographic areas that are creating most of the new jobs are shown to be different from those that are losing jobs through plant closure. This corroborates what Hammerman and others have noted, that people are being locked in geographically to regions that are declining economically. Although some analysts believe that more jobs are being created than destroyed and that people need only relocate to growing regions, Harris notes that people in the hard-hit regions are going to need greater assistance than they are getting now because the impact on many people is too intense to cope with adequately.

In the debate over deindustrialization and plant closure, one of the most critical issues is the long-term impact on workers. Researchers have interpreted existing data to back up both pro and con arguments over the need for legislation. In an attempt to develop a more accurate data base from which to evaluate impacts on displaced workers, the Bureau of Labor Statistics conducted a special survey in 1984 of workers who lost jobs between 1979 and 1983. The final article in this section, by Paul O. Flaim and Ellen Sehgal, summarizes and evaluates the resulting data. It is an important addition to the literature on plant closure because it focuses on people and how they are affected by shutdowns. It also provides the kind of detail that allows meaningful conclusions to be drawn about certain categories of displaced workers—for instance, that blacks, Hispanics, and older workers have greater difficulty regaining employment after layoff from a plant closure. Paul O. Flaim is Chief of the Division of Data Development and Users' Services, Office of Employment and Unemployment Statistics, U.S. Bureau of Labor Statistics. Ellen Sehgal is Senior Economist in the same division.

# 6
# Jobs, Income, and Health

*Barry Bluestone*
*Bennett Harrison*

A t a minimum, almost any kind of capital mobility produces some short-term, or frictional, unemployment. A new machine is brought into a plant, an old one is shipped out, and two workers are subsequently let go. If the two readily find comparable jobs, or better ones, the income loss to them and their families and the productivity loss to society are inconsequential. At the other extreme, as usually occurs when a plant like the Campbell Works in Youngstown closes down, the displacement can be devastating for some workers. The consequences are especially severe if the shutdown occurs during a recession when the competition for other jobs is fierce, or if it occurs in a small or remote community where no other jobs exist. The serious problems begin with unemployment, but they seldom stop there.

Evidence from a broad array of case studies suggests that long-term unemployment is the result of plant closings for at least one-third of those directly affected. The family income loss that accompanies it is nearly universal and often substantial. In a report prepared for the Federal Trade Commission, C & R Associates reviewed twelve case studies of factory shutdowns and reported that in all the studies reviewed the impact on the employees was severe. This is not exactly a new story. In 1961, ten months after Mack Truck abandoned its 2,700-employee assembly plant in Plainfield, New Jersey, 23 percent of its work force were still without jobs, well after unemployment benefits were exhausted. A similar proportion remained unemployed two years after the 1956 Packard plant shutdown involving 4,000 workers. Another one-third of the Packard work force found jobs after the closing but lost these within the first twenty-four months of the original plant shutdown. Having lost all their seniority, often amounting to ten years or more, these workers were vulnerable to layoffs on their new jobs.

More recent research corroborates the evidence from these earlier studies. Writing at Cornell University, Professors Robert Aronson and Robert McKersie

Barry Bluestone and Bennett Harrison, *The Deindustrialization of America: Plant Closings, Community Abandonment, and the Dismantling of Basic Industry* (New York: Basic Books, 1982), pp. 51–66.

surveyed workers who lost their jobs in upstate New York when Westinghouse, Brockway Motors, and then GAF shut down operations in their communities between November 1976 and July 1977. The researchers found that nearly 40 percent of the work force experienced unemployment of forty or more weeks, while one-quarter of the 2,800 affected workers spent a year or more without work. Their vulnerability to unemployment continued even beyond this point. Two years after the closings, 10 percent of the original sample group were still unemployed.

A plant closing during a recession is likely to be even more devastating in terms of re-employment possibilities because of the absence of jobs in other sectors. Thus when Armour and Company closed its Oklahoma City meat-packing plant during the 1960–61 economic downturn and laid off 400 of its work force, 50 percent remained unemployed for at least six months. More recently, the closing of a chemical company branch plant in Fall River, Massachusetts, during the 1975 recession resulted in unemployment that lasted on the average nearly sixty weeks, with some workers idled as long as three years. Thirty-nine percent of the workers in the sample found jobs only after their unemployment compensation had long been exhausted. Ironically, despite the obvious economic distress after the closing of Youngstown Sheet and Tube, conditions would have been much worse if employment in the automobile industry had not expanded to partially fill the void in steel. A temporary boom in output at the Lordstown complex of the General Motors Corporation cushioned the immediate blow to the Youngstown community.

These particular findings are necessarily based on a handful of plant closings, because only a tiny fraction of all shutdowns have been surveyed by social scientists. Such sobering evidence might therefore be ignored or at least criticized as unrepresentative. However, there is evidence from a nationwide sample of approximately 4,000 men, all over the age of forty-five, that confirms these findings as anything but atypical.

Herbert Parnes and Randy King followed this group over a seven-year period from 1966 to 1973. They found that almost one in twenty of this national sample, who had worked for the same employer for at least five years, experienced permanent involuntary separation during the survey period. Even among this experienced group, 20 percent remained unemployed for at least six months before finding another job. Moreover, the subsequent employment record of those separated from their jobs never regained its previous stability. At the time of the 1973 survey, at least two years after initial separation, 6 percent of the total group of displaced workers were unemployed, as compared with only 1 percent of a control group with the same demographic characteristics, but with no record of permament job termination. Seventy-six percent of the control group, but only 66 percent of the displaced group, worked all fifty-two weeks in 1973. These results are remarkably consistent with the findings of the individual plant closing studies.

The Parnes and King analysis reveals some surprises. For example, no one appears to be immune to job loss, no matter how well placed. Popular conceptions notwithstanding, displacement respects neither educational attainment nor occupational status. There is virtually no difference in educational background between those displaced and the total population at risk. Similarly, there were no substantial differences among professionals, clerical workers, operatives, and service workers in the chances of being displaced. In the words of these Ohio State researchers: "Apparently the risk of displacement from a job after reasonably long tenure is insensitive to conventional measures of human capital and to the particular occupations in which men are employed."[1] When a plant shuts down, or operations are permanently curtailed so that some workers receive layoffs without recall, engineers lose their jobs along with janitors.

This does not imply that all groups are equally vulnerable to dislocation related to capital mobility. In the Cornell study mentioned earlier, women were found to be twice as likely as men to be unemployed for longer than a year. As secondary earners, it is possible that some married women had the "luxury" of an extended job search, but for most, the problem was finding comparable employment. The latter is certainly true for nonwhite minorities, as Gregory Squires of the U.S. Civil Rights Commission notes.

Blacks are especially hard-hit because they are increasingly concentrated within central cities and in those regions of the country where plant closings and economic dislocation have been most pronounced. While blacks constituted 16 percent of all central city residents in 1960, before the recent spate of primarily northern-based shutdowns, they accounted for 22 percent of the urban population in 1975. Similarly, blacks and other people of color did not share in the suburban housing and business boom of this period. In spite of fifteen years of civil rights legislation, they were at best able to increase their share of the suburban population from 4.8 percent to 5.0 percent by 1978. Moreover, as the number of jobs grew more rapidly in the South, whites moved in to take the overwhelming majority of them.

To add to the inequity of burden, nonwhite minorities also tend to be concentrated in industries that have borne the brunt of recent closings. This is particularly true in the automobile, steel, and rubber industries. One Washington bureaucrat remarked during the hearings on the Chrysler loan guarantee that it should have been named the Coleman Young bail-out bill and filed under one of the titles of the Civil Rights Act (Young is the black mayor of Detroit). In August 1979 virtually 30 percent of Chrysler's national employment was made up of black, Hispanic, and other minorities, while over one-half of its Detroit work force was nonwhite. These groups are also at greater risk because they are more dependent than whites on wages and salaries as sources of family income. Eighty percent of minority earnings are derived from wages and salaries, compared to only 75 percent for whites.

How capital mobility can have a discriminatory impact, either intentionally or not, is shown clearly in two examples provided in Squires' work. When a laundry located in St. Louis began to decentralize in 1964, its work force was 75 percent black. By 1975 after it had opened up thirteen suburban facilities and reduced its downtown operations, its black work force was down to 5 percent. In 1976 a Detroit manufacturer relocated production to its facility in a rural county just over the state line in Ohio. Salaried employees, most of whom were white, were offered assistance in finding new jobs. Hourly employees, most of whom were black, received no such aid. As a result, minorities, who had constituted 40 percent of the work force at the Detroit facility, comprised barely 2 percent at the Ohio plant.

The nearly immutable code of "last hired, first fired," combined with entrenched patterns of housing segregation, have left minorities at a real disadvantage when manufacturing plants close down, retail shops move out, and economic activity spreads to the suburbs and beyond. The dream of jobs with high wages and decent fringe benefits that once lured blacks to the North has turned into a nightmare for those who now face termination in the once bustling factories of the industrial Midwest.

## Income Loss and Underemployment

The incidence of job loss and the duration of unemployment are, of course, only two measures of the personal costs associated with economic dislocation. Not only do workers lose jobs, but the new jobs they eventually get do not provide as much income or status.

Again, the Parnes and King national sample confirms what had been found in the case studies. Almost *three-fifths* of the displaced workers experienced a decline in occupational status, in contrast to only one-fifth of the control group. What is most revealing about this result is that the downward occupational mobility was most acute among professional and managerial workers. In the initial 1966 survey, 27 percent of the workers eventually displaced, and an identical proportion of the control group, were in these top occupational categories. Seven years later, the proportion of displaced workers in these categories had fallen to 18 percent, while the proportion among the control group had actually increased to more than one-third.

In essence, the data indicate that permanently laid-off workers do not suffer merely a temporary loss. Many appear to make no complete occupational recovery even after a number of years, and some victims never recover. One example, provided by Parnes and King, is the forty-eight-year-old accountant who in 1966 had served with his employer for twenty-seven years. His annual earnings were $18,500 at the time of his layoff. After being unemployed for over a year, he finally found a job as a salesman. By 1973 he had worked for only eighteen weeks at that job, earning $4,000—implying an annual salary of somewhat under $12,000. In another example, in 1966 a fifty-seven-year-old metal roller with an eighth grade education and forty years of service in his job earned $9,000 per year. After his separation from this job, he was unemployed for thirteen weeks before finding a job as a gardener at an annual salary of $3,000.

Thanks to the efforts of researchers at the Public Research Institute of the Center for Naval Analyses in Virginia, there now are some statistical estimates of income loss that apply to entire industries. Using Social Security data, Louis Jacobson and his colleagues have been able to calculate the earnings losses of permanently displaced, prime-age male workers in a number of key industries. To do this, Jacobson calculates the actual earnings of workers in a given industry who remain continuously employed in that sector. This earnings trajectory is then compared with the earnings records of workers who experience permanent layoffs from the same industry. For most cases there is an immediate drop in income subsequent to termination, followed by a rise in earnings as those displaced find new employment in other firms. Of course, some job losers are affected quite adversely, with their earnings falling to zero, while others find comparable work almost immediately. The "actual earnings profile" reflects the *average* earnings of the full cohort of displaced workers.

Jacobson's estimates listed in table 6-1 indicate that in the first two years following involuntary termination, the average annual earnings loss ranges from less than 1 percent for workers formerly employed in the production of TV receivers to more than 46 percent in steel. Even after *six years,* workers in some industries continued to suffer as much as an 18 percent shortfall. Those displaced from the better-paying, unionized industries such as meat-packing, flat glass, automobile, aerospace, steel, and petroleum refining experienced the greatest reduction in income, but even in the low-wage sector including women's apparel, shoes, toys, and rubber footwear, six or more years elapsed before displaced workers caught up with those who had the good fortune to hold on to their jobs.

Presumably, many, if not most, of these workers received unemployment insurance benefits (UIB) when they first lost their jobs. However, during this period, as is true of the present, UIB had a twenty-six-week maximum. Clearly this could only compensate for a small portion of the earnings loss in the first two years and only a minuscule fraction during the full six. Furthermore, workers who found another job immediately following a layoff, even one that paid well below their previous wage, were not eligible for unemployment benefits at all. On the other hand, workers in a small number of industries such as auto manufacturing may have been employed by firms that paid supplemental unemployment benefits (SUB). A few displaced workers might also have qualified for trade readjustment assistance (TRA), but these "better off" workers are by far the exception, not the rule.

**Table 6-1**
**Long-Term Earnings Losses of Permanently**
**Displaced Prime-Age Male Workers**

| Industry | Average Annual Percentage Loss | |
|---|---|---|
| | First Two Years | Subsequent Four Years |
| Automobiles | 43.4 | 15.8 |
| Steel | 46.6 | 12.6 |
| Meat packing | 23.9 | 18.1 |
| Aerospace | 23.6 | 14.8 |
| Petroleum refining | 12.4 | 12.5 |
| Women's clothes | 13.3 | 2.1 |
| Electronic components | 8.3 | 4.1 |
| Shoes | 11.3 | 1.5 |
| Toys | 16.1 | -2.7 |
| TV receivers | 0.7 | -7.2 |
| Cotton weaving | 7.4 | -11.4 |
| Flat glass | 16.3 | 16.2 |
| Men's clothing | 21.3 | 8.7 |
| Rubber footwear | 32.2 | -.9 |

*Source:* Louis S. Jacobson, "Earnings Losses of Workers Displaced from Manufacturing Industries," in William G. Dewald, ed., *The Impact of International Trade and Investment on Employment*, A Conference of the U.S. Department of Labor (U.S. Government Printing Office, 1978), and Louis S. Jacobson, "Earnings Loss Due to Displacement," (Working Paper CRC-385, The Public Research Institute of the Center for Naval Analyses, April 1979).

This is clearly evident in a recent study of job loss among blue-collar women in the apparel and electrical goods industries conducted by Ellen Rosen. Here it was found that even after including UIB and TRA on top of any reemployment income, 92 percent of the affected workers suffered an annual income loss, with over 20 percent losing more than $3,000. On average, these women ended up losing over one-fifth of their normal yearly income following termination. This figure is nearly identical to the 18 percent cut in median family income found by Aronson and McKersie in their upstate New York study.

Other research reveals similar degrees of economic loss. When the Mathematica Policy Research Center interviewed approximately 1,500 displaced workers, over 900 of whom received TRA in addition to regular unemployment compensation, it was found that in the first year after layoff, total household income for those in the Mathematica TRA sample dropped by about $1,700 in real terms. Three and one-half years after layoff, those who were never recalled by their initial employers still had, on average, lower real weekly earnings than before. Moreover, about 38 percent of the original sample lost their health insurance coverage sometime during the initial unemployment stint.

Workers stand to lose all or part of their pension rights as well. Many Packard workers were left without a penny after paying into their retirement fund for years. A study commissioned by the Federal Reserve Bank of Boston showed that, during the early 1970s, of all the Massachusetts shoe workers displaced by shutdowns, 62 percent lost the pension benefits that they otherwise would have enjoyed had their plants remained in business. Enactment of the Employee Retirement Income Security Act (ERISA) in 1974 curbed many of the earlier abuses of pension rights, but corporate termination of a particular plant or division (what the U.S. Department of Labor, Labor-Management Services Administration calls a "partial termination") still can result in some benefits not being paid.

When the Diamond-Reo Corporation went out of business in 1975, 2,300 UAW members, including 500 retirees and 750 vested active employees, were affected. Vested active employees are those who are presently working and have sufficient seniority to qualify for company pensions when they retire. Because the pension plan was seriously underfunded, before ERISA all 750 vested actives would have lost their pension rights and the retirees would have suffered severe cutbacks averaging nearly 80 percent of their monthly pension benefits. Because of ERISA, about four out of five of the retirees suffered only relatively minor cutbacks in their expected pensions. But, for the one in five who chose nonguaranteed early retirement supplements, there were drastic cutbacks. An even more serious loss was borne by the 750 vested employees who had not yet retired.

How workers fare after a permanent layoff, if they do not drop out of the labor market altogether, depends on the types of jobs they find when they go to look for new work. Our own research, based on the Social Security Longitudinal Employer-Employee Data file (LEED), the same one used by Jacobson, reveals that postseparation earnings are largely determined by which market segment a displaced worker eventually enters. By way of example, consider what happened to prime-age workers in the New England aircraft industry when that sector went into a tailspin as a result of slackened procurement for the Vietnam war. Those who ended up in "secondary" jobs did considerably worse than those who remained in the aircraft industry or were able to transfer to other companies in the "primary" segment of the labor market.

Aircraft workers who were able to retain their jobs earned 21.1 percent more in 1972 than they did in 1967 (see table 6–2). Those who left the region following the severe layoffs that began in 1968, to search out aircraft industry jobs elsewhere, did not fare quite as well, averaging only a 16.5 percent earnings gain. Those who left the industry altogether but found work in other parts of the primary labor market, had higher earnings in the later year, but their wage gain was only one third as large as that achieved by those who were able to keep their aircraft jobs.

By contrast, those forced to take menial jobs in restaurants, hospitals, and various other sectors where they could not use their aircraft industry skills lost,

**Table 6–2**

**What Happens to the Earnings of Workers Who Leave the Aircraft Industry**

*(in current dollars)*

| 1972 Industry/Region | Number of Workers | Percentage in Category | 1967 Average Earnings | 1972 Average Earnings | Percentage Change in Average Earnings |
|---|---|---|---|---|---|
| Aircraft/inside New England | 37,700 | 64.7 | $9,575 | $11,595 | +21.1 |
| Aircraft/outside New England | 600 | 1.0 | 9,829 | 11,455 | +16.5 |
| Other "primary"[a] industries | 11,900 | 20.4 | 8,733 | 9,345 | +7.0 |
| Other "secondary"[b] industries | 2,100 | 3.6 | 6,054 | 4,468 | −26.2 |
| Not in jobs covered by Social Security | 3,700 | 6.3 | 6,175 | 0 | — |
| Disabled, deceased, unknown | 2,300 | 3.9 | — | — | — |
| Total | 58,300 | 100.0 | | | |

*Source:* Special tabulations of Social Security LEED File prepared by Alan Matthews and Barry Bluestone, Social Welfare Research Institute, Boston College, September 1979.

[a]"Primary" industries include most durable manufacturing, wholesale trade, public utilities, and some services.

[b]"Secondary" industries include most nondurable manufacturing, retail trade, and lower-skill requirement, higher-turnover personal services.

on average, 26 percent of their former income levels (even before accounting for inflation). An additional 6 percent of the original sample had no "covered" earnings in 1972, despite the fact that they neither retired nor became disabled. Those in this category either had no wage income at all during the year or worked in jobs not covered by social security, mainly in the federal government. Similar results to those obtained in aircraft were found for the metalworking machinery industry, although those who ended up in secondary jobs were not affected quite so adversely.

To the extent that the relative earnings losses represent real losses in productivity, table 6–2 suggests that one-third of the entire New England aircraft labor force experienced a productivity loss ranging from 4.6 percent for those who left the region for aircraft jobs elsewhere to 47.3 percent among those who ended up taking secondary jobs. For those who found no job at all after leaving the industry, the social efficiency loss was, mathematically speaking, infinite.

Leaving the aircraft industry has even worse consequences for women than for their male counterparts. Men who ended up in secondary jobs experienced a 14 percent earnings loss on average, while women lost a whopping 40 percent. Adding to this disparity is the difference in the probability that a woman would end up in the secondary segment. Only one out of fifty men employed in aircraft in 1967 was located in a secondary job in 1972. In sharp contrast, nearly one in eight women experienced such an industrial demotion. For women, getting into a high productivity, high-wage manufacturing job is a real victory; being forced out involves a real defeat.

Even these numbers present too sanguine a picture of income loss. Unlike other studies, these particular tabulations do not distinguish between voluntarily leaving and involuntary layoffs. Consequently, we cannot tell whether the workers who left aircraft or metalworking did so voluntarily, were the subjects of layoff, or experienced a plant closing. Because people who quit often leave because they have found better jobs, the earnings statistics in table 6–2 almost certainly underestimate the losses of those who left involuntarily.

## Loss of Family Wealth

Families who fall victim to brief periods of lost earnings are frequently able to sustain their standards of living through unemployment insurance and savings. Unfortunately for the victims of plant closings, the consequences are often much more severe, ranging from a total depletion of savings to mortgage fore-closures and reliance on public welfare. Families sometimes lose not only their current incomes, but their total accumulated assets as well.

During the Great Depression, the waves of plant closings that spread across the country drove millions of families into poverty. A study completed in 1934 of Connecticut River Valley textile workers showed that two years after the mills closed down, 75 percent of the families affected were living in poverty, compared with 11 percent before the shutdown. More than one in four families was forced to move in order to find lower rents. Some families lost their houses when they fell behind on mortgages. Thirty-five percent reported no new purchases of clothing, and the consumption of other items was reduced significantly.

This experience did not die out with the end of the Depression. A similar fate is faced by workers and their families who suffer permanent layoff today. When the Plainfield, New Jersey, Mack Truck facility shut down in 1960, workers had to reduce their food and clothing consumption substantially, and they turned to borrowing and installment credit for other necessities. Aircraft workers in Hartford County, Connecticut, the jet engine capital of the world, responded to the loss of their jobs in the mid-1970s by sharply reducing their expenditures on food, clothing, and medical care in addition to a long list of "luxury" items such as recreation and house repair. Out of the eighty-one workers interviewed in a study by Rayman and Bluestone, three of these dis-placed jet engine workers lost their houses to foreclosure. Among participants in the upstate New York study conducted by Aronson and McKersie, 11 percent reported cutting back on housing expenses, 16 percent reduced their food consumption, 31 percent bought less clothing, and 43 percent spent less on recreation. In what could lead to a mortgaging of their families' health, one in seven reduced their expenditures for medical care. These figures are remark-ably close to those found in the Hartford County research.

Such grave consequences are found in other case studies as well. In his study of the Wickwire shutdown (Colorado Fuel and Iron Corporation), Feli-cian Foltman found that workers were often forced to sell their automobiles and personal possessions in order to qualify for relief payments. Even more recently, the closing of a chemical company in Massachusetts forced families to rely on food stamps and housing assistance after their savings were used up. In the wake of permanent layoffs in the mid-1970s at the RCA Mountaintop semiconductor complex near Wilkes-Barre, Pennsylvania, researchers reported a major increase in young males on state welfare—a condition only permitted by law once someone virtually exhausts all of his savings and cashes in all of his assets.

Somewhat of an embarrassment to the research community is the fact that so little is known about the loss in assets suffered by workers when they lose their jobs. Saving for a rainy day is a normal part of a family's security umbrella, but for many it is never enough to cover the damage caused by the downpour from a plant closing.

## Impacts on Physical and Mental Health

The loss of personal assets places families in an extraordinarily vulnerable position, because when savings run out, people lose the ability to respond to short-run crises. The first unanticipated financial burden that comes along—an unexpected health problem, a casualty or fire loss, or even a minor automobile accident—can easily hurl the family over the brink of economic solvency. The trauma associated with this type of loss extends well beyond the bounds of household money matters.

Medical researchers have found that acute economic distress associated with job loss causes a range of physical and mental health problems, the magnitudes of which are only now being assessed. Simply measuring the direct employment and earnings losses of plant closings therefore tends to seriously underestimate the total drain on families caught in the midst of capital shift.

Dr. Harvey Brenner of Johns Hopkins University, along with Sidney Cobb at Brown University and Stanislav Kasl at Yale University, have done careful studies in this area. Writing in *Psychometric Medicine,* Kasl and Cobb report high or increased blood pressure (hypertension) and abnormally high cholesterol and blood-sugar levels in blue collar workers who lost their jobs due to factory closure. These factors are associated with the development of heart disease. Other disorders related to the stress of job loss are ulcers, respiratory diseases, and hyper-allergic reactions. Higher levels of serum glucose, serum pepsinogen, and serum uric acid found in those experiencing job termination relative to levels in a control group of continuously employed workers suggest unduly high propensities to diabetes and gout. Compounding these problems is the fact that economically deprived workers are often forced to curtail normal health care and suffer from poorer nutrition and housing.

The Kasl and Cobb findings are by no means unique. Aronson and McKersie write that two-fifths of their sample reported deterioration in their physical and emotional well-being since their termination. Headaches, upset stomachs, and feelings of depression were the most widely-reported health problems. Aggressive feelings, anxiety, and alcohol abuse were the observed psychological consequences of the Youngstown steel closings. Similar conditions were widely reported among the aircraft workers in the Hartford County study. In most of these cases, the factor of time seems to be essential. Those who need much of it to find another job suffer the most.

Workers generally lose health benefits when they lose their jobs. According to Don Stillman of the United Auto Workers, fewer than 30 percent of the unemployed have any health insurance at all. Those who do have to spend 20 to 25 percent of their unemployment benefits merely to continue their former coverage—if continuation is available at all. Premiums for nongroup coverage average twice those for group plans, yet the benefits are lower. There are so many deductibles that nongroup health insurance covers an average of only 31 percent of a family's incurred medical costs.

Brenner's work gives evidence of yet a much broader and deeper problem. Making use of correlation and regression analysis, he has been investigating the statistical linkages between so-called economic stress indicators and seven indices of pathology. The economic stress indicators include per capita income, the rate of inflation, and the unemployment rate. The indices of pathology are:

Age and sex specific mortality rates

Cardiovascular-renal disease mortality rates

Suicide mortality rates

Homicide mortality rates

Mental hospital admission rates

Imprisonment rates

Cirrhosis of the liver mortality rates

Using national data for the period 1940–73, Brenner found that unemployment plays a statistically significant role in affecting several forms of "social trauma." In particular, he concludes that a 1 percent increase in the aggregate unemployment rate sustained over a period of six years has been associated with approximately:

37,000 total deaths (including 20,000 cardiovascular deaths)

920 suicides

650 homicides

500 deaths from cirrhosis of the liver

4,000 state mental hospital admissions

3,300 state prison admissions

These results, of course, do not directly address the question of unemployment caused by deindustrialization *per se,* but it is likely that permanent layoffs cause even more "social trauma" than unemployment arising from other causes. For example, in the aftermath of the Federal Mogul Corporation closing of its roller-bearing plant in Detroit, eight of the nearly two thousand affected workers took their own lives. This macabre statistic is unfortunately not unusual. In their study of displaced workers, Cobb and Kasl found a suicide rate "thirty times the expected number."[2]

Of course, suicide is only the most extreme manifestation of the severe emotional strain caused by job loss. Family and social relationships are nearly always strained by protracted unemployment. Richard Wilcock and W.H. Franke, in their now famous work on permanent layoffs and long-term unemployment, suggest that social, medical, and psychological costs may even outweigh direct economic costs in severity. They note:

> Perhaps the most serious impact of shutdowns, particularly for many of the long-term unemployed, was a loss of confidence and a feeling of uselessness. . . . The unemployed worker loses his daily association with fellow workers. This loss means not only disappearance of human relationships built up over a period of years, but also the end of a meaningful institutional relationship. When he is severed from his job, he discovers that he has lost, in addition to the income and activity, his institutional base in the economic and social system.[3]

Loss of a work network removes an important source of human support. As a result, psychosomatic illnesses, anxiety, worry, tension, impaired interpersonal relations, and an increased sense of powerlessness arise. As self-esteem decreases, problems of alcoholism, child and spouse abuse, and aggression increase. Unfortunately, these tragic consequences are often overlooked when the costs and benefits of capital mobility are evaluated.

Special psychological problems arise when a plant closing occurs in a small community, especially when the establishment was the locality's major employer. Writing about the closing of a plant in southern Appalachia, Walter Strange notes that the people

> lost the central focus which had held the community together—its reason for existence—a focus which was held in common as community property, one which provided not only for economic needs but . . . a structural framework which gave coherence and cohesion to their lives.[4]

These effects typically lessen or disappear following successful reemployment. Yet, "stressful situations" caused by a plant closing can linger long after the final shutdown has occurred. Moreover, feelings of lost self-esteem, grief, depression, and ill health can lessen the chances of finding reemployment; this failure, in turn, can exacerbate the emotional distress, generating a cycle of destruction. Ultimately a debilitating type of "blaming the victim" syndrome can evolve, causing dislocated workers to feel that the plant closing was their own fault. Strange argues "that those feelings of self-doubt can create fear of establishing a new employment relationship or complicate the adjustment process to a new job." As the sociologist Alfred Slote put it, in his seminal work on job termination:

> The most awful consequence of long-term unemployment is the development of the attitude, "I couldn't hold a job even if I found one," which transforms a man from unemployed to unemployable.[5]

## Notes

1. Herbert S. Parnes and Randy King, "Middle-Aged Job Losers," *Industrial Gerontology* 4, no. 2 (spring 1977), p. 82.

2. Sidney Cobb and Stanislaw Kasl, "Termination: The Consequences of Job Loss," Public Health Service, Center for Disease Control, National Institute for Occupational Safety and Health, U.S. Department of Health, Education, and Welfare, Washington, D.C. (June 1977), p. 134.

3. Richard Wilcock and W.H. Franke, *Unwanted Workers: Permanent Layoffs and Long-Term Unemployment* (New York: Glencoe Free Press, 1963), pp. 166, 185.

4. Walter Strange, "Job Loss: A Psychosocial Study of Worker Reactions to a Plant Closing in a Company Town in Southern Appalachia," National Technical Information Service (NTIS), 1977, as quoted in C & R Associates, "Community Costs of Plant Closings: Bibliography and Survey of the Literature," report prepared for the Federal Trade Commission under Contract No. L0362 (July 1978), p. 39.

5. Alfred Slote, *Termination: The Closing of Baker Plant* (Indianapolis, Ind.: Bobbs-Merrill, 1969), p. xix.

# 7
# Five Case Studies

*Herbert Hammerman*

I t is generally agreed that technological change in the United States has had long-term beneficial effects in terms of greater productivity, faster economic growth, more jobs, and higher wages and employee benefits. However, the short-term cost of such industrial progress to individual workers displaced from their jobs is not always fully recognized.

Between April 1962 and May 1963, the Bureau of Labor Statistics conducted five case studies of the effects of plant shutdowns or large-scale layoffs. These studies reveal that even under favorable labor market conditions, many workers, once displaced, were unable to find new jobs. Others had long periods of unemployment and experienced considerable hardship. This was particularly true of older workers, women, and workers with the least education or the lowest levels of skill. In most cases, displaced workers got little help from their former employers and relied mainly on personal contacts with friends and relatives in locating work.

Sizable numbers stopped seeking employment and most of those who were reemployed earned less—substantially less in many cases—and had lower benefits; many had jobs of lower skill, and all but a small number lost their seniority protections.

Interplant transfers were offered in but two cases, and only under union contract requirements. Even in the one case in which a transfer with seniority was offered, no more than one out of five accepted. Among the obstacles to mobility were the high costs of relocation, home ownership, the secondary role of the wife's job in the family, children in school, family and social ties, and fear of future layoffs.

## The Cases

These studies included plants in five manufacturing industries: petroleum refining, automotive equipment, glass jars, floor covering, and iron foundries.

Herbert Hammerman, "Five Case Studies of Displaced Workers," *Monthly Labor Review* 87, no. 6 (June 1964), pp. 663–670.

The plants were located in six areas, from the East Coast to the Mountain States, most of them in the Midwest. The number of workers displaced totaled close to 3,000 and ranged from about 100 to over 1,000.

The layoffs took place between July 1960 and June 1962. Surveys were conducted between April 1962 and May 1963. The period elapsing between layoffs and surveys varied from 6 to 21 months.

### Labor Market Conditions

All areas were substantially industrialized and highly diversified. The smallest had a labor force of a little under 50,000; the largest, well over 500,000. Five of the six areas were standard metropolitan statistical areas. In five areas, unemployment rates at the time of the layoffs were in excess of or close to the "relatively substantial unemployment" level of 6 percent. Conditions improved subsequently and by the time the surveys were conducted, unemployment had declined substantially in each area.

### Role of Technological Change

In each case, technological change was a factor directly or indirectly in the shutdown or layoff. One plant installed new labor saving production processes; another lost business because of a change in customer demand to a product using a different material. In some instances, the building and equipment were old.

In all but one case, however, other factors played a significant role as well, such as a sharp decrease in consumer demand for a particular product; the dislocation resulting from a shift from long-run operations for one customer to short-run operations for many customers; and labor-management conflict.

The experiences of the Bureau in attempting to select cases of worker displacement due to technological change clearly indicate that such cases are difficult to isolate. In most instances, more than one factor influenced the ultimate decision to close the plant, and it was extremely difficult to determine which factor or factors were decisive.

## Measures to Prevent Displacement

Layoffs may be prevented or minimized by various means. One is attrition, in which workers who quit or retire are not replaced. Early retirement of older workers, spreading available work by measures such as the elimination of overtime, and timing the layoff to take place during a period of business expansion are also sometimes possible. In only one case, the oil refinery, was any attempt made beforehand to reduce the extent of the layoff. At this company, more than half of the projected employment reduction was achieved by attrition. No new employees had been hired for over three years before the first group was laid off. The firm also induced workers over age 51 who were not scheduled for layoff under the seniority regulations to retire early by offering them a substantial "age allowance" separation payment in addition to their regular severance pay and annuity. A maximum "age allowance" of $4,800 was paid at age 58, with the amount scaled down toward age 51 and age 65, respectively. Those accepting early retirement accounted for one-sixth of the displaced workers.

## Measures to Find Jobs

The nature and extent of assistance given displaced workers in their efforts to secure reemployment varied considerably from case to case, depending on employers' attitudes, the history of labor-management relationships, and union contract provisions. In four of the cases, the majority of the workers were represented by unions affiliated with the AFL-CIO; in the fifth case, there was an independent union. Types of assistance used or attempted in one or more of the case studies included early notice of the impending layoff, placement services, interplant transfers, employer-sponsored retraining programs, and in-plant reassignments and transfers. Generally, these efforts were of limited help.

### Early Notice

One firm ceased new hiring three years before the layoffs began, notified the workers well in advance of termination, and phased out the layoffs over a period of a year. In all other cases, notice was considerably shorter. The longest advance notice was six months; the shortest, little more than two months.

### Placement Services

By far, the most effective source of assistance in locating jobs appears to have been personal contacts. "Friends or relatives" were credited with finding the jobs of from one-half to two-thirds of the reemployed workers in the five cases studied.

The only substantial assistance in securing jobs for displaced workers was given by the oil company and, in another case, by the union. The company assisted displaced workers through its own employment office, contacting about 600 firms in the area, and also entered into an agreement to pay the placement fees of two private employment agencies. The company stated that it had assisted one-third of the workers who found employment in locating their jobs, although one-half of that number were no longer working on these particular jobs at the time of the survey.

In the automotive equipment case, the union invited all displaced members to fill out a job referral form that was circulated among companies with which it had contracts. As a result of these efforts, approximately 200 of the more than 1,000 displaced workers were employed by companies under contract to the union. The largest proportion was hired by an expanding firm that accepted displaced workers up to age 55. This age limit was higher than most, but union representatives stated that this company was willing to hire older workers because of its need for workers with "instant skills," who could adapt to new jobs with a minimum of retraining. The union found that it was not able to place men over age 55 readily, or women at any age.

*Interplant Transfers*

Although all five companies were multiplant firms, only two offered interplant transfers, in both instances under union negotiated plans. In neither instance were relocation allowances granted. In the case of the automotive equipment firm, the contract with the major union representing production and maintenance workers provided that, if the company shut down a plant and transferred its operations to another plant, the employees would be given an opportunity to transfer to the other plant with their jobs. They would be credited with full seniority for purposes of layoffs, recall, and economic benefits such as pensions and vacation. On the basis of this agreement, about 1 out of 5 displaced workers transferred to another plant of the company in an adjacent state. The agreement also gave displaced employees preferential hiring rights at other plants. Workers exercising such rights would start as new employees for purposes of layoff, but would carry seniority with them for economic benefits. About 3 percent of all displaced workers were transferred under this provision. A craft union representing some skilled workers in the plant had not negotiated an interplant transfer provision and its members were not given an opportunity to transfer after the plant closed.

In the second case, the floor covering plant, the contract required the transfer of economic benefits, but no job security benefits. Not more than 1 out of 8 workers took advantage of this provision. The importance of job security to those displaced is indicated by the fact that a much larger proportion of workers accepted transfer in the automotive equipment case than in the floor-covering case, despite the fact that the distance was twice as far. Some effects of these provisions are discussed in the section on mobility and reemployment below.

*Employer Retraining Programs*

None of the five employers adopted programs to retrain displaced workers for jobs elsewhere. The automotive equipment company publicly announced establishment of a $100,000 retraining fund shortly after announcement of the impending shutdown. About 30 percent of the displaced workers registered for training. The program was abandoned when the company concluded that, because of age, inadequate schooling, or low scores on aptitude tests, there would have been no reasonable prospect for job placement for most trainees. It stated that many were not willing to train for service jobs paying much lower wages than they had been receiving, and few were willing to give up unemployment compensation and supplementary unemployment benefits for which they would have been disqualified under existing regulations while engaged in a full-time training program. In the case of the oil refinery, about 1 out of 10 laid-off workers reported that earlier training given by their employers for jobs in the plant helped them in getting jobs outside; all were employed at the time of the survey.

### Inplant Reassignments and Transfers

In the only case that did not involve a plant shutdown, the layoff was on the basis of plantwide seniority. This procedure left numerous vacancies which were filled by reassignment, transfer, and retraining of the remaining employees.

## Measures to Maintain Income

### Unemployment Insurance

The most important source of income for the displaced workers was unemployment insurance. In four cases, the proportion receiving such benefits ranged from 69 to 94 percent of the total, for an average of from 18 to 27 weeks. Even in the fifth case, with relatively low unemployment, close to one-half of the workers received benefits.

### Supplementary Unemployment Benefits

In only one case—the automotive equipment plant—had supplementary unemployment benefits been negotiated. By the time of the survey, benefits had been received by two-thirds of the displaced workers of this company for an average of 22 weeks.

### Severance Pay

Some form of severance pay was obtained by displaced workers in four of the five cases, but in only one instance was it sufficiently great to be of substantial assistance in a period of protracted unemployment. That plan provided a "service allowance" based on the worker's wage rate and length of service. The lowest amount paid under the formula was in excess of $600.

*Early Retirement Pay*

Although all pension programs provided for early retirement at age 60 or sooner, in only two cases did more than a very small proportion of the displaced workers benefit from these provisions. The automotive equipment case involved a substantial number of older workers and, in the 14 months after the shutdown was announced, pensions were paid to 375 workers, 283 of them in the major bargaining unit. While some of these were for normal and disability retirement, the bulk of them were for early retirement. With the payment of these pensions, the fund was not sufficient to cover the vested rights of the younger workers. Therefore, in accordance with contract provisions, deferred pensions were substantially reduced for those in the 50–59 age group and were eliminated for those under 50. At the oil refinery, a contributory plan provided immediate or deferred annuities regardless of age. Ninety percent of those 55 and over received immediate annuities and another 5 percent got deferred annuities; 1 out of 3 displaced workers under 55 received immediate or deferred annuities.

In all other cases, most workers, including many with long service, lost their pension rights entirely.

## Characteristics of the Displaced Workers

The typical displaced worker was a white male in his late forties. He was married, owned his home, had two dependents, and some high school education.

Women accounted for two out of five workers in one case, one out of five in another, and insignificant proportions in all others. In no case did nonwhites exceed 7 percent of the total. Although most workers were age 45 or over, the proportion in this category varied from 21 percent in one case to 94 percent in another. In all cases, a majority were married and owned their homes. The number of dependents tended to be fewer among the older groups. While in four cases a majority of the workers had some high school education, in all but one instance the proportion of graduates was relatively small.

## The Search for Employment

Job hunting was a difficult experience for many displaced workers. For a large number, it was fruitless. Most of the displaced workers had accumulated long years of service in a particular line of work and many were ill-prepared for the strenuous efforts of job hunting.

*Employment and Unemployment*

At the time of the surveys, only two out of three of all displaced workers in these case studies were employed. One out of ten was retired or for other

**Table 7–1**
**Unemployment of Displaced Workers**

| | Period between Layoff and Survey (months) | Percent Unemployed | | Among Displaced Workers at Time of Survey |
| | | In Labor Market Area | | |
| | | At Time of Layoff | At Time of Survey | |
|---|---|---|---|---|
| Petroleum refinery | 6–18[a] | 4.6–9.9 (average 7.8)[a] | 4.7 | 8 |
| Automotive equipment plant | 10 | 6.8 | 4.9 | 27 |
| Glass jar plant | 8 | 7.5 | 5.1 | 39 |
| Floor covering plant | 16 | 5.9 | 3.0 | 17 |
| Foundries | 21 and 13[b] | 2.8[b] and 6.9 | 3.1[b] and 3.6 | 28[b] |

[a]Gradual layoff over a period of twelve months.
[b]Rate of each of the two areas in which a foundry was closed; areas have been combined in data for displaced workers.

reasons not seeking employment, while close to one out of four was unemployed and seeking employment. The rate of unemployment varied from 8 to 39 percent. As shown in table 7–1, it was substantially higher than the unemployment rate in the labor market areas as a whole. In all but one case, the unemployment figures for displaced workers were at least five times greater than the unemployment rate in the labor market areas as a whole.

There was substantial long-term unemployment (see figure 7–1). In four cases, over one-half of the displaced workers had been unemployed at least 16 weeks; in two of these cases, the proportion was two-thirds or more. In the same four cases, those unemployed at least one-half year ranged from over two-fifths to more than one-half.

A substantial proportion of the displaced workers had held no jobs at all after their layoff. However, a considerable number, ranging from one out of eight to about three out of eight, had held more than one job.

*Age and Reemployment*

Reemployment was markedly higher among workers below age 45 than among older workers. In four of the cases shown in table 7–2, below, the older groups contained a greater proportion of workers who were unemployed and seeking work. In the fifth case, the great majority of the older workers laid off had retired voluntarily, although by the time of the survey many were reemployed and a small percentage were seeking jobs. In all instances, a larger proportion of older workers were not seeking employment.

In two cases, where narrower age breakdowns were feasible, by far the highest unemployment rates were found in the 55–59 age groups. A majority of workers in the 60 and over age groups were not seeking work. Although a considerable number of these had taken early retirement benefits, many others

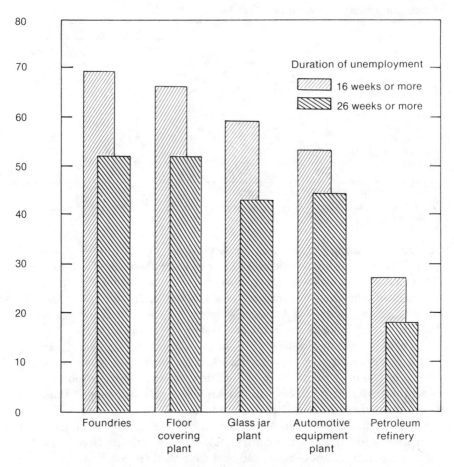

**Figure 7–1. Long-Term Unemployed as a Percentage of Total Displaced Workers**

may have been discouraged from looking for work in the face of age discrimination. More displaced workers volunteered comments on the subject of age discrimination than on any other matter. Most were workers in their fifties or above, but many were younger, a number in their early forties.

*Education and Reemployment*

Displaced workers who completed high school had substantially lower unemployment rates than those who did not. In three cases, the graduate's unemployment rate was less than one-half that of the nongraduate. The differences in unemployment rates between those who had no high school education and those who had some were much smaller.

**Table 7-2**
**Age Compared to Unemployment**

| | Percent of Workers in Each Age Group at Time of Survey | | | |
| | Unemployed | | Not Seeking Employment | |
| | Less Than 45 Years | 45 Years and Over | Less Than 45 Years | 45 Years and Over |
|---|---|---|---|---|
| Petroleum refinery | 8 | 5 | 1 | 39 |
| Automotive equipment plant | 10 | 29 | 12 | 14 |
| Glass jar plant | 35 | 41 | 3 | 16 |
| Floor covering plant | 12 | 19 | 0 | 15 |
| Foundries | 25 | 32 | 0 | 19 |

**Table 7-3**
**Unemployment by Sex**

| | Percent Unemployed | | | |
| | Automotive Equipment Plant | | Glass Jar Plant | |
| | Men | Women | Men | Women |
|---|---|---|---|---|
| *Age* | | | | |
| Less than 35 years | 5 | 22 | 10 | 58 |
| 35–44 years | 7 | 25 | 22 | 63 |
| 45–54 years | 17 | 67 | 19 | 70 |
| 55–59 years | 32 | 62 | 51 | 71 |
| *Education* | | | | |
| No high school | 26 | 67 | 32 | 65 |
| Some high school | 25 | 64 | 23 | 59 |
| High school graduate | 11 | 31 | 11 | 65 |

The older worker with higher education was more likely to be reemployed. Among older workers, high school graduates fared better than nongraduates, and workers with some high school had lower unemployment rates than those with no high school. Workers not seeking employment were found for the most part among the less educated. Suffering from the combined handicap of inadequate education and older age, many withdrew from the labor market before they normally would have retired.

*Women and Reemployment*

The rates of unemployment among women were 56 and 61 percent, respectively, or almost three times the rate among men in the two cases where meaningful comparisons were possible. Only one-fourth of the women were employed in each case; the others were not seeking work. In one case, almost seven out of eight women had been out of work one-half year or more, compared with one out of three men. In the other case, the ratio was two out of three women as against one out of four men.

As shown in table 7–3, displaced women had a much higher unemployment rate than men at each age group under 60 and at each educational level.

The impact of age discrimination would seem to have been felt earlier by women than by men. The highest level of unemployment was reached by men at age 55–59 in both cases. However, among women, unemployment reached its peak at age 45–54 in one case and virtually its peak (within 1 percentage point) in the other case.

## Skill Level and Reemployment

In each case studied, a higher unemployment rate was found among less skilled workers. Unemployment ranged from none to 33 percent for maintenance workers, from 8 to 39 percent for machine operators, and from 20 to 59 percent for laborers. A similar pattern was revealed when hourly earnings were used as an approximate measure of skill; the highest unemployment rates were found at the lowest earnings levels.

## Industries Providing Jobs

Table 7–4 showing industries in which displaced workers obtained employment reveals that few were able to find jobs in the same industry. In four cases, such workers constituted no more than 21 percent of the total reemployed. Of those obtaining employment in the same industry, most in two cases and all in a third had to move to other areas.

Most reemployed workers secured jobs in manufacturing industries. However, substantial proportions, ranging from one out of four to almost three out of five, were employed in nonmanufacturing industries.

## Mobility and Reemployment

The role of mobility in obtaining employment is indicated by the fact that greater proportions of employed workers than the unemployed had sought

**Table 7–4**
**Industries Providing Reemployment**

|  | Percent Reemployed in | | |
| --- | --- | --- | --- |
|  | *Same Industry* | *Other Manu-facturing* | *Nonmanu-facturing* |
| Petroleum refinery | 7 | 57 | 36 |
| Automotive equipment plant | 47[a] | 28 | 25 |
| Glass jar plant | 21 | 32 | 47 |
| Floor covering plant | 17[b] | 46 | 37 |
| Foundries | 8 | 34 | 58 |

[a]Includes 33 percent who transferred under union contract provisions to another plant of the same company in another area. The remaining 14 percent were employed by other companies in the same industry.

[b]All 17 percent transferred under union contract provisions to another plant of the same company in another area.

work outside their home cities. Moreover, in all but one case, more than twice the proportion of reemployed workers than workers still unemployed indicated that they had looked for work further than 50 miles from home.

The two cases involving interplant transfers cast some light on inducements and obstacles to worker mobility. Only the guarantee of transfer with full seniority rights was sufficient to induce substantial numbers of displaced workers to relocate. Even in that case, some four out of five did not accept relocation. Relatively few workers were willing to transfer with accumulated rights to pensions, vacation, and other economic benefits, but with no seniority rights on layoffs. A study of the characteristics of the transferees indicates that other inducements to relocate were the need to conserve rights to pensions and other employee benefits, fear of age discrimination, and the economic pressures of larger families.

Obstacles to mobility included the secondary role of the wife's job in the family, home ownership, family and social ties, children in school, fear of future layoffs, and the high cost of transfer. Apart from costs of relocation, many transferred workers found it necessary or expedient to maintain two homes and to commute between areas on weekends, at least in the first year after transfer. Some complained that tax laws worked in favor of the companies which could write off the cost of their move, while transferring workers received no deductions whatsoever.

### Training and Reemployment

Only a very small number of displaced workers, ranging from 2 to 7 percent, had taken any training courses, other than on-the-job, after displacement. Nevertheless, a large majority indicated that they would be interested in taking such courses if they did not have to pay for them. Workers manifested much variety of interest. Many men were interested in learning special skills such as welding, electronics, auto mechanics, and machine repair. Women emphasized office and clerical occupations and nursing.

## Job Effects of Displacement

In addition to unemployment, displaced workers suffered other job losses: lower earnings, work of lower skill, loss of employee benefits, loss of seniority protection, and premature withdrawal from the labor force. Other groups also make contributions to the social costs of displacement. Labor unions suffer a decline in membership; businesses and the community lose the income derived from the displaced workers' wages; and governments lose tax revenues and often have to increase their relief payments.

### Effects on Earnings

The great majority of those who were reemployed received lower hourly earnings. In each of the five cases, more than one-half of the reemployed workers had lower earnings, with the ratio almost as high as four out of five in one case (see figure 7-2). Moreover, many workers took a cut of 20 percent or more in earnings. These constituted at least one out of every four reemployed workers, and in one case amounted to more than one-half of the total. In contrast, only small proportions achieved higher earnings. In the two cases involving inter-plant transfers—the automotive equipment plant and the floor covering plant—relatively high proportions (32 and 18 percent, respectively) were reemployed at the same earnings levels.

When older workers obtained employment, they had to accept a much greater decline in hourly earnings than did younger workers. Table 7-5 shows the percent by age group of workers who took at least a 20 percent decrease in hourly earnings. In three of the four cases for which such tabulations were feasible, the proportion of workers whose earnings had dropped at least 20 percent increased substantially after age 45. Also by the same measure, the least educated workers took the sharpest cuts in wages (table 7-6).

## Effects on Employee Benefits

Displaced workers frequently complained, often bitterly, of the loss of employee benefits. This was one of the most serious hardships resulting from worker displacements because benefits had been counted on, for greater security for themselves and their families in old age and in illness. Moreover, since many

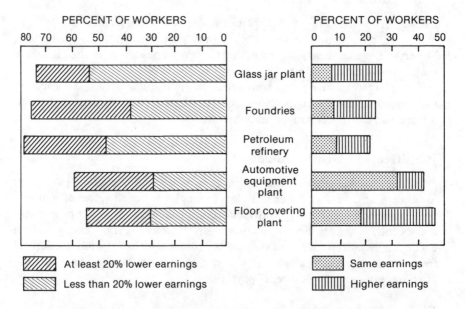

**Figure 7-2. Change in Earnings of Reemployed Workers**

**Table 7–5**
**Earnings Decrease and Age**

| | Percent of Workers in Age Group Whose Hourly Earnings Decreased at Least 20 Percent | | | |
|---|---|---|---|---|
| | Less Than 35 Years | 35–44 Years | 45–54 Years | 55 Years and Over |
| Petroleum refinery | 24 | 35 | 31[a] | — |
| Automotive equipment plant | 17 | 18 | 30 | 33 |
| Glass jar plant | 31 | 35 | 72 | 74 |
| Floor covering plant | 11 | 13 | 35 | 57 |

[a]45 years and over.

**Table 7–6**
**Education and Earnings Decrease**

| | Percent of Workers at Educational Level Whose Hourly Earnings Decreased at Least 20 Percent | | |
|---|---|---|---|
| | No High School | Some High School | High School Graduate |
| Petroleum refining | — | 43 | 31 |
| Automotive equipment plant | 39 | 28 | 23 |
| Glass jar plant | 70 | 55 | 32 |
| Floor covering plant | 31 | 18 | 22 |

benefits are based upon length of service, workers obtaining other employment had to start anew in accumulating rights. Most of the reemployed workers indicated that fringe benefits on their current jobs were less liberal than on their previous jobs.

*Changes in Type of Job*

Many of the displaced workers experienced a downgrading of skills. This was more true of semiskilled than of skilled occupations. While the change in jobs for a majority of workers in the more skilled maintenance occupations generally meant no change in occupational group, in no case did as many as one-third of the machine operators obtain jobs in the same occupational group. Substantial proportions of the operators who were reemployed were working as laborers or custodial workers.

*Effects on Union Membership*

The layoffs had a serious effect on membership in labor unions. Prior to displacement, nine out of ten were union members. By the time of the surveys, membership in unions was reduced to no more than one out of three in two cases and in no event higher than two out of three. The highest proportion was found among displaced workers of the automotive equipment plant where substantial numbers were either transferred to another area under union contract or obtained jobs at other unionized plants with assistance of the union. Even when consideration is limited only to those workers who had found jobs, the figures still show a substantial drop in union membership, to a range of from two-fifths to three-fifths of the total. This decline could be accounted for in part by the fact that many new jobs were in unorganized industries or plants.

### Effect on Seniority

In most union contracts, seniority protects the longer service worker in case of layoffs. It is often a factor in promotions, generally determines eligibility for and the size of certain employee benefits, and may confer such advantages as choice of shift. With this seniority gone, the displaced worker starts a new job as a new employee, having the least security and lowest employee benefits in the plant. The bulk of the displaced workers had over 10 years of seniority. In some plants, substantial proportions had longer service. Workers with at least 20 years of seniority accounted for one out of four in one plant, one out of three in a second, and nine out of ten in a third.

### Early Withdrawals from the Labor Force

Sizable numbers, ranging from 9 to 14 percent of the total, indicated that they were no longer seeking employment. Such withdrawals from the labor force represented substantial proportions of workers in the 60–64 age group. It seemed clear that many found themselves compelled to end their careers as wage earners earlier than they had previously planned. At best, their withdrawal meant early retirement with pensions below the amount which would have been due them at normal retirement and a lower level of living than had been anticipated. In many instances, older workers without pensions withdrew because of their inability to obtain jobs; this was particularly true in the case of women.

# 8

# Magnitude of Job Loss

*Candee S. Harris*

T he priority accorded job-related issues at all levels of government has risen dramatically in recent years. The most apparent and immediate stimulus for this increased concern is the recent recession, during which unemployment rates reached a postdepression high of 10.8 percent in December of 1982. The length and depth of this recession reflected the convergence of several factors, including a long-term secular decline in the rate of growth, a cyclical downturn exacerbated by restrictive macroeconomic policies, and an acceleration of the structural reorientation of the economy, resulting in very uneven distributions of growth and decline both regionally and industrially.

These economic conditions have engendered a number of difficult political and economic questions. Is America deindustrializing, or is this simply a deep trough in the business cycle? Is the economy experiencing a decline in manufacturing and other goods-producing industries, either absolutely or relatively? What are the implications of these changing conditions for the American economy as a whole and for the regions, industries, and workers comprising the economy?

The mounting public debate about these issues has elicited apparently conflicting evidence. On one hand, social scientists analyzing the U.S. economy from a very broad perspective largely dismiss fears of deindustrialization. A recent examination of the U.S. economy by Robert Lawrence found that the manufacturing sector performed rather well compared to other advanced industrial economies during the decade of the 1970s. Between 1970 and 1980 aggregate employment in manufacturing grew 4.7 percent, compared to net losses in most other countries. Manufacturing output and productivity also increased during the decade. However, the small positive trend indicated by these aggregate figures masks serious decline in particular manufacturing industries.

Candee S. Harris, "The Magnitude of Job Loss from Plant Closings and the Generation of Replacement Jobs: Some Recent Evidence," *The Annals,* American Academy of Political and Social Science 475 (Sept. 1984), pp. 15–27.

Perception of the problem is also colored by the cyclicality in the economic performance of the manufacturing sector. During the recent recession, the number of manufacturing jobs fell by 2.5 million. More than 95 percent of this reduction was among production workers. The manufacturing sector has always been more sensitive to cyclical fluctuations than the rest of the economy, but the outlook for the recovery to replace all manufacturing production jobs lost in this recession is not optimistic.

Furthermore, unemployment in this recession diverges from the pattern in the three previous recessions in terms of intensity—there were record levels of unemployment—and in terms of composition. A much higher proportion of job losers in 1981 and 1982 were permanently separated from their employers compared to the recessions of the 1970s, 53 percent versus 37 percent. In 1982 there were 1.4 million fewer production jobs in manufacturing than in 1970.

Beyond cyclical factors, the proportion of total economic activity in manufacturing industries is declining in response to changes in demand and productivity. Over the next 20 years the share of the work force that will be employed in the manufacturing sector is expected to drop from 22 percent to a range of 5 to 14 percent.

A seemingly endless stream of closing manufacturing plants flows across the pages of the press, vividly documenting the serious negative effects on many workers and communities touched by the processes of change. Some basic manufacturing industries in the United States, such as steel and automobile production, are being supplanted by competitors in the international market. Technological advances that contribute to improved competitiveness are rendering obsolete many of the better-paying, skilled and semiskilled occupations associated with heavy manufacturing. As employment in these industries declines, displaced workers are shifting into other sectors of the economy or onto the unemployment rolls. However, the profiles of the typical worker and the typical job in these declining industries differ from those in expanding sectors of the economy, such as the much touted, high technology industries. Judgments of the desirability of an economy concentrated in goods-producing versus service industries aside, workers and communities with a stake in the older economic constellation are ill prepared for an intense pace of disinvestment and for the changing focus of new investment.

On the broadest level, implementation of effective macroeconomic policies to fortify the current recovery and to encourage the long-term growth and international competitiveness of American industry is essential. Downturns in the economy accentuate all instances of job loss and slow the processes by which lost employment opportunities are replaced through the expansion of existing firms and the creation of new business establishments.

Even if successful, broad macroeconomic policies to stimulate recovery will not sufficiently improve conditions in the pockets of serious distress produced by secular shifts in the composition of economic activity. Policies targeted at specific problem areas will be needed to facilitate the transition. Many such policies already exist and are embodied in programs whose objectives range from providing training to workers displaced by trade liberalization to granting tax incentives to foster business development in areas with high unemployment. The accelerating pace of economic change has been accompanied by the call for additional legislative and administrative efforts that will encourage the progressive development of the United States and will provide assistance to those victimized by the development process.

Among the policies sought are those that will regulate the process of plant closings and compensate workers and communities for economic losses associated with closings. Such legislation has been enacted in a few states and has been proposed in many more. National legislation has been presented to Congress in several forms attempting to address some of the problems arising from the decisions of firms to close or relocate facilities. Some of the measures put forward include requiring companies to provide (1) prior notification of radical reductions in employment or closing, (2) severance pay and retraining programs for displaced workers, and (3) in certain cases financial compensation to the community.

In addition to the political obstacles blocking enactment and implementation of plant-closing legislation, policy formulation has been impeded by the lack of empirical information on the magnitude and geographic concentration of the problem. In response to the growing pressure for and against plant-closing legislation, data collection efforts and statistical estimates of the impact of closing on employment loss have increased, but a comprehensive assessment of the problem has been lacking.

Using business data from the U.S. Establishment and Enterprise Micro-data (USEEM), developed at the Brookings Institution, this analysis examines the magnitude of employment loss associated with plant closings and general business turnover in the United States between 1976 and 1982. Beyond merely counting the number of jobs lost through closings, the propensity of different types of business enterprise to regenerate employment through the dissolution of establishments is considered. How are these processes of job loss and job creation linked to shifts in the composition of the economy in terms of industrial and business size characteristics? The dynamics of employment behavior in the manufacturing sector are contrasted with those in other industries, particularly the service sector. Differences in the job-replacement behavior of small and large firms are discussed in relation to secular growth trends in manufacturing and other industries. Finally, the geographic mapping of closings and job replacement is considered.

## Plant Closings: An Operational Definition

The normal functioning of the U.S. economy is the consequence of many offsetting factors, the product of a turbulent process of the formation and dissolution of large numbers of business establishments. An analysis of plant closings should attempt to minimize the statistical noise generated by this usual business turnover in a dynamic economy. Our concern is focused on massive layoffs and closings, which displace large numbers of workers and seriously affect the economic health of the communities in which they are located. Business dissolutions, or establishments that cease operation, include businesses that close for any reason, including bankruptcy, retirement of the owner, financial failure, insufficient profitability, and so on. In an effort to eliminate most of the usual turnover, this study limits the definition of plant closings to those dissolving establishments that are affiliates of business enterprises with 100 or more employees. A more refined definition further requires that the establishment be a subordinate of a multiestablishment firm, that is, a branch or subsidiary.

## Recent Evidence on Plant Closings

Plant closings, as narrowly defined in this article, constituted one-fourth of all dissolutions of manufacturing establishments between 1978 and 1982, but accounted for over two-thirds of jobs lost in these dissolutions. These figures for plant closings translate into a loss of 3.5 to 4 million jobs, or one out of every four jobs in large manufacturing branch plants. Reflecting the dominance of large firms in the manufacturing sector, the overall industry rates of employment loss from dissolutions of all subordinate establishments closely resemble those of plant closings in large firms. These overall rates are not significantly affected by the higher rates of loss in small firms (see table 8–1).

The most striking contrast presented in table 8–1 is between the very high closing rates in subordinate establishments in the upper portion of the table and the low rates in nonsubordinate establishments. Within the same firm size classes, based on the number of employees at a firm, closing rates range from two to five times higher for the subordinate establishments. Clearly, for multiestablishment firms, closing a branch operation is a policy option frequently exercised within the context of a broad management strategy. As Frederick M. Sherer has pointed out,

> under certain cost conditions, production cutbacks can be accomplished more economically by shutting down one or more whole plants in an integrated network than by reducing output at each of many independent plants.[1]

### Cyclicality

The propensity of large firms to exercise their option to close branches is demonstrated in the dramatic increase in closing rates of subordinate establishments in the second period (1980–82) over the first period (1978–80). These

**Table 8–1**
**Manufacturing Dissolutions and Employment Loss by Establishment Type, 1978–82**
*(percentage)*

|  | Number of Employees at a Firm | | All Firms |
|---|---|---|---|
|  | *Under 100* | *100 or More* | *All Firms* |
| Subordinate establishments*a* | | | |
| *1978–1980* | | | |
| Closing rate | 18.9 | 12.3 | 13.6 |
| Employment loss | 17.8 | 7.9 | 8.1 |
| *1980–1982* | | | |
| Closing rate | 38.1 | 27.2 | 29.3 |
| Employment loss | 24.8 | 16.1 | 16.2 |
| Increase in rate of | | | |
|     employment loss | 39.3 | 103.8 | 100.0 |
| Nonsubordinate establishments*b* | | | |
| *1978–1980* | | | |
| Closing rate | 9.6 | 3.0 | 9.3 |
| Employment loss | 7.8 | 3.2 | 5.3 |
| *1980–1982* | | | |
| Closing rate | 10.3 | 5.1 | 10.1 |
| Employment loss | 9.4 | 5.9 | 7.5 |
| Increase in rate of | | | |
|     employment loss | 20.5 | 84.4 | 41.5 |

*Source:* Unpublished tabulations, U.S. Establishment and Enterprise Microdata Base, Brookings Institution, Washington, D.C.
*a*Branch and subsidiary establishments.
*b*Independent firms and owners of multiestablishment firms.

two periods represent rather distinct portions of the business cycle. In the first period, the real gross national product increased by about 2.5 percent, whereas during the second period it declined by slightly more than that amount. Unemployment increased throughout the four years, but grew at a much higher rate during the second two-year period. Manufacturing employment growth, as measured in the USEEM data base, was about 6.0 percent between 1978 and 1980, but was –5.2 percent between 1980 and 1982. In sum, the first period is generally expansionary, and the second, recessionary.

Closing rates for subordinate establishments doubled in every size class in the second period, while nonsubordinate establishments registered more moderate increases. Particularly notable is the doubling of the rate of employment loss due to subordinate plant closings from about 8 percent to over 16 percent. Rates of employment loss in dissolutions in subordinates of smaller firms increased by only 39 percent. Similarly, employment loss rates in large nonsubordinate firms increased 84 percent, while in small nonsubordinate firms employment loss rates grew only 20 percent.

Closing rates in small independent firms exhibit much less variability between the two periods. In fact, rates of employment loss in dissolutions in firms with fewer than 20 employees declined slightly in the second period, possibly indicating a countercyclical tendency for this size class in aggregate. Employment in manufacturing branches of large firms is extremely volatile. Closings of such plants accounted for more than two-thirds of all manufacturing jobs lost in dissolutions between 1978 and 1982. Consistent with the results of other analyses, these findings indicate that large manufacturing firms are much more responsive to cyclical fluctuations in the economy than are small firms.

## Rates of Job Replacement, 1976–82

The relative importance of the jobs lost in plant closings depends in large part on whether and how they are replaced. Although detailed data on job replacement differentiating subordinate and nonsubordinate establishments are not currently available, we can examine the generation of new employment opportunities in different employment size classes of firms. The two extreme firm size classes presented in table 8–2, "under 20 employees" and "100 or more employees," represent the vast majority of independents and of subordinates, respectively. For the remainder of this analysis, the latter firm size class will be used to represent closings.

**Table 8–2**
**Employment Losses and Job Replacement Rates, 1976–82**

|  | All Industries | Manufacturing | Services |
|---|---|---|---|
| **1976–82** | | | |
| Employment loss (in thousands) | | | |
| Closings[a] | 16,177 (33.4) | 5,439 (30.9) | 3,471 (27.3) |
| Turnover[b] | 8,605 (11.3) | 1,259 (5.8) | 1,449 (7.9) |
| Job replacement rates | | | |
| All firms | 1.51 | 1.08 | 2.12 |
| Firms with under 20 employees | 2.37 | 2.61 | 3.87 |
| Firms with 20–90 employees | 1.49 | 1.33 | 2.13 |
| Firms with 100 or more employees | 1.25 | 0.91 | 1.72 |
| Interim changes | | | |
| Closings—employment loss (in thousands) | | | |
| 1976–78 | 4,511 (6.0) | 1,760 (8.1) | 864 (4.7) |
| 1978–80 | 3,912 (7.5) | 1,216 (5.4) | 858 (4.1) |
| 1980–82 | 7,754 (9.1) | 2,464 (10.2) | 1,749 (9.1) |
| Replacement rates | | | |
| 1976–78 | 1.65 | 1.26 | 2.61 |
| 1978–80 | 2.06 | 1.80 | 2.72 |
| 1980–82 | 1.09 | 0.57 | 1.52 |

*Source:* Unpublished tabulations, U.S. Establishment and Enterprise Microdata Base.

*Note:* Numbers in parentheses represent employment loss in terms of percentage of 1976 base employment.

[a]Closings are dissolutions of establishments in firms with 100 or more employees.

[b]Turnover covers establishments of firms with fewer than 100 employees.

Jobs lost due to dissolutions can be replaced in two ways. First, new businesses can be created or new branch plants opened. Second, employment in some existing establishments can expand sufficiently to offset both contracting employment in other establishments and losses due to dissolutions. Tables that follow present gross figures for employment losses associated with dissolutions of establishments, employment gains associated with formations of new establishments, the sum of formation gains and the net expansion in existing establishments, and job replacement rates. Table 8–2 presents job replacement rates for all industries, for the manufacturing sector, and for the nongovernmental service sector.

Within each industrial sector, the replacement rates are consistently lower for larger firms. Replacement rates in these large firms decline abruptly in the 1980–82 period. During the recent recession, large firms in the manufacturing sector managed to replace only two out of every ten jobs lost in dissolutions. The poor employment growth performance of large firms accounted for all of the net decline in manufacturing employment between 1980 and 1982 (–5.2 percent). The smallest firm size class, under 20 employees, retained its healthy performance in all three observation periods, creating two to three new jobs for every one lost in dissolutions. Surprisingly, the middle sized class (20 to 99 employees), actually improved its generation of replacement jobs during the downturn of the second time period, from 1.3 jobs to 2.3. Unfortunately, the predominance of large firm employment is reflected in the overall performance of the industry, which registered net losses between 1980 and 1982, despite the substantial growth in the smaller size classes. More detailed industry breakouts might reveal that the smaller firms are engaged in different aspects of manufacturing than are the large firms. Small or mid-sized manufacturing firms might be concentrated in those industries that are less vulnerable to cyclical fluctuations.

In order to compare the differential job replacement performance of large and small firms in manufacturing to that in an expanding industrial sector, the job replacement in the service industries was examined. The smallest service firms also demonstrated the highest job replacement ratios and the least cyclicality in aggregate. In contrast to the manufacturing sector, one would expect service sector businesses of different sizes to be more equally distributed across narrowly defined industrial categories and, therefore, to be equally affected by recessionary conditions. This was not, in fact, the case. The large firms suffered the greatest deterioration in job replacement rates in the second period, but the behavior of the mid-range firms in this industrial sector more closely resembled that of the large firms.

Both the rates of employment loss and the job replacement rates reflect the reorientation of the economy. As shown in table 8–2, closing rates are lower for the service industries than for manufacturing industries in all periods and across all size classes. Reinforcing the industrial shift, replacement rates are almost twice as high in services as in manufacturing.

## Regional Aspects of Plant Closings and Job Replacement

Regional employment losses associated with large firm closings in all industries, along with job replacement ratios, are presented in table 8–3 for the whole period 1976–82 and in tables 8–4 and 8–5 for the interim period between 1972 and 1982. Differences in regional rates of job loss due to closings are somewhat surprising. The rates of loss are higher not in the declining regions as one would expect, but in the regions experiencing better than average net employment growth. For example, between 1976 and 1980 the region with the greatest

**Table 8–3**
**Employment Loss in Closings and Job Replacement Rates: All Industries by Region, 1976–82**

| Region | Employment Closings[a] (in thousands) | Loss in Percentage[b] | Replacement Rates |
|---|---|---|---|
| U.S. total | 16,177 | (33.4) | 1.50 |
| New England | 852 | (28.8) | 1.49 |
| Mid-Atlantic | 2,696 | (29.6) | 1.17 |
| East North Central | 3,077 | (29.7) | 1.23 |
| West North Central | 958 | (29.2) | 1.68 |
| South Atlantic | 2,639 | (35.9) | 1.51 |
| East South Central | 947 | (34.0) | 1.29 |
| West South Central | 1,808 | (38.0) | 1.93 |
| Mountain | 688 | (40.2) | 2.15 |
| Pacific | 2,512 | (41.2) | 1.70 |

*Source:* Unpublished tabulations, U.S. Establishment and Enterprise Microdata Base.
[a]Closings in firms with 100 or more employees.
[b]Employment loss as a percentage of 1976 base employment.

**Table 8–4**
**Employment Loss in Dissolutions of Manufacturing Establishments by Region**
*(percentage)*

| Region | 1972–74[a] | 1974–76[a] | 1978–80[b] | 1980–82[b] |
|---|---|---|---|---|
| New England | 8.0 | 8.1 | 7.4 | 11.4 |
| Mid-Atlantic | 7.9 | 11.0 | 7.1 | 11.5 |
| East North Central | 5.4 | 8.9 | 6.4 | 10.8 |
| West North Central | 7.0 | 9.9 | 5.6 | 10.5 |
| South Atlantic | 8.1 | 11.6 | 7.3 | 13.7 |
| East South Central | 7.6 | 10.3 | 6.0 | 13.7 |
| West South Central | 7.8 | 11.1 | 7.3 | 12.6 |
| Mountain | 9.2 | 14.1 | 7.1 | 14.7 |
| Pacific | 8.6 | 10.9 | 9.0 | 14.0 |

[a]David Birch, "The Job Generation Process" (Final report to the Economic Development Administration, U.S. Department of Commerce, grant no. OER-608-G78-7, June 1979).
[b]Unpublished tabulations, U.S. Establishment and Enterprise Microdata Base.

**Table 8–5**
**Manufacturing Job Replacement Ratios for All Establishments**
**by Region**

| Region | 1972–74[a] | 1974–76[a] | 1978–80[b] | 1980–82[b] |
|---|---|---|---|---|
| U.S. | 1.60 | 0.67 | 1.85 | 0.57 |
| New England | 1.42 | 0.45 | 1.58 | 0.52 |
| Mid-Atlantic | 1.06 | 0.28 | 1.45 | 0.52 |
| East North Central | 1.85 | 0.57 | 1.33 | 0.22 |
| West North Central | 2.15 | 0.65 | 2.21 | 0.57 |
| South Atlantic | 1.65 | 0.61 | 2.16 | 0.69 |
| East South Central | 1.97 | 0.71 | 1.83 | 0.32 |
| West South Central | 1.86 | 1.16 | 2.94 | 1.05 |
| Mountain | 1.70 | 1.01 | 3.01 | 0.88 |
| Pacific | 1.60 | 1.47 | 2.10 | 0.86 |

[a]David Birch, "The Job Generation Process."
[b]Unpublished tabulations, U.S. Establishment and Enterprise Microdata Base.

percentage loss was the Pacific, which lost 2.5 million jobs, or 41 percent of its 1976 work force.

Clearly, the rate of employment loss from closings is not as important as the rate at which those jobs are replaced. Of the four regions in the North, only the West North Central achieved a better-than-average rate of job replacement, whereas replacement rates in all regional divisions of the West were well above average.

Regional shifts in manufacturing employment are even more pronounced, as reflected in the closing and replacement rates for that sector in tables 8–4 and 8–5. In each of the interim periods noted between 1972 and 1982, manufacturing dissolution rates were generally higher in the expanding regions. Again job replacement rates flagged the geographic undercurrents of change in the distribution of manufacturing employment in the United States. All regions of the North have had persistently lower than average rates of job creation since 1974, which has helped effect the shift toward the South and West.

It must be noted, however, that what might be called the winning regions have historically had relatively low shares of the nation's industry, so these recent trends have operated to offset existing discrepancies. Nonetheless, this seemingly equitable redistribution at the macroeconomic level does little to ameliorate the negative impact on workers and communities in the losing regions.

## Replacement Rates in Large Firms

Because it is predominantly the changes in employment in large, multiestablishment firms that determine replacement rates and net manufacturing employment growth, location decisions of these companies largely determine regional performance. The declining manufacturing employment in the North indicates that the regional distribution of new plant openings is not the same as

**Table 8-6**
**Manufacturing and Business Service Establishments of Large Firms:**
**Distribution of Employment in Formations and Dissolutions, 1976-80**
*(percentage)*

| Region | Employees in Subordinates (1976) | Share of Employment (1976) | Share of Employees in Dissolutions | Share of Employees in Formations |
|---|---|---|---|---|
| United States | | 100.0 | 100.0 | 100.0 |
| Northeast | 51.7 | 26.5 | 24.9 | 19.6 |
| North Central | 59.7 | 30.6 | 28.0 | 27.0 |
| South | 64.9 | 29.4 | 31.4 | 35.9 |
| West | 56.3 | 13.4 | 15.7 | 17.8 |

*Source:* Unpublished tabulations, U.S. Establishment and Enterprise Microdata Base.

that of the plant closings, resulting in localized economic decline. For the period 1976–80, table 8-6 presents the regional shares of employment, employment loss in dissolutions, and employment gains from formations of manufacturing and business service establishments in large firms. The regional distribution of employment losses from dissolutions is much more similar to the distribution of total employment than is the distribution of employment gains from formations. The North Central and the Northeast, in particular, received notably lower shares of new jobs from formations than shares of jobs lost in dissolutions. The opposite was true for the West and South, which together garnered about 338,000 more jobs from new formations than were lost in dissolutions in large manufacturing firms.

Given these findings concerning the greater volatility of jobs in subordinate establishments, one would expect those regions with disproportionately high shares of such establishments to experience more fluctuation in rates of employment gain and loss. Comparison of the figures in the first column of table 8-6 with those in table 8-5 supports this surmise. The two regions with substantially higher shares of manufacturing and business service employment in branch and subsidiary establishments—the North Central and the South—did experience the greatest deterioration in job replacement rates during the downturns in 1974–76 and 1980–82. The employment growth benefits derived from attracting such establishments are mitigated by increased vulnerability to cyclical fluctuations. The job replacement rates in the East North Central, for example, show steady decline since 1972. That region's replacement rate dropped from 0.57 in the 1974–76 downturn to only 0.22 in the current recessionary period (see table 8-5). This mismatch between the locus of job creation results in intense regional inequities. In 1980 the unemployment rate in Houston was only 4.7 percent, while in Detroit it was over 14 percent.

## Jobs Lost versus Jobs Created

In the restructuring of the economy, there is another impetus toward inequities—the mismatch between the skills of the displaced workers and the requirements of new employment opportunities. Between 1976 and 1980, about 42 percent of the net growth in manufacturing employment and 26 percent of the employment gains from formations of new manufacturing establishments were in high technology manufacturing industries. This shift within the manufacturing sector to more technically advanced industries complicates the process of job replacement in the manufacturing sector.

It is often assumed that the new employment opportunities created by the expansion of high technology industries are better paying, less hazardous, and more personally satisfying. In 1980 high technology's average rates of compensation of manufacturing production workers in particular and all employees in general were higher than those in low technology manufacturing industries. High technology production workers earned about $.50 more per hour and the average high technology employee made $3500 more per year than their counterparts in the low technology manufacturing sector.

As with most aggregate statistics, these figures disguise discrepancies in the occupational structure and employee compensation patterns within the industries. Scientific, engineering, and technical workers make up about 18 percent of total employment in the high technology industries. A majority of high technology jobs are for assemblers and other production workers. How do these production workers compare with those in the declining industries that are experiencing a majority of the plant closings?

Demographic characteristics of the high technology production workers suggest that new jobs created in these industries may not absorb the blue-collar production workers being displaced in declining manufacturing industries. For example, in two of the fastest growing high technology industries—the electronic computing industry and the communication equipment industry—about 35 percent and 40 percent of the employees, respectively, are women. In the steel and automobile industries, they account for less than 15 percent. In addition, the high technology industries have lower unionization rates and lower proportions of minority production workers.

In sum, the new jobs being created in the manufacturing sector do not very well match those being eliminated. Although not conclusive, such evidence should be taken into account when considering development programs targeted at high technology industries, particularly in regions with high concentrations of declining manufacturing industries.

## Summary and Conclusion

Closings of large firms eliminated over 16 million jobs between 1976 and 1982. Almost one-third of these were in the manufacturing sector. Although small manufacturing firms (those with fewer than 100 employees) registered annual net employment growth rates around 6 percent between 1976 and 1982, larger firms contracted their employment. Rates of employment loss due to closings of manufacturing branches doubled in the 1980–82 period over the 1978–80 period, combining with lower replacement rates to produce a net decline of 5.2 percent in manufacturing. By contrast, the service sector performed strongly throughout the six-year period, with lower rates of employment loss and much higher rates of job replacement.

The mismatch between the regional distribution of employment losses from closings and the distribution of formations and net change in existing establishments, especially in large manufacturing firms, produced uneven effects across the nation. The resultant regional redistribution of manufacturing activity especially worsened the position of the northern regions. Continuing a long-term shift of activity, the regions experiencing the highest rates of growth in employment and in population were those that previously had the smallest shares of each, namely, the South and the West. Formations of new businesses and net employment are apparently following patterns of population migration. Unfortunately, displaced manufacturing workers tend to be concentrated in the North, where low business formation and job replacement rates exacerbate current regional distress.

Since the mid-1970s, plant closings have eliminated over 900,000 manufacturing jobs every year. Although this represents less than 10 percent of the manufacturing labor force, these displaced workers are concentrated in particular industries and, more important, in particular regions with relatively low rates of job creation. Despite some success in replacing lost jobs, remaining inequities in the regional distribution of business closings and business formations and the mismatch between displaced workers and new employment opportunities in manufacturing have combined to create important adjustment problems. Trying to halt or reverse the reorientation of the manufacturing sector is neither feasible nor desirable, but the problems associated with the transition are significant and deserve national attention.

## Note

1. Frederick M. Sherer, *Industrial Market Structure and Economic Performance,* 2nd ed (Chicago, Ill.: Rand McNally, 1980), p. 103.

# 9
# Reemployment and Earnings

*Paul O. Flaim*
*Ellen Sehgal*

**W**hat happens to workers when recessions close their plants or severely curtail operations? What happens to those who lose their jobs because of structural problems of the type that have recently affected some of our key manufacturing industries? How many of these workers manage to return to the same or similar jobs as economic conditions improve? How many remain without jobs or eventually settle for different and usually lower paying jobs?

In an attempt to obtain answers to these questions in connection with the 1980–81 and 1982–83 recessions, two agencies of the U.S. Department of Labor arranged for a special household survey in January 1984. Among the principal findings:

1. A total of 11.5 million workers 20 years of age and over lost jobs because of plant closings or employment cutbacks over the January 1979–January 1984 period. Those who had worked at least 3 years on their jobs, the focus of this study, numbered 5.1 million.

2. About one-half of the 5.1 million workers reported they had become displaced because their plants or businesses closed down or moved. Two-fifths reported job losses due to "slack work" (or insufficient demand), and the rest said their shifts or individual jobs had been abolished.

3. About 3.5 million of the displaced workers had collected unemployment insurance benefits after losing their jobs. Nearly one-half of these reported they had exhausted their benefits.

4. Many no longer had health insurance coverage, including some who subsequently found work.

5. Of the 5.1 million displaced workers, about 3.1 million had become reemployed by January 1984, but often in different industries than in the ones

Paul O. Flaim and Ellen Sehgal, "Displaced Workers of 1979–83: How Well Have They Fared?" *Monthly Labor Review* 108, no. 6 (June 1985), pp. 3–16.

they had previously worked. About 1.3 million were looking for work, and the remaining 700,000 had left the labor force.

6.  Of the 3.1 million displaced workers who were reemployed, about one-half were earning as much or more in the jobs they held when surveyed than in the ones they had lost. However, many others had taken large pay cuts, often exceeding 20 percent.

7.  Blacks accounted for about 600,000 of the 5.1 million displaced workers, and Hispanics made up 300,000. The proportion reemployed as of January 1984 was relatively small for both of these groups—42 percent for blacks and 52 percent for Hispanics. Conversely, the proportions looking for work were relatively high—41 percent for blacks and 34 percent for Hispanics.

These data are discussed in detail below, as are the concepts of displacement and how they were applied in this special survey.

## Concept and Measurement

Concern over displaced workers began to grow during the early 1980s when it was feared that a large part of the employment cutbacks taking place in some industries might be permanent, leaving many of the affected workers with little hope of reemployment in the same industry. The steel industry and the auto industry were prime examples of this type of situation. And many other manufacturing industries, particularly in the hard goods sector, were similarly affected by a combination of cyclical factors and such deep-seated structural problems as plants that were no longer competitive in the face of foreign imports.

Given this situation, it was feared that a large number of workers who had spent many years in relatively high-paying jobs would suddenly find themselves without work and with little hope of finding similar employment. These are the persons generally referred to as displaced (or dislocated) workers. While there has never been a precise definition of such workers, the term is generally applied to persons who have lost jobs in which they had a considerable investment in terms of tenure and skill development and for whom the prospects of reemployment in similar jobs are rather dim.

Because there were only widely different estimates of a rather speculative nature as to the number of such workers as of late 1983, the Employment and Training Administration contracted with the Bureau of Labor Statistics to design a special survey to identify and count them. The survey was planned as a supplement to the Bureau of the Census' Current Population Survey (which provides the monthly estimates of unemployment). It was first of all decided to identify all adult workers who had lost a job over the 1979–83 period because of "a plant closing, an employer going out of business, a layoff from which . . . (the worker in question) was not recalled, or other similar reasons." For these workers, a series of questions would then follow to determine the precise reason for the job loss, the nature of the job in terms of industry and occupation, how long the workers had held the job, how much they had been earning, and whether they had been covered by group health insurance. Other questions focused on the period of unemployment which might have followed the job loss, including the receipt and possible exhaustion of unemployment insurance benefits, and the possible loss of health insurance coverage. If the worker in question was again employed at the time of the interview, additional information was sought on the earnings on the current job.

This sequence of questions yielded information that allowed much flexibility in deciding who among these workers could properly be considered as "displaced." Different cutoffs could be made in terms of the years of tenure on the job lost, the period of unemployment resulting, the extent of the cut in wages incurred in taking a new job, and other possible factors.

In publishing the preliminary results of the survey, and in conducting the more detailed analysis discussed in this article, the only cutoffs that were made were those deemed absolutely necessary in order not to stray too far from the general consensus as to who is and who is not a displaced worker. Thus, an exclusion was first made with regard to workers whose job losses could not be categorized definitively as displacements—those attributed either to seasonal factors or to a variety of miscellaneous reasons that could not be easily classified. An additional exclusion was made with regard to all workers with less than three years in the jobs they had lost.

Summarizing the results of the survey, a total of 13.9 million workers twenty years of age and over were initially identified as having lost a job over the January 1979–January 1984 period because of plant closings, employers going out of business, or layoffs from which they had not been recalled. Further probing disclosed that about 2.4 million of this total had lost their jobs because of seasonal causes or a variety of other reasons which could not be easily classified. These were dropped from the universe to be examined.

Of the remaining 11.5 million workers, a large proportion had only been at their jobs for a relatively short time before they were dismissed. For example, 4.4 million had been at their jobs a year or less. To focus only on workers who had developed a rather firm attachment to their jobs, the universe to be studied was limited to those with at least three years of tenure on the jobs they lost. As noted, these numbered 5.1 million. Had a more liberal cutoff of two years been used as a parameter, the count of displaced workers would have been raised to 6.9 million. On the other hand, the imposition of a five-year cutoff would have lowered the total to 3.2 million.

Not all of the 5.1 million workers deemed to have been displaced should be regarded as having suffered serious economic consequences. While a great majority were indeed either still unemployed or had taken jobs entailing a drop in pay, or had left the labor force, there were also many for whom the job loss had been only a temporary setback. Some had apparently been out of work for only a very short period and, as already noted, many were actually earning more when surveyed than in the jobs they had lost. In short, while all of the 5.1 million workers had clearly been displaced from a job at some point over the 1979–83 period, not all could be properly regarded as being still "displaced" when surveyed in January 1984. And even among the majority for whom the "displaced" label was still applicable when surveyed, there were many who probably found suitable employment in subsequent months.

## Who Were the Displaced?

A large number of the 5.1 million workers who had been displaced from their jobs fit the conventional description. They were primarily men of prime working age, had lost typical factory jobs, were heavily concentrated in the Midwest and other areas with heavy industry, and, if reemployed, were likely to have shifted to other industries. However, the universe also included persons from practically all industry and occupational groups, a large number of whom were women.

### Age, Sex, Race, and Hispanic Origin

As shown in table 9–1, men 25 to 54 years of age accounted for nearly 2.6 million of the displaced workers, or slightly more than one-half. There were 200,000 men age 20 to 24, about 460,000 men 55 to 64, and 90,000 in the 65 and over group. The younger the workers, the more likely they were to have found new jobs after their displacement. As shown in table 9–1, the proportion reemployed as of January 1984 ranged from a high of 72 percent for men age 20 to 24 to a low of 17 percent for those 65 years of age and over. Most of the men in the latter age group had apparently retired after losing their jobs.

**Table 9–1**
**Employment Status of Displaced Workers by Age, Sex, Race, and Hispanic Origin, January 1984**
*(percentage)*

| Characteristic | Number (thousands)[a] | Total | Employed | Unemployed | Not in the Labor Force |
|---|---|---|---|---|---|
| *Total* | | | | | |
| Total, 20 years and over | 5,091 | 100.0 | 60.1 | 25.5 | 14.4 |
| 20 to 24 years | 342 | 100.0 | 70.4 | 20.2 | 9.4 |

| | | | | | |
|---|---|---|---|---|---|
| 25 to 54 years | 3,809 | 100.0 | 64.9 | 25.4 | 9.6 |
| 55 to 64 years | 748 | 100.0 | 40.8 | 31.8 | 27.4 |
| 65 years and over | 191 | 100.0 | 20.8 | 12.1 | 67.1 |
| *Men* | | | | | |
| Total, 20 years and over | 3,328 | 100.0 | 63.6 | 27.1 | 9.2 |
| 20 to 24 years | 204 | 100.0 | 72.2 | 21.7 | 6.1 |
| 25 to 54 years | 2,570 | 100.0 | 68.2 | 26.8 | 5.0 |
| 55 to 64 years | 461 | 100.0 | 43.6 | 34.1 | 22.3 |
| 65 years and over | 92 | 100.0 | 16.8 | 12.9 | 70.3 |
| *Women* | | | | | |
| Total, 20 years and over | 1,763 | 100.0 | 53.4 | 22.5 | 24.2 |
| 20 to 24 years | 138 | 100.0 | 67.8 | 18.0 | 14.2 |
| 25 to 54 years | 1,239 | 100.0 | 58.0 | 22.6 | 19.4 |
| 55 to 64 years | 287 | 100.0 | 36.3 | 28.0 | 35.7 |
| 65 years and over | 99 | 100.0 | 24.6 | 11.3 | 64.1 |
| *White* | | | | | |
| Total, 20 years and over | 4,397 | 100.0 | 62.6 | 23.4 | 13.9 |
| Men | 2,913 | 100.0 | 66.1 | 25.1 | 8.8 |
| Women | 1,484 | 100.0 | 55.8 | 20.2 | 24.1 |
| *Black* | | | | | |
| Total, 20 years and over | 602 | 100.0 | 41.8 | 41.0 | 17.1 |
| Men | 358 | 100.0 | 43.9 | 44.7 | 11.4 |
| Women | 244 | 100.0 | 38.8 | 35.6 | 25.6 |
| *Hispanic origin* | | | | | |
| Total, 20 years and over | 282 | 100.0 | 52.2 | 33.7 | 14.1 |
| Men | 189 | 100.0 | 55.2 | 35.5 | 9.3 |
| Women | 93 | 100.0 | 46.3 | 30.0 | 23.6 |

*Note:* Detail for the above race and Hispanic-origin groups will not sum to totals because data for the "other races" group are not presented and Hispanics are included in both the white and black population groups.

[a]Data refer to persons with tenure of three years or more who lost or left a job between January 1979 and January 1984 because of plant closings or moves, slack work, or the abolishment of their positions or shifts.

The women who had been displaced from their jobs numbered nearly 1.8 million, with 1.2 million of them in the 25 to 54 age group. As indicated by table 9–1, these women were less likely than the displaced men to have returned to work as of January 1984 and were far more likely to have left the labor force regardless of their age.

About 600,000 of the displaced workers were black, and less than one-half of them were reemployed when interviewed (42 percent). The proportion unemployed was almost as large (41 percent). Hispanic workers accounted for about 280,000 of the displaced. For them, the proportion reemployed (52 percent) was higher than for blacks but considerably lower than for whites. Of the whites who had been displaced, over three-fifths were reemployed and less than a quarter were unemployed.

*Industry and Occupation*

Nearly 2.5 million of the displaced workers, or almost one-half of the total, had lost jobs in manufacturing, an industry group that now accounts for less than one-fifth of total employment. Some of the key durable goods industries which were most severely affected by the recessionary contractions of demand as well as by more fundamental structural problems figured most prominently as the sources of displacements. There were, for example, about 220,000 workers who had lost jobs in the primary metals industry, 400,000 who had worked in machinery (except electrical), and 350,000 had been in the transportation equipment industry, with autos accounting for 225,000 of the latter (see table 9–2).

Reflecting primarily the long-lasting nature of the problems of the steel industry and of the areas where its plants are (or were) located, less than one-half (46 percent) of the workers who had been displaced from primary metal jobs were reemployed when surveyed. About 39 percent were unemployed, and 16 percent had left the labor force. However, the reemployment percentage for workers displaced from jobs in the nonelectrical machinery industry (62 percent) and the transportation equipment industry (63 percent) was considerably higher. But even among these workers, many were now working in different industries, and usually at lower wages.

While these troubled durable goods industries figured most prominently as sources of workers' displacements, it should be noted that other industries, both within and outside the manufacturing sector, had also contributed heavily to the problem. For example, 800,000 workers had been displaced from jobs in the various nondurable goods industries, 500,000 had been in retail sales, another 500,000 in services, and 400,000 in construction.

In terms of their occupational distribution, a large number of displaced workers (1.8 million) had lost jobs as operators, fabricators, and laborers—the typical jobs on a factory floor. But all occupational groups had contributed to the displacement problem. There were, for example, 700,000 persons who had

**Table 9-2**
**Employment Status of Displaced Workers by Industry and Class of Worker of Lost Job, January 1984**
*(percentage)*

| Industry | Number (thousands)[a] | Total | Employed | Unemployed | Not in the Labor Force |
|---|---|---|---|---|---|
| Total, workers 20 years and over[b] | 5,091 | 100.0 | 60.1 | 25.5 | 14.4 |
| Nonagricultural private wage and salary workers | 4,700 | 100.0 | 59.8 | 25.8 | 14.4 |
| Mining | 150 | 100.0 | 60.4 | 31.0 | 8.6 |
| Construction | 401 | 100.0 | 55.0 | 30.7 | 14.3 |
| Manufacturing | 2,483 | 100.0 | 58.5 | 27.4 | 14.1 |
| Durable goods | 1,675 | 100.0 | 58.2 | 28.9 | 12.9 |
| Lumber and wood products | 81 | 100.0 | 67.9 | 19.1 | 13.0 |
| Furniture and fixtures | 65 | 100.0 | c | c | c |
| Stone, clay, and glass products | 75 | 100.0 | 47.5 | 30.5 | 22.0 |
| Primary metal industries | 219 | 100.0 | 45.7 | 38.7 | 15.6 |
| Fabricated metal products | 173 | 100.0 | 62.0 | 32.2 | 5.8 |
| Machinery, except electrical | 396 | 100.0 | 62.3 | 27.4 | 10.3 |
| Electrical machinery | 195 | 100.0 | 48.2 | 34.5 | 17.3 |
| Transportation equipment | 354 | 100.0 | 62.6 | 26.0 | 11.4 |
| Automobiles | 224 | 100.0 | 62.9 | 24.0 | 13.1 |
| Other transportation equipment | 130 | 100.0 | 62.1 | 29.4 | 8.5 |
| Professional and photo-graphic equipment | 54 | 100.0 | c | c | c |
| Other durable goods industries | 62 | 100.0 | c | c | c |
| Nondurable goods | 808 | 100.0 | 59.1 | 24.2 | 16.7 |
| Food and kindred products | 175 | 100.0 | 52.5 | 32.6 | 15.0 |
| Textile mill products | 80 | 100.0 | 59.8 | 26.2 | 13.9 |
| Apparel and other finished textile products | 132 | 100.0 | 63.0 | 14.2 | 22.8 |
| Paper and allied products | 60 | 100.0 | c | c | c |
| Printing and publishing | 103 | 100.0 | 58.0 | 22.9 | 19.1 |
| Chemical and allied products | 110 | 100.0 | 64.0 | 27.3 | 8.7 |
| Rubber and miscellaneous plastics products | 100 | 100.0 | 62.8 | 18.3 | 18.8 |
| Other nondurable goods industries | 49 | 100.0 | c | c | c |
| Transportation and public utilities | 336 | 100.0 | 57.9 | 26.8 | 15.3 |
| Transportation | 280 | 100.0 | 58.8 | 30.5 | 10.7 |
| Communication and other public utilities | 56 | 100.0 | c | c | c |
| Wholesale and retail trade | 732 | 100.0 | 61.4 | 21.6 | 16.9 |
| Wholesale trade | 234 | 100.0 | 69.6 | 22.0 | 8.4 |
| Retail trade | 498 | 100.0 | 57.6 | 21.5 | 20.9 |
| Finance, insurance, and real estate | 93 | 100.0 | 78.5 | 12.4 | 9.1 |
| Services | 506 | 100.0 | 65.0 | 20.5 | 14.5 |
| Professional services | 187 | 100.0 | 64.0 | 19.8 | 16.1 |

*Table 9–2 continued*

| Industry | Number (thousands)$^a$ | Total | Employed | Unemployed | Not in the Labor Force |
|---|---|---|---|---|---|
| Other service industries | 318 | 100.0 | 65.6 | 20.9 | 13.5 |
| Agricultural wage and salary workers | 100 | 100.0 | 69.9 | 22.9 | 7.2 |
| Government workers | 248 | 100.0 | 63.3 | 18.7 | 18.0 |
| Self-employed and unpaid family workers | 25 | 100.0 | $^c$ | $^c$ | $^c$ |

$^a$Data refer to persons with tenure of three years or more who lost or left a job between January 1979 and January 1984 because of plant closings or moves, slack work, or the abolishment of their positions or shifts.

$^b$Total includes a small number who did not report industry or class of worker.

$^c$Data not shown where base is less than 75,000.

lost managerial and professional jobs, 1.2 million who had been in technical, sales, and administrative jobs, and slightly over 1 million who had been in precision production, craft, and repair jobs (see table 9–3).

In general, the more skilled the occupation the more likely was the displaced worker to be reemployed. Thus, about 75 percent of those who had been in managerial and professional jobs were back at work when interviewed. In contrast, among the workers who had lost low-skill jobs as handlers, equipment cleaners, helpers, and laborers, less than one-half were working in January 1984.

*Regional Distribution*

Although displaced workers were found in all regions of the country, a particularly large number (about 1.2 million) was found to reside in the East North Central area, which includes the heavily industrialized states of the Midwest. (See table 9–4 for regional data and area definitions.) Another large concentration of such workers (800,000) was found in the Middle Atlantic area, which consists of New Jersey, New York, and Pennsylvania.

The severity of the job losses incurred in these two areas during 1979–83 was denoted not only by the relatively large numbers of displaced workers found within them in January 1984, but also by the fact that the proportion that had managed to return to work, either in their former jobs or entirely new ones, barely exceeded 50 percent. As a further indication of the seriousness of the displacement problem in the East North Central area, this region was found to contain nearly one-third of the displaced workers who were unemployed in January 1984 (400,000 out of 1.3 million), and almost one-half of them were reported as having been jobless six months or more.

**Table 9–3**
**Employment Status of Displaced Workers by Occupation of Lost Job, January 1984**
*(percentage)*

| Occupation | Number (thousands)[a] | Total | Employed | Unemployed | Not in the Labor Force |
|---|---|---|---|---|---|
| Total, workers 20 years and over[b] | 5,091 | 100.0 | 60.1 | 25.5 | 14.4 |
| Managerial and professional specialty | 703 | 100.0 | 74.7 | 16.6 | 8.8 |
| Executive, administrative, and managerial | 444 | 100.0 | 75.7 | 15.6 | 8.7 |
| Professional specialty | 260 | 100.0 | 72.9 | 18.2 | 8.9 |
| Technical, sales, and administrative support | 1,162 | 100.0 | 60.6 | 21.1 | 18.3 |
| Technicians and related support | 122 | 100.0 | 67.9 | 25.3 | 6.8 |
| Sales occupations | 468 | 100.0 | 66.7 | 14.6 | 18.7 |
| Administrative support, including clerical | 572 | 100.0 | 54.1 | 25.5 | 20.5 |
| Service occupations | 275 | 100.0 | 51.0 | 24.1 | 24.9 |
| Protective service | 32 | 100.0 | c | c | c |
| Service, except private household and protective | 243 | 100.0 | 53.0 | 23.6 | 23.4 |
| Precision production, craft, and repair | 1,042 | 100.0 | 61.6 | 26.1 | 12.3 |
| Mechanics and repairers | 261 | 100.0 | 61.3 | 29.3 | 9.4 |
| Construction trades | 315 | 100.0 | 63.2 | 23.8 | 13.0 |
| Other precision production, craft, and repair | 467 | 100.0 | 60.8 | 25.8 | 13.4 |
| Operators, fabricators, and laborers | 1,823 | 100.0 | 54.6 | 31.6 | 13.7 |
| Machine operators, assemblers, and inspectors | 1,144 | 100.0 | 56.0 | 27.5 | 16.5 |
| Transportation and material moving occupations | 324 | 100.0 | 63.8 | 28.7 | 7.5 |
| Handlers, equipment cleaners, helpers, and laborers | 355 | 100.0 | 41.8 | 47.6 | 10.6 |
| Construction laborers | 55 | 100.0 | c | c | c |
| Other handlers, equipment cleaners, helpers, and laborers | 300 | 100.0 | 42.0 | 47.0 | 11.0 |
| Farming, forestry, and fishing | 68 | 100.0 | c | c | c |

[a]Data refer to persons with tenure of three years or more who lost or left a job between January 1979 and January 1984 because of plant closings or moves, slack work, or the abolishment of their positions or shifts.
[b]Total includes a small number who did not report occupation.
[c]Data not shown where base is less than 75,000.

*Tenure on Jobs Lost*

Many of the displaced workers had been at their jobs for many years. Of the 5.1 million total—all of whom had worked at least three years on the jobs they had

**Table 9-4**
**Employment Status and Area of Residence in January 1984 of Displaced Workers by Selected Characteristics**
(*in thousands*)

| Characteristic | Total[a] | New England | Middle Atlantic | East North Central | West North Central | South Atlantic | East South Central | West South Central | Mountain | Pacific |
|---|---|---|---|---|---|---|---|---|---|---|
| *Workers who lost jobs* | | | | | | | | | | |
| Total | 5,091 | 260 | 794 | 1,206 | 426 | 664 | 378 | 484 | 211 | 667 |
| Men | 3,328 | 155 | 530 | 772 | 282 | 428 | 236 | 347 | 152 | 427 |
| Women | 1,763 | 105 | 264 | 434 | 145 | 236 | 143 | 137 | 59 | 241 |
| *Reason for job loss* | | | | | | | | | | |
| Plant or company closed down or moved | 2,492 | 118 | 410 | 556 | 208 | 339 | 204 | 231 | 103 | 323 |
| Slack work | 1,970 | 106 | 269 | 513 | 164 | 236 | 132 | 211 | 83 | 256 |
| Position or shift abolished | 629 | 36 | 115 | 138 | 54 | 89 | 42 | 42 | 26 | 88 |
| *Industry of lost job* | | | | | | | | | | |
| Construction | 481 | 16 | 68 | 88 | 36 | 81 | 34 | 63 | 30 | 63 |
| Manufacturing | 2,514 | 158 | 414 | 658 | 210 | 296 | 189 | 215 | 58 | 315 |
| Durable goods | 1,686 | 94 | 260 | 514 | 137 | 175 | 107 | 142 | 40 | 218 |
| Nondurable goods | 828 | 64 | 154 | 145 | 73 | 122 | 82 | 73 | 18 | 97 |
| Transportation and public utilities | 352 | 14 | 61 | 83 | 34 | 34 | 33 | 41 | 19 | 32 |
| Wholesale and retail trade | 740 | 41 | 100 | 182 | 68 | 132 | 40 | 54 | 32 | 90 |
| Finance and service industries | 648 | 22 | 122 | 133 | 45 | 70 | 32 | 54 | 39 | 132 |
| Public administration | 84 | 2 | 10 | 22 | 5 | 13 | 4 | 8 | 5 | 16 |
| Other industries[b] | 272 | 5 | 20 | 40 | 28 | 38 | 45 | 49 | 27 | 19 |

| Employment status in January 1984 | | | | | | | | | | |
|---|---|---|---|---|---|---|---|---|---|---|
| Employed | 3,058 | 171 | 428 | 621 | 276 | 461 | 209 | 344 | 148 | 399 |
| Unemployed | 1,299 | 48 | 225 | 400 | 96 | 117 | 113 | 85 | 33 | 181 |
| Percent less than 5 weeks | 22.1 | c | 24.1 | 21.2 | 13.0 | 29.4 | 17.3 | 25.4 | c | 18.4 |
| Percent 27 weeks or more | 38.8 | c | 36.8 | 47.2 | 47.5 | 25.5 | 51.7 | 29.8 | c | 28.0 |
| Not in the labor force | 733 | 41 | 141 | 185 | 54 | 85 | 56 | 55 | 30 | 86 |

*Note:* The following list shows the states which make up each of the geographical divisions used in this table: New England—Connecticut, Maine, Massachusetts, New Hampshire, Rhode Island, and Vermont; Middle Atlantic—New Jersey, New York, and Pennsylvania; East North Central—Illinois, Indiana, Michigan, Ohio, and Wisconsin; West North Central—Iowa, Kansas, Minnesota, Missouri, Nebraska, North Dakota, and South Dakota; South Atlantic—Delaware, District of Columbia, Florida, Georgia, Maryland, North Carolina, South Carolina, Virginia, and West Virginia; East South Central—Alabama, Kentucky, Mississippi, and Tennessee; West South Central—Arkansas, Louisiana, Oklahoma, and Texas; Mountain—Arizona, Colorado, Idaho, Montana, Nevada, New Mexico, Utah, and Wyoming; Pacific—Alaska, California, Hawaii, Oregon, and Washington.

[a] Data refer to persons with tenure of three years or more who lost or left a job between January 1979 and January 1984 because of plant closings or moves, slack work, or the abolishment of their positions or shifts.

[b] Includes a small number who did not report industry.

[c] Data not shown where base is less than 75,000.

lost—nearly one-third had spent at least ten years in their jobs. Another one-third had been at their jobs from five to nine years. The remaining one-third had lost jobs at which they had worked either three or four years. Not surprisingly, the older the displaced workers the more likely they were to report a relatively longer period of service in the jobs they had lost. This is clearly shown in table 9-5, which gives the percent distribution of the displaced by age and years of tenure on the lost job. As shown, while the overall median job tenure for the entire 5.1 million total was 6.1 years, median tenure for those 55 to 64 years of age was 12.4 years. Nearly one-third of the workers in this age group reported they had lost jobs in which they had spent 20 years or more.

## Displacements and Their Aftermath

Various questions concerning the reasons for the displacements and what occurred in their aftermath were also asked as part of the January 1984 survey. The data obtained through these questions are the focus of the following sections.

### Reasons for Dismissals

About one-half of the 5.1 million displaced workers reported they had lost their jobs because their plant or business had closed down or moved. Another two-fifths cited "slack work" as the reason (an answer which may be translated as insufficient demand for the products or services of the employer). The remainder reported simply that their individual jobs, or the entire shift on which they had been working, had been abolished (see table 9-6).

Older workers were most likely to have lost their jobs due to plant closings. Evidently, while their seniority protected their jobs in the face of such problems as "slack work," it afforded little protection against the shutdown of their plants or the folding of their companies. The younger displaced workers, however, were about as likely to have lost their jobs due to slack work as due to plant closings.

**Table 9-5**
**Tenure on Lost Job**

| Age | Total | 3 to 4 Years | 5 to 9 Years | 10 Years or More | 20 Years or More | Median Years of Tenure |
|---|---|---|---|---|---|---|
| Total, 20 years and over | 100.0 | 36.2 | 33.6 | 30.2 | 8.8 | 6.1 |
| 25 to 54 years | 100.0 | 37.9 | 36.9 | 25.1 | 4.7 | 5.8 |
| 55 to 64 years | 100.0 | 15.5 | 23.2 | 61.3 | 27.9 | 12.4 |
| 65 years and over | 100.0 | 14.6 | 31.1 | 54.2 | 30.0 | 11.9 |

## Table 9–6
## Displaced Workers by Reason for Job Loss and by Age, Sex, Race, and Hispanic Origin
*(percentage)*

| Characteristic | Number (thousands)[a] | Total | Plant or Company Closed Down or Moved | Slack Work | Position or Shift Abolished |
|---|---|---|---|---|---|
| *Total* | | | | | |
| Total, 20 years and over | 5,091 | 100.0 | 49.0 | 38.7 | 12.4 |
| 20 to 24 years | 342 | 100.0 | 47.1 | 47.1 | 5.8 |
| 25 to 54 years | 3,809 | 100.0 | 46.3 | 41.0 | 12.7 |
| 55 to 64 years | 748 | 100.0 | 57.8 | 28.2 | 14.0 |
| 65 years and over | 191 | 100.0 | 70.8 | 18.1 | 11.1 |
| *Men* | | | | | |
| Total, 20 years and over | 3,328 | 100.0 | 46.0 | 42.9 | 11.1 |
| 20 to 24 years | 204 | 100.0 | 39.5 | 59.6 | .9 |
| 25 to 54 years | 2,570 | 100.0 | 43.9 | 44.8 | 11.3 |
| 55 to 64 years | 461 | 100.0 | 55.6 | 30.5 | 14.0 |
| 65 years and over | 92 | 100.0 | 68.7 | 15.7 | 15.5 |
| *Women* | | | | | |
| Total, 20 years and over | 1,763 | 100.0 | 54.6 | 30.8 | 14.6 |
| 20 to 24 years | 138 | 100.0 | 58.3 | 28.7 | 12.9 |
| 25 to 54 years | 1,239 | 100.0 | 51.1 | 33.3 | 15.6 |
| 55 to 64 years | 287 | 100.0 | 61.4 | 24.5 | 14.1 |
| 65 years and over | 99 | 100.0 | 72.8 | 20.3 | 6.9 |
| *White* | | | | | |
| Total, 20 years and over | 4,397 | 100.0 | 49.6 | 37.9 | 12.5 |
| Men | 2,913 | 100.0 | 46.0 | 42.6 | 11.4 |
| Women | 1,484 | 100.0 | 56.7 | 28.7 | 14.6 |
| *Black* | | | | | |
| Total, 20 years and over | 602 | 100.0 | 43.8 | 44.7 | 11.6 |
| Men | 358 | 100.0 | 44.9 | 46.4 | 8.8 |
| Women | 244 | 100.0 | 42.2 | 42.2 | 15.7 |
| *Hispanic origin* | | | | | |
| Total, 20 years and over | 282 | 100.0 | 47.4 | 45.2 | 7.3 |
| Men | 189 | 100.0 | 48.1 | 43.8 | 8.1 |
| Women | 93 | 100.0 | 46.2 | 48.1 | 5.7 |

*Note:* Detail for the above race and Hispanic-origin groups will not sum to totals because data for the "other races" group are not presented and Hispanics are included in both the white and black population groups.

[a]Data refer to persons with tenure of three years or more who lost or left a job between January 1979 and January 1984 because of plant closings or moves, slack work, or the abolishment of their positions or shifts.

## Notification of Dismissal

More than one-half of the displaced workers reported that they had received an advance notice of their dismissal, or that they had expected it. However, only one in ten of these had apparently left their jobs before the actual dismissal occurred (see table 9–7).

Workers who reported that they lost their jobs because the plant or company closed or moved (61 percent) were more likely than workers who reported other reasons for job loss (52 percent) to respond that they received advance notice or had expected a dismissal. But even among those whose plants had closed, only a little more than one-tenth reported that they had left their jobs before they ended.

Of the displaced workers who did leave their jobs before they were to be laid off, a substantially higher proportion were reemployed in January 1984 (79 percent) than was the case among those who were informed but stayed on (60 percent). The evidence here, therefore, adds some support for policies to encourage firms to provide early notification of layoffs; but, as noted, most workers remained on their jobs even with the advance notification.

## Moving to Another Area

Only a small minority of the 5.1 million displaced workers (680,000) moved to a different city or county to look for work or to take a different job. However, of those who did move, a higher proportion were reemployed in January 1984— almost three in four, in contrast to three in five of the nonmovers (see table 9–8). Men were more likely to move than women, and of the male movers, proportionately more were reemployed (77 percent) than was the case for their women counterparts (60 percent). Relatively few older workers relocated—only 6 percent among those 55 and over. However, even among them, about three-fifths of those who moved were working again, a substantially higher proportion than for nonmovers.

Although the data point up the employment benefits of relocation, it should be recognized that there are important reasons for the reluctance of workers to move. Many have established community ties; they may own homes which are particularly hard to sell if located in a depressed area; and there may be family members who are still employed locally, thereby adding to the costs of a move. They may also not have sufficient information about job opportunities in other areas. Finally, it has been found that a sizable proportion of workers who do relocate are likely to return.

A recently published guidebook for employers on managing plant closings estimates that only about 20 percent or fewer workers in a plant would consider relocating as part of their "reemployment strategy." The authors mention, for example, that only 20 percent of laid-off steelworkers from a Youngstown steel

Table 9-7
## Displaced Workers by Age, Whether They Received Advance Notice or Expected Layoff, Selected Reason for Job Loss, and Employment Status, January 1984[a]
*(in thousands)*

| Characteristic | Total Who Lost Jobs | | | | Plant or Company Closed Down or Moved | | | | All Other Reasons | | | |
|---|---|---|---|---|---|---|---|---|---|---|---|---|
| | | Employment Status in January 1984 | | | | Employment Status in January 1984 | | | | Employment Status in January 1984 | | |
| | Total | Employed | Unemployed | Not in the Labor Force | Total | Employed | Unemployed | Not in the Labor Force | Total | Employed | Unemployed | Not in the Labor Force |
| *All persons 20 years and over* | | | | | | | | | | | | |
| Total[a] | 5,091 | 3,058 | 1,299 | 733 | 2,492 | 1,547 | 509 | 437 | 2,599 | 1,512 | 791 | 296 |
| Received advance notice or expected layoff | 2,870 | 1,715 | 709 | 446 | 1,525 | 945 | 297 | 283 | 1,346 | 770 | 412 | 163 |
| Left before job ended | 318 | 250 | 23 | 45 | 185 | 151 | 7 | 27 | 133 | 99 | 16 | 18 |
| Did not leave before job ended | 2,532 | 1,450 | 683 | 399 | 1,331 | 787 | 290 | 254 | 1,202 | 664 | 393 | 145 |
| Did not receive advance notice or expect layoff | 2,221 | 1,343 | 590 | 287 | 967 | 602 | 211 | 154 | 1,253 | 741 | 378 | 134 |
| *20 to 34 years* | | | | | | | | | | | | |
| Total | 2,034 | 1,330 | 504 | 200 | 885 | 615 | 184 | 86 | 1,148 | 715 | 320 | 114 |
| Received advance notice or expected layoff | 1,160 | 771 | 274 | 114 | 550 | 393 | 100 | 58 | 609 | 379 | 174 | 56 |
| Left before job ended | 146 | 117 | 11 | 17 | 74 | 61 | 3 | 9 | 72 | 57 | 7 | 8 |
| Did not leave before job ended | 1,004 | 643 | 264 | 97 | 470 | 325 | 96 | 48 | 534 | 319 | 167 | 48 |
| Did not receive advance notice or expect layoff | 874 | 558 | 230 | 85 | 335 | 222 | 84 | 28 | 539 | 336 | 146 | 57 |

Table 9-7 continued

| Characteristic | Total Who Lost Jobs | | | | Plant or Company Closed Down or Moved | | | | All Other Reasons | | | |
|---|---|---|---|---|---|---|---|---|---|---|---|---|
| | Total | Employment Status in January 1984 | | | Total | Employment Status in January 1984 | | | Total | Employment Status in January 1984 | | |
| | | Employed | Unemployed | Not in the Labor Force | | Employed | Unemployed | Not in the Labor Force | | Employed | Unemployed | Not in the Labor Force |
| **35 to 54 years** | | | | | | | | | | | | |
| Total | 2,118 | 1,384 | 534 | 200 | 1,039 | 714 | 203 | 122 | 1,079 | 670 | 331 | 78 |
| Received advance notice or expected layoff | 1,183 | 784 | 284 | 115 | 626 | 439 | 115 | 71 | 557 | 345 | 169 | 43 |
| Left before job ended | 137 | 112 | 10 | 15 | 85 | 73 | 3 | 9 | 52 | 40 | 7 | 6 |
| Did not leave before job ended | 1,040 | 668 | 272 | 100 | 541 | 367 | 112 | 62 | 499 | 302 | 160 | 37 |
| Did not receive advance notice or expect layoff | 935 | 599 | 250 | 85 | 413 | 274 | 87 | 51 | 522 | 325 | 163 | 34 |
| **55 years and over** | | | | | | | | | | | | |
| Total | 939 | 345 | 261 | 334 | 568 | 218 | 122 | 229 | 371 | 127 | 139 | 105 |
| Received advance notice or expected layoff | 528 | 160 | 151 | 217 | 349 | 113 | 82 | 154 | 179 | 47 | 69 | 63 |
| Left before job ended | 35 | 21 | 2 | 12 | 26 | 18 | — | 9 | 9 | 3 | 2 | 4 |
| Did not leave before job ended | 489 | 139 | 148 | 203 | 320 | 95 | 82 | 143 | 169 | 44 | 66 | 59 |
| Did not receive advance notice or expect layoff | 412 | 186 | 109 | 117 | 219 | 105 | 40 | 75 | 192 | 80 | 70 | 42 |

aData refer to persons with tenure of three years or more who lost or left a full-time wage and salary job between January 1979 and January 1984 because of plant closings or moves, slack work, or the abolishment of their positions or shifts.

**Table 9-8**
**Displaced Workers by Whether They Moved to a Different City or County to Find or Take Another Job, by Age, Sex, and Current Employment Status, January 1984**
*(in thousands)*

| Age and Sex | Nonmovers | | | | Movers | | | |
|---|---|---|---|---|---|---|---|---|
| | | Employment Status in January 1984 | | | | Employment Status in January 1984 | | |
| | Total | Employed | Unemployed | Not in the Labor Force | Total | Employed | Unemployed | Not in the Laobr Force |
| *Total* | | | | | | | | |
| Total, 20 years and over[a] | 4,374 | 2,537 | 1,157 | 680 | 682 | 500 | 134 | 48 |
| 25 to 54 years | 3,234 | 2,044 | 859 | 332 | 556 | 413 | 108 | 34 |
| 25 to 34 years | 1,370 | 864 | 365 | 141 | 318 | 221 | 71 | 26 |
| 35 to 44 years | 1,055 | 706 | 267 | 81 | 158 | 125 | 26 | 6 |
| 45 to 54 years | 809 | 473 | 227 | 109 | 80 | 67 | 11 | 2 |
| 55 years and over | 880 | 312 | 246 | 321 | 53 | 32 | 12 | 9 |
| *Men* | | | | | | | | |
| Total, 20 years and over | 2,784 | 1,700 | 800 | 284 | 519 | 401 | 96 | 21 |
| 25 to 54 years | 2,114 | 1,399 | 609 | 107 | 440 | 342 | 78 | 19 |
| 25 to 34 years | 936 | 616 | 270 | 50 | 262 | 191 | 55 | 16 |
| 35 to 44 years | 671 | 459 | 189 | 23 | 117 | 98 | 18 | 2 |
| 45 to 54 years | 507 | 324 | 150 | 33 | 61 | 54 | 5 | 2 |
| 55 years and over | 510 | 191 | 155 | 164 | 38 | 24 | 12 | 2 |
| *Women* | | | | | | | | |
| Total, 20 years and over | 1,590 | 837 | 357 | 397 | 163 | 99 | 38 | 27 |
| 25 to 54 years | 1,120 | 645 | 250 | 225 | 116 | 71 | 30 | 15 |
| 25 to 34 years | 434 | 249 | 94 | 91 | 56 | 30 | 15 | 11 |
| 35 to 44 years | 384 | 247 | 78 | 58 | 41 | 27 | 9 | 5 |
| 45 to 54 years | 303 | 149 | 77 | 76 | 19 | 13 | 6 | — |
| 55 years and over | 369 | 121 | 92 | 157 | 14 | 8 | — | 7 |

[a]Data refer to persons with tenure of three years or more who lost or left a job between January 1979 and January 1984 because of plant closings or moves, slack work, or the abolishment of their positions or shifts.

plant had moved out of the area; that only 20 percent of enrollees in the Job Search and Relocation Assistance Pilot Program of the U.S. Department of Labor, and only 6 percent of enrollees for Trade Adjustment Assistance, used the relocation assistance which was offered them.

### How Long Without Work?

On average, the displaced workers had spent nearly 6 months without work after they had lost their jobs. That is, the median period without work, which need not have been a continuous spell and could have included time spent outside the labor force, was 24.1 weeks. However, it should also be noted that about one-fourth of these 5.1 million workers were still jobless when surveyed. For many of them, the period of unemployment would obviously extend beyond the January 1984 survey period.

As has historically been the case for the unemployed in general, older workers were without work longer than their younger counterparts. For workers 55 years and over, the median period without a job was 30 weeks, while for workers 25 to 34 it was 22 weeks.

Workers who were no longer in the labor force in January 1984 had been without work many more weeks, on average, than those who were still looking for work (57 versus 32 weeks), while workers who were reemployed had spent far fewer weeks without a job (13) (see table 9–9).

### Receipt of Unemployment Insurance

The economic difficulties of most of the displaced workers were alleviated by their receipt of unemployment insurance benefits. Yet, while 3.5 million of the 5.1 million displaced workers had received such benefits, almost one-half had exhausted them by January 1984 (see table 9–10). Understandably, the probability of exhausting one's benefits was closely tied to the length of one's period of unemployment, being very high for workers reporting more than 6 months (27 weeks) without work and much lower for those with only a short spell of joblessness.

A larger percentage of the workers who were unemployed in January 1984 had received unemployment insurance benefits—80 percent—than their counterparts who were either reemployed or had left the labor force—65 percent for both. Of the workers who had received benefits, the proportion that had exhausted them by January 1984 was about 50 percent for those still unemployed, 40 percent for those reemployed, and 70 percent for those no longer in the labor force.

### Loss of Health Insurance

Because a large proportion of the displaced workers had held relatively "good" jobs in terms of pay and other benefits, a large majority of them had participated in a group health insurance program on these jobs. As shown in table 9–11, many of them no longer were covered under any plan when surveyed in January 1984.

Of the 3.1 million persons who were working again in January 1984, 2.5 million had been covered by group health insurance coverage on their lost jobs. Even among these, about 1 in 4 were no longer covered under a health plan in January 1984.

For the 1.3 million displaced workers who were jobless in January 1984 and who previously had been covered by group health insurance, 60 percent no longer had any coverage at the time of the survey. For black unemployed workers previously covered, the uncovered proportion was 75 percent when surveyed.

In general, women were less likely than men to be left without any health insurance coverage after displacement, even if unemployed. This is probably

**Table 9–9**
**Displaced Workers by Weeks Without Work, Age, and Employment Status, January 1984**[a]

| | Weeks without work | | | | | |
|---|---|---|---|---|---|---|
| Characteristic | Less Than 5 | 5 to 14 | 15 to 26 | 27 to 52 | More than 52 | Median without work |
| *Total* | | | | | | |
| Age 20 and over | 1,173 | 912 | 707 | 983 | 1,211 | 24.1 |
| 25 to 54 years | 856 | 729 | 538 | 745 | 871 | 23.1 |
| 25 to 34 years | 399 | 347 | 214 | 349 | 359 | 21.9 |
| 35 to 44 years | 268 | 228 | 200 | 220 | 278 | 22.3 |
| 45 to 54 years | 189 | 154 | 125 | 177 | 234 | 25.8 |
| 55 years and over | 203 | 109 | 122 | 179 | 302 | 29.8 |
| *Employed* | | | | | | |
| Age 20 and over | 910 | 657 | 453 | 590 | 393 | 13.1 |
| 25 to 54 years | 705 | 540 | 364 | 486 | 334 | 13.4 |
| 25 to 34 years | 322 | 252 | 147 | 222 | 129 | 12.5 |
| 35 to 44 years | 223 | 185 | 134 | 150 | 130 | 15.4 |
| 45 to 54 years | 160 | 103 | 83 | 114 | 74 | 15.3 |
| 55 years and over | 119 | 65 | 52 | 63 | 41 | 12.4 |
| *Unemployed* | | | | | | |
| Age 20 and over | 166 | 201 | 201 | 264 | 447 | 32.2 |
| 25 to 54 years | 124 | 156 | 142 | 185 | 348 | 32.6 |
| 25 to 34 years | 64 | 75 | 57 | 81 | 153 | 33.8 |
| 35 to 44 years | 40 | 37 | 50 | 57 | 106 | 30.9 |
| 45 to 54 years | 21 | 43 | 35 | 46 | 90 | 32.5 |
| 55 years and over | 25 | 31 | 50 | 65 | 88 | 33.3 |
| *Not in the labor force* | | | | | | |
| Age 20 and over | 98 | 55 | 53 | 130 | 370 | 56.8 |
| 25 to 54 years | 27 | 34 | 33 | 74 | 189 | 57.6 |
| 25 to 34 years | 14 | 20 | 10 | 46 | 77 | 53.0 |
| 35 to 44 years | 6 | 7 | 17 | 13 | 42 | 54.7 |
| 45 to 54 years | 8 | 7 | 7 | 16 | 69 | 96.2 |
| 55 years and over | 59 | 14 | 19 | 51 | 173 | 61.2 |

[a]"Displaced" refers to persons whose jobs were lost because of plant closings or moves, slack work, or the abolishment of their positions or shifts.

# Table 9-10
## Workers Who Lost Jobs in Past Five Years by Duration of Joblessness, Receipt of Unemployment Insurance, Whether Benefits Exhausted, Weeks Without Work, and Employment Status, January 1984[a]
*(in thousands)*

| Weeks Without Work and Employment Status | Lost a Job in Last 5 Years | | | Plant or Company Closed Down or Moved | | | All Other Reasons | | |
|---|---|---|---|---|---|---|---|---|---|
| | Total | Received Unemployment Benefits | Exhausted Benefits | Total | Received Unemployment Benefits | Exhausted Benefits | Total | Received Unemployment Benefits | Exhausted Benefits |
| **Both sexes, all persons** | | | | | | | | | |
| Total[a] | 5,091 | 3,497 | 1,670 | 2,492 | 1,589 | 755 | 2,599 | 1,908 | 915 |
| Less than 5 weeks | 1,173 | 298 | 44 | 665 | 144 | 21 | 508 | 155 | 23 |
| 5 to 14 weeks | 912 | 687 | 59 | 419 | 297 | 19 | 494 | 391 | 40 |
| 15 to 26 weeks | 707 | 604 | 165 | 325 | 270 | 63 | 381 | 334 | 102 |
| 27 to 51 weeks | 656 | 583 | 316 | 309 | 270 | 157 | 347 | 313 | 160 |
| 52 weeks or more | 1,538 | 1,273 | 1,064 | 724 | 584 | 482 | 814 | 689 | 582 |
| *Employed* | | | | | | | | | |
| Total | 3,058 | 1,973 | 802 | 1,547 | 904 | 357 | 1,512 | 1,068 | 445 |
| Less than 5 weeks | 910 | 182 | 18 | 546 | 98 | 8 | 364 | 84 | 9 |
| 5 to 14 weeks | 657 | 499 | 44 | 313 | 225 | 16 | 343 | 274 | 28 |
| 15 to 26 weeks | 453 | 389 | 111 | 204 | 171 | 43 | 249 | 218 | 69 |
| 27 to 51 weeks | 368 | 342 | 182 | 190 | 169 | 98 | 178 | 172 | 84 |
| 52 weeks or more | 615 | 533 | 436 | 269 | 228 | 186 | 346 | 305 | 251 |
| *Unemployed* | | | | | | | | | |
| Total | 1,299 | 1,043 | 541 | 509 | 390 | 203 | 791 | 653 | 338 |
| Less than 5 weeks | 166 | 69 | 9 | 61 | 15 | 2 | 105 | 54 | 7 |
| 5 to 14 weeks | 201 | 167 | 11 | 75 | 59 | 3 | 126 | 108 | 8 |
| 15 to 26 weeks | 201 | 174 | 38 | 88 | 75 | 12 | 113 | 99 | 26 |
| 27 to 51 weeks | 199 | 176 | 93 | 72 | 64 | 34 | 127 | 112 | 59 |
| 52 weeks or more | 512 | 447 | 387 | 206 | 174 | 151 | 306 | 273 | 236 |
| *Not in the labor force* | | | | | | | | | |
| Total | 733 | 481 | 327 | 437 | 294 | 195 | 296 | 187 | 132 |
| Less than 5 weeks | 98 | 48 | 17 | 58 | 30 | 10 | 40 | 18 | 7 |
| 5 to 14 weeks | 55 | 22 | 3 | 30 | 13 | — | 24 | 9 | 3 |
| 15 to 26 weeks | 53 | 40 | 16 | 33 | 24 | 8 | 20 | 17 | 8 |
| 27 to 51 weeks | 89 | 65 | 41 | 47 | 37 | 25 | 42 | 28 | 16 |
| 52 weeks or more | 411 | 294 | 241 | 249 | 182 | 145 | 162 | 112 | 96 |

[a]Data refer to persons with tenure of three years or more who lost or left a job between January 1979 and January 1984 because of plant closings or moves, slack work, or the abolishment of their positions or shifts.

**Table 9-11**
**Displaced Workers by Health Insurance Coverage and Employment Status, January 1984**
*(in thousands)*

| Characteristic | Total[a] | Covered by Group Health Insurance on Lost Job | | | Not Covered on Lost Job |
|---|---|---|---|---|---|
| | | Total | Not Covered under Any Plan in January 1984 | | |
| | | | Number | Percent | |
| *Total* | | | | | |
| Total, 20 Years and Over | 5,091 | 3,977 | 1,381 | 34.7 | 1,033 |
| Employed | 3,058 | 2,454 | 573 | 23.4 | 554 |
| Unemployed | 1,299 | 1,037 | 612 | 59.0 | 236 |
| Not in the labor force | 733 | 486 | 196 | 40.3 | 242 |
| *Men* | | | | | |
| Total, 20 Years and Over | 3,328 | 2,757 | 985 | 35.7 | 507 |
| Employed | 2,117 | 1,780 | 413 | 23.2 | 301 |
| Unemployed | 903 | 743 | 469 | 63.1 | 139 |
| Not in the labor force | 307 | 235 | 102 | 43.6 | 67 |
| *Women* | | | | | |
| Total, 20 Years and Over | 1,763 | 1,220 | 396 | 32.4 | 526 |
| Employed | 941 | 675 | 160 | 23.7 | 253 |
| Unemployed | 396 | 294 | 142 | 48.4 | 98 |
| Not in the labor force | 426 | 251 | 93 | 37.2 | 175 |
| *White* | | | | | |
| Total, 20 Years and Over | 4,397 | 3,433 | 1,118 | 32.6 | 902 |
| Employed | 2,754 | 2,203 | 516 | 23.4 | 509 |
| Unemployed | 1,031 | 822 | 452 | 55.0 | 192 |
| Not in the labor force | 613 | 408 | 150 | 36.7 | 201 |
| *Black* | | | | | |
| Total, 20 Years and Over | 602 | 468 | 239 | 51.0 | 117 |
| Employed | 252 | 208 | 50 | 23.9 | 38 |
| Unemployed | 247 | 193 | 144 | 74.5 | 44 |
| Not in the labor force | 103 | 67 | 45 | 66.7 | 34 |
| *Hispanic origin* | | | | | |
| Total, 20 years and over | 282 | 193 | 66 | 34.2 | 83 |
| Employed | 147 | 111 | 29 | 25.6 | 32 |
| Unemployed | 95 | 60 | 33 | 55.5 | 33 |
| Not in the labor force | 40 | 22 | 5 | 20.5 | 17 |

[a]Data refer to persons with tenure of three years or more who lost or left a job between January 1979 and January 1984 because of plant closings or moves, slack work, or the abolishment of their positions or shifts.

because many of them had spouses who were working, and thus were likely to have been covered under the spouse's plan.

Among the previously covered displaced workers who were out of the labor force when surveyed, about 40 percent were not covered under any plan in January 1984. Again, for blacks the proportion who had lost all coverage was much larger—67 percent.

Some additional information on this topic is provided by a University of Michigan survey conducted in 1983 in the Detroit area. This survey found that, of those persons who had been without work for only three months or less, about 30 percent had no health insurance coverage. In contrast, the uncovered proportion among those without work for more than two years was 55 percent. Almost four-fifths of those workers had previously had health insurance when employed. The male workers were more likely than their female counterparts to be without health insurance at the time of the survey.

## New Jobs

Of the 5.1 million displaced workers, 2.8 million who had been displaced from full-time wage and salary jobs were reemployed in January 1984. Among them, 2.3 million were again working at full-time wage and salary jobs, about 220,000 were in other types of full-time employment (mainly self-employment), and about 360,000 were holding part-time jobs (see table 9–12).

Many reemployed workers were in occupations different from those they previously had held. For example, among the workers who were employed in January 1984, about 525,000 had been in managerial and professional specialty occupations at their lost jobs. Of these, only about one-half were reemployed in such jobs. Similarly, about 640,000 had been in precision production, craft, and repair work at their lost jobs; among them only 360,000 were working again in these occupations in January 1984 (see table 9–13).

Reemployed workers not only were working in different occupations, but also in different industries. For example, of the 980,000 displaced workers who had been in durable goods manufacturing, only about 40 percent were reemployed in these industries in January 1984. Similarly, about 35 percent of 493,000 workers were reemployed in nondurable goods manufacturing. In wholesale and retail trade, 50 percent of 455,000 were reemployed and in service industries, 46 percent of 347,000. Table 9–14 shows the percentage reemployed by key industry group.

As shown, even among the nearly one-half million reemployed who had been displaced from nondurable goods industries, only about one-third were again working in this industry group in January 1984. In fact, generally more than one-half of the displaced workers who were reemployed in January 1984 were no longer in the industry group from which they had been displaced.

Understandably, the workers who had been displaced from high wage industries were most likely to have suffered a drop in earnings in taking a new job. For example, as seen in table 9–15, for the 980,000 who had previously been in durable goods manufacturing, the median weekly earnings on the old jobs had been $344. In contrast, the median for the jobs they held in January 1984 was only $273. And it should be noted that these numbers, which are

**Table 9–12**
**Displaced Full-Time Workers by Industry, by Reemployment in January 1984, and by Comparison of Earnings between New and Old Jobs**
(in thousands)

| Industry of Lost Job | Total Reemployed January 1984 | Part-Time Job | Full-time Wage and Salary Job | | | | | | Self Employment or Other Full-time Job |
|---|---|---|---|---|---|---|---|---|---|
| | | | Total[a] | Earnings Relative to Those of Lost Job | | | | | |
| | | | | 20 Percent or More Below | Below, but Within 20 Percent | Equal or Above, but Within 20 Percent | 20 Percent or More Above | | |
| Displaced after 3 years or more on job[b] | 2,841 | 357 | 2,266 | 621 | 320 | 571 | 533 | | 218 |
| Construction | 253 | 26 | 199 | 48 | 30 | 47 | 61 | | 28 |
| Manufacturing | 1,418 | 151 | 1,200 | 366 | 171 | 286 | 247 | | 67 |
| Durable goods | 954 | 106 | 797 | 281 | 102 | 181 | 155 | | 51 |
| Primary metals industries | 98 | 14 | 77 | 40 | 5 | 22 | 5 | | 7 |
| Steel[c] | 78 | 14 | 59 | 33 | 3 | 14 | 5 | | 4 |
| Other primary metals | 20 | — | 18 | 7 | 2 | 9 | — | | 2 |
| Fabricated metal products | 102 | 12 | 81 | 30 | 6 | 21 | 16 | | 9 |
| Machinery, except electrical | 244 | 17 | 215 | 77 | 34 | 39 | 40 | | 12 |
| Electrical machinery | 94 | 10 | 84 | 26 | 12 | 14 | 22 | | — |
| Transportation equipment | 219 | 30 | 174 | 66 | 22 | 42 | 34 | | 14 |
| Automobiles | 141 | 19 | 115 | 43 | 16 | 21 | 26 | | 7 |
| Other transportation equipment | 77 | 11 | 59 | 23 | 6 | 21 | 8 | | 7 |
| Nondurable goods | 464 | 45 | 403 | 85 | 69 | 105 | 92 | | 16 |
| Transportation and public utilities | 191 | 15 | 154 | 40 | 22 | 44 | 27 | | 22 |
| Wholesale and retail trade | 399 | 72 | 296 | 61 | 41 | 79 | 85 | | 31 |
| Finance and service industries | 378 | 58 | 270 | 59 | 35 | 83 | 74 | | 50 |
| Public administration | 48 | 4 | 42 | 11 | 5 | 7 | 18 | | 2 |
| Other industries[d] | 153 | 31 | 104 | 36 | 16 | 24 | 22 | | 18 |

[a] Includes 221,000 persons who did not report earnings on lost job.
[b] Data refer to persons who lost or left a full-time wage and salary job between January 1979 and January 1984 because of plant closings or moves, slack work, or abolishment of their positions or shifts.
[c] Includes blast furnaces, steelworks, rolling and finishing mills, and iron and steel foundries.
[d] Includes a small number who did not report industry.

## Table 9–13
## Reemployed Workers by Occupation in January 1984 and by Occupation of Job Lost in Preceding Five Years
(in thousands)

| Occupation on Job Lost | Total Employed[a] | Occupation on Job Held in January 1984 | | | | | | | | | | |
| --- | --- | --- | --- | --- | --- | --- | --- | --- | --- | --- | --- | --- |
| | | Managerial and Professional Specialty | | Technical, Sales, and Administrative Support | | | Service Occupations | Precision Production, Craft, and Repair | Operators, Fabricators, and Laborers | | | Farming, Forestry, and Fishing |
| | | Executive, Administrative, and Managerial | Professional Specialty | Technicians and Related Support | Sales Occupations | Administrative Support, including Clerical | | | Machine Operators, Assemblers, and Inspectors | Transportation and Material Moving Occupations | Handlers, Equipment Cleaners, Helpers, and Laborers | |
| Total, 20 years and over | 3,058 | 282 | 194 | 73 | 359 | 364 | 320 | 621 | 387 | 223 | 183 | 52 |
| Managerial and professional specialty | 525 | 153 | 116 | 16 | 62 | 79 | 31 | 38 | 11 | 11 | 6 | 2 |
| Executive, administrative, and managerial | 336 | 141 | 26 | 10 | 43 | 57 | 12 | 27 | 7 | 7 | 3 | 2 |
| Professional specialty | 189 | 12 | 91 | 6 | 18 | 22 | 19 | 11 | 4 | 4 | 3 | — |
| Technical, sales, and administrative support | 704 | 70 | 38 | 41 | 197 | 188 | 56 | 50 | 27 | 19 | 16 | 3 |
| Technicians and related support | 83 | 3 | 10 | 39 | 4 | 4 | 6 | 6 | 6 | 1 | 6 | — |
| Sales occupations | 312 | 34 | 15 | — | 159 | 27 | 18 | 30 | 10 | 11 | 6 | 2 |
| Administrative support, including clerical | 309 | 34 | 13 | 2 | 34 | 157 | 32 | 14 | 11 | 7 | 4 | 1 |
| Service occupations | 140 | 1 | 6 | 2 | 10 | 8 | 81 | 18 | 4 | 5 | 5 | — |
| Precision production, craft, and repair | 642 | 33 | 19 | 4 | 28 | 25 | 35 | 359 | 64 | 27 | 40 | 9 |
| Operators, fabricators, and laborers | 995 | 18 | 14 | 10 | 58 | 64 | 118 | 145 | 277 | 159 | 107 | 26 |
| Machine operators, assemblers, and inspectors | 640 | 6 | 10 | 8 | 37 | 44 | 94 | 98 | 248 | 35 | 50 | 9 |
| Transportation and material moving occupations | 207 | 4 | 2 | 1 | 14 | 7 | 6 | 19 | 12 | 107 | 24 | 9 |
| Handlers, equipment cleaners, helpers, and laborers | 148 | 7 | 2 | 1 | 8 | 13 | 16 | 28 | 16 | 16 | 33 | 8 |
| Farming, forestry, and fishing | 47 | 5 | — | — | 3 | 0 | 0 | 9 | 4 | 4 | 9 | 13 |

[a]Data refer to persons with tenure of three years or more who lost or left a job between January 1979 and January 1984 because of plant closings or moves, slack work, or the abolishment of their positions or shifts.

**Table 9–14**
**Key Industry Group Reemployment**

|  | Durable | Non-durable | Trade | Services |
|---|---|---|---|---|
| Durable goods | 40 | 14 | 9 | 8 |
| Nondurable goods | 6 | 35 | 6 | 4 |
| Wholesale trade | 5 | 4 | 10 | 5 |
| Retail trade | 12 | 9 | 40 | 15 |
| Service | 16 | 19 | 17 | 46 |
| Other industries | 22 | 19 | 18 | 22 |

**Table 9–15**
**Earnings of Reemployed Workers**

| Industry of Lost Jobs | Reemployed Workers (in thousands) | Median Weekly Earnings | |
|---|---|---|---|
| | | Lost Job | Job Held in January 1984 |
| Durable goods | 980 | $344 | $273 |
| Primary metals | 100 | 407 | 246 |
| Transportation equipment | 222 | 399 | 319 |
| Nondurable goods | 493 | 264 | 254 |
| Textile mill products | 48 | 181 | 187 |
| Apparel and other finished textile products | 83 | 202 | 197 |

shown below for a few illustrative industries, understate the actual loss in purchasing power as they are stated in "current" dollars, that is, they do not take into account the effects of inflation. Workers who had been displaced from jobs in nondurable goods manufacturing (made up primarily of lower paying industries) showed only slight declines, if any, between their earnings on their new and old jobs. For example, the median weekly earnings on their lost jobs were $202 for workers in apparel and other finished textile products, while their earnings on their new jobs were $197; for workers in textile mill products, their median earnings on their lost jobs were $181, and on their new jobs, $187.

Among the individual displaced workers who had previously been in full-time jobs in durable goods industries and who were again working full time in January 1984, about 40 percent had seen their weekly earnings drop by 20 percent or more. Yet, as seen in table 9–12, for those who had been displaced from jobs in other industries, the earnings in the new jobs compared more favorably with those in the old jobs.

Of the entire universe of about 2 million workers who were in full-time wage and salary jobs both before displacement and when surveyed, and who reported the earnings both for their old and new jobs, more than one-half (55 percent) were making as much or more in January 1984 than before displacement. These workers could, therefore, be seen as having readjusted rather well after their initial job losses. However, among these 2 million workers, there were also 900,000 who had taken some pay cuts, and for about 600,000 of these the cut was in the range of 20 percent or more.

In addition to the workers who had taken pay cuts although they were again working in full-time jobs, there were also, as already noted, a considerable number—about 360,000—who had gone from a full-time to a part-time job. Needless to say, these workers were even more likely to have suffered a considerable drop in weekly earnings after their displacement. When these are added to our universe, we can conclude that at least one-half of the displaced workers who were reemployed in January 1984 were earning less than in the jobs they had lost.

Among the findings from other studies on displacement which have dealt with earnings differences between the displaced workers' old and new jobs, are the following:

Older workers and workers with less education are more likely to experience earnings losses.

Because there are fewer job opportunities available, earnings losses are larger in areas of high unemployment and in small labor markets.

Earnings losses are particularly large for workers displaced from well-paying unionized industries such as autos and industrial chemicals.

A special assessment of Department of Labor funded programs in six local areas that provided training and other services to displaced workers in 1982–83, found that for the program participants who were reemployed, the average wages at their new jobs had dropped substantially from their prelayoff wages: The mean hourly wage at the new jobs was in the $7 to $8 range, whereas the mean wage at layoff ranged from approximately $9 to $11 an hour. In addition to the losses in wages, there were obviously some losses of fringe benefits relative to those enjoyed on the previous jobs.

## Steel and Automobile Workers

Much of the public discussion about workers' displacements in recent years has focused on the steel and auto industries. This is probably because any plant shutdowns or mass layoffs in these two industries have a particularly large impact on the geographic areas where they are concentrated, as well as a large multiplier effect on the other sectors of the economy. Moreover, the two industries were not only hard hit by the recessions of the early 1980s, but also had to retrench and alter their production methods because of foreign competition and other structural factors. These developments led to large reductions in employment, with the payrolls in both of these industries being considerably lower in January 1984, even after some rapid recovery from the latest recession, than they had been 5 years earlier. Specifically, over this 5-year period, employment had dropped by about 400,000 (or nearly one-third) in the primary metals industry and by about 200,000 (or one-fifth) in the motor vehicles industry. Of course, many other durable goods industries also underwent large reductions in employment over this period, but because their plants are generally not as concentrated in certain areas, nor as dominant in the local economies as are steel and automobile plants, their cutbacks received less nationwide publicity.

## Steel Workers

Of the 5.1 million displaced workers in January 1984, about 220,000 had worked in primary metals industries (largely steel). Forty percent of them reported they lost their jobs because their plants had closed down, and most of the others cited slack work as the reason for job loss. Reflecting the deep-seated problems of this industry and the generally depressed conditions of some of the areas where its plants are (or were) located, less than one-half (46 percent) of these displaced workers were working again in January 1984. Nearly 40 percent were still looking for work, while 16 percent were no longer in the labor force. Among those who had lost their jobs because of plant closings, almost one-fourth had left the labor force. Thus, the employment status of the workers displaced from primary metals jobs was far worse than that for the entire universe of displaced workers.

Not surprisingly, of the former steel (and other primary metals) workers who were again employed when surveyed, most had left the primary metals industry. Only 25,000 of them were working in durable-goods industries in January 1984. Of the others, some 20,000 were in services industries, 15,000 in construction, and another 15,000 in retail trade. Having had to find work in generally new fields, the displaced workers who had previously held jobs in primary metals industries reported a larger decline in earnings at their new jobs (40 percent) than workers from any other industry group. As already indicated, median earnings of these reemployed workers were $246 at their new jobs versus $407 at their old ones. Such earnings losses must have caused substantial changes in the consumption pattern of these workers and their families.

## Automobile Workers

About 225,000 auto workers had been displaced from their jobs during the January 1979–January 1984 survey period. Of these, 44 percent reported they had lost their jobs because their plants had closed, while 46 percent reported slack work as the reason for job loss. Reflecting partly the fact that the industry had enjoyed a substantial recovery by January 1984, nearly two-thirds of these workers were again employed when surveyed. However, while some automobile workers had gone back to their former jobs, many others had apparently switched to different—and generally lower paying—jobs in other industries. As indicated above, for all those who were reemployed, the median weekly earnings for the jobs they held in January 1984 were substantially lower than the median for the auto industry jobs they had lost.

It is also important to note that 25 percent of the displaced auto workers were still looking for work in January 1984 and that 13 percent had left the labor force. For those who lost their jobs because their plant closed, the proportions unemployed or out of the labor force in January 1984 were even a bit higher.

Of course, an additional number of automobile workers were recalled to their jobs during 1984. Employment in the motor vehicles and equipment industry increased from about 850,000 (seasonally adjusted) in January 1984 to about 900,000 by the year's end. So, the displacement problem in this industry was likely to have been alleviated considerably during the year following the survey.

## Other Studies of Displaced Workers

In addition to the data from the January 1984 survey, special case studies evaluating the effectiveness of Department of Labor programs for displaced workers, particularly displaced auto and steel workers, are another valuable source of information on this topic.

In order to obtain information on the effectiveness of various types of assistance which might be provided to displaced workers, the Department of Labor funded a series of pilot projects in 1980–83. One project, the Downriver Community Conference Economic Readjustment Program, served laid-off automotive workers from the Detroit metropolitan area. Among the findings from this demonstration study are the following:

1. The displaced workers were predominantly men, aged 25 to 44, and married. Most had graduated from high school; however, when tested in the program, one-fifth scored below a sixth-grade literacy level. They had, on average, worked more than 10 years on the lost job—and they had earned about $10 an hour.

2. Depending upon the particular plant from which they had been laid off, the workers were found to have received either unemployment insurance benefits, unemployment insurance coupled with company-funded supplemental unemployment benefits, or, in some cases, both of these benefits

as well as trade adjustment assistance, which was paid to those whose jobs were deemed to have been lost because of imports. Therefore, some of the workers had their prelayoff earnings almost entirely replaced by benefits, at least for a time.

3. Although resources were made available to the workers for job search and relocation outside their area, only 8 percent of the program enrollees relocated. About 20 percent of those who relocated subsequently returned.

4. Two years after the job loss, only about 50 percent of the workers in the program had found another job. The reemployment rate declined the longer the workers remained in the program, and this reflected in part the worsening labor market conditions in the Detroit area during that particular period.

5. On average, the earnings of participants who became reemployed were more than 30 percent below their prelayoff earnings.

The Department of Labor had also funded a pilot program in Buffalo, New York (among other sites), the aim of which was to assist displaced workers, largely from auto and steel jobs. In this demonstration, it was found that the reemployed workers were placed in jobs paying a mean wage of about $6.50 an hour, a decline from a mean prelayoff hourly wage of more than $10 an hour. The program participants were primarily men, between their mid-20s and mid-40s, most with a high school education. Nearly 70 percent of the participants were reemployed at the time of the project's termination, with the younger workers being slightly more likely to be placed in jobs than were the others.

Some additional data on displaced workers are available from a sample of 379 workers from a population of about 11,000 workers on indefinite layoff from a major automobile manufacturer in 1983. The survey, which was funded by the Department of Commerce, was conducted by the University of Michigan from November 1983 to January 1984. Among the findings are the following:

Auto workers who were recalled to jobs with their previous employer reported a mean hourly wage of $12.26, with a weekly gross pay of $490.42. In contrast, the other reemployed workers cited a mean hourly wage of $7.42 and an average weekly gross pay of $314.70.

Of the 379 respondents, 30 percent had been recalled to their old jobs at the time of the survey, 25 percent were employed elsewhere, about 35 percent were looking for work, and 10 percent were no longer in the labor force.

Compensation payments (for example, unemployment insurance and trade adjustment assistance benefits) had covered, on average, about 30 percent of the displaced workers' income loss since they had been laid off. The proportion of lost income offset by such benefits was lower the longer the layoff period, dropping from about 55 percent for workers laid off less than one year to about 13 percent for those laid off more than two years.

Workers with more than ten years' seniority at their old jobs had received benefits that replaced larger proportions of their lost wages. However, these workers also reported relatively lower earnings when they were reemployed.

## Summary

The two recessions of the early 1980s, coupled with more deep-seated structural problems affecting certain industries, took a heavy toll among American workers. About 5.1 million who had worked at least three years on their jobs found themselves without employment over the 1979–83 period due to plant closings, payroll curtailments, or companies going out of business. In some cases, these job losses were only temporary, entailing little sacrifice in terms of unemployment and lost income. In many other cases, the readjustment to the job loss has been much more painful.

Some of the workers displaced from their jobs over this five-year period had returned to work after a relatively short time, and their earnings when surveyed in January 1984 were as high or higher than they had been before the job loss. Many others had found different jobs, but frequently at much lower wages than in the jobs from which they had been displaced. About one-fourth were still unemployed when surveyed, though some may have been employed during part of the period since their displacement. Finally, about 15 percent had left the labor force.

Given the resiliency of the U.S. economy and the rapid advances which it posted during most of 1984, it is quite likely that many of the displaced who were still jobless in January 1984 were either recalled to their old jobs or managed to find new ones during the year. But even as the year came to a close, some industries—steel being a prime example—were still plagued by serious structural problems. This, in turn, was reflected by the still high jobless rates in some geographic areas where the displacement problem had taken a particularly large toll. For many of the workers displaced from long-held jobs in these areas, the prospects of reemployment were obviously not very bright—unless they were willing to relocate to new areas and to search in new fields.

# Part III
# Management, Union, and Public Policies

Although there is controversy over the causes of deindustrialization, nearly everyone agrees its impact on affected workers is severe, as we have seen in Part II. Controversy returns with this group of readings; there is widespread disagreement concerning what should be done to alleviate the problems brought about by plant closure.

At the root of the controversy is whether management, union, or government policies should predominate. Should the burden of adjusting to plant closure be left mainly in management's hands? After all, because companies make the decisions to close in the first place, it is their mess; let them clean it up. Or should the unions play a primary role? Because most of the workers affected by plant closure are unionized, isn't it the unions' job to protect the people they represent? On the other hand, what if management and unions are not getting the job done? Government policies are the alternative, and these vary from state to state. Because there is no real federal industrial policy in the United States, thus far there is no specific federal legislation directed at plant closure. This is a subject for the following section, which examines various legislative proposals. In this section we are concerned with how management and union policies are working out and whether public policies need strengthening.

The initial selection, by Arnold R. Weber and David P. Taylor, is chosen because it is a thoughtful early piece that focuses on one of the key issues of policy on plant closure: advance notification. If there is a single issue around which the policy controversy centers, it is this one. Even the staunchest advocate of the free-market approach would appear to have sympathy for the impact of closure on workers. Although advance notice is not the only feature that legislative advocates are seeking, it is perhaps their strongest card. The reason, as Weber and Taylor indicate, is that advance notice does much to alleviate the impact of a closure by giving workers adequate time to adjust psychologically and seek alternatives in the labor market. Moreover, advance notice does not lead to increased turnover, at least according to Weber and Taylor's research. Arnold R. Weber, for many years a professor at the Graduate School of Business, University of Chicago, and former Assistant Secretary of Labor, is currently president of Northwestern University. David P. Taylor is Deputy Director General of the International Labour Organization in Geneva, Switzerland.

In the following selection, Edward J. Blakely and Philip Shapira are critical of both conservative and liberal plant closure policies. They advocate greater community involvement, as have other observers, but break new ground by suggesting policies that have yet to be applied in the United States. For example, they recommend changing the tax code to eliminate incentives for firms to close plants and would provide for more participation of workers and community leaders in plant closure decisions. Some of their ideas might not be easily implemented, given the current political climate, but they are fresh and innovative nonetheless. Edward J. Blakely is a professor of city and regional planning at the University of California, Berkeley, and Philip Shapira is a Fellow in the Office of Technology Assessment of the United States Congress.

Richard B. McKenzie is probably the leading exponent of the free-market approach to plant closure policy. The next reading is excerpted from his principal book, *Fugitive Industry*. McKenzie's arguments can be summarized as follows: (1) the free-market economy has done reasonably well in providing jobs, and attempts to restrict business would be more harmful than closures themselves; (2) workers in industries affected by plant closure, such as steel, are compensated for job displacement through higher wages; (3) restrictions on plant closure hamper the competitive process, deter investment, and reduce economic growth; and (4) no firm would close a truly profitable plant to gain higher rates of return elsewhere. Although each of his premises can be disputed, there is no way to prove conclusively whether the free-market or the government involvement approach is better. Perhaps the key reason McKenzie's arguments have gained currency is their comportment with the wave of conservatism that has swept the United States and the free-market, pro-business, anti-union policies of the Reagan administration. Richard B. McKenzie is a professor of economics at Clemson University and Senior Fellow at the Heritage Foundation.

The next excerpt, by Archie B. Carroll, is included because of its appeal for greater voluntary exercise of social responsibility on the part of management. Carroll's ideas may seem attractive to those of every political belief because nearly everyone would agree that an enlightened sense of management responsibility and action can do much to alleviate the impact of plant closure. As Carroll suggests, it may even be possible to avoid the closure entirely through an employee buyout. He cites the actions of Brown & Williamson Tobacco Corporation as a positive approach to plant closure. Although the end result at that company was a loss of jobs, workers felt fairly treated, and detrimental effects were minimized. Archie B. Carroll is a professor of management at the University of Georgia.

In the next reading, Dale Yoder and Paul D. Staudohar raise the question of who should ultimately be responsible for plant closings—the firm, the government, or both? Using a case study approach, they compare how General Motors and Ford initiated plant shutdowns based on their respective management policies. Whereas GM created an environment of uncertainty and anxiety among its employees, Ford generated a cooperative, understanding atmosphere. The authors also discuss the role government should play in determining what kind of public policy legislation should be passed to ensure employee, community, and company protection. Dale Yoder, a well-known analyst of management issues, is professor emeritus at the Graduate School of Business, Stanford University.

In the following essay, Anne T. Lawrence presents a typology of union responses to plant closure. Included in the typology are collective bargaining, political action, and direct action strategies. She examines the prevalence of these strategies and discusses their effectiveness for reducing membership loss and preventing plant closures. Anne T. Lawrence is a Postdoctoral Fellow in the Research Training Program on Organizations and Mental Health at Stanford University.

In the concluding selection, William Foote Whyte examines programs for employee ownership of businesses. Such programs are an alternative to plant closure and also may provide a way of revitalizing a company's labor relations so that plant closure need not be considered in the first place. Worker ownership takes various forms, as Whyte points out through case examples. His analysis suggests that although programs for sharing ownership are not always successful, they have helped save thousands of jobs in American industry. William Foote Whyte is a professor at the New York State School of Industrial and Labor Relations, Cornell University.

# 10
# Problems of Advance Notice

*Arnold R. Weber*
*David P. Taylor*

I f management officials universally felt that advance notice of plant shutdown was a costless procedure, there would be every reason to expect that
the duration of the period of advance notice would be coextensive with the
permissible period. Some managements, however, have been reluctant to give
maximum advance notice for fear the efficiency of the work force would be
reduced and the orderly shutdown of the plant impaired. To bring these problems into perspective, this discussion will deal with both the preconceived and
actual problems of advance notice.

A leading issue in determining the duration of the advance notice period is
the concern that the announcement of the plant closing will reduce employee
morale and productivity. There is some belief that once the employees know
the plant is to be closed, they will become apathetic about their work or, worse,
engage in sabotage. All reports indicate that this problem has not developed.
The announcement of an impending plant shutdown, to be sure, does cause an
initial shock; few workers feel any pleasure at the prospect of permanent layoff.
This reaction has not, however, resulted in any perceptible decrease in productivity. Even in the soap company case, where there was considerable resent-
ment over the manner in which the announcement was made, productivity
remained at prior levels. Paradoxically, in the can company case, an official
reported that output per man-hour actually increased after the announcement.
This was interpreted as an attempt by the workers to persuade management to
reverse its decision and keep the plant open.

A related aspect of the productivity problem is the concern that a prolonged period of advance notice will lead to early quits and a crippling reduction in the work force before final production requirements have been met.
Again, management officials report that this problem has failed to materialize.
Where there have been some early quits in the industrial cases studied, the
manpower losses have been minor and have not limited the efficient operation
of the plants involved.

Arnold R. Weber and David P. Taylor, "Procedures for Employee Displacement: Advance
Notice of Plant Shutdown," *Journal of Business* 36, no. 3 (July 1963), pp. 312-315.

Two factors have acted to keep the quit rate low. First, almost all companies studied had some kind of severance pay plan in effect. Under the terms of these plans employees who quit forfeit severance pay. In addition, the most skilled employees, whose loss would have had the greatest effect on productivity, tended to be those with the most seniority—those who had the greatest benefits to realize under the severance pay plans. Thus the employees who did quit early were the less skilled, junior workers who were easy to replace on a temporary basis. Indeed, some firms studied were so unconcerned with the effect the loss these workers would have on productivity that they were allowed to quit early without penalty. Interplant transfer programs have the same holding effect as severance pay plans. Because these require the preservation of employee status, the worker cannot leave his original plant before he is terminated.

Second, labor-market factors often serve to reduce the likelihood of early quits. That is, when the plant is located in a small or depressed labor market, which has often been the case, alternative job opportunities are generally limited. Consequently, early quits have little attraction to an employee if such action appears likely to extend his period of unemployment.

Management's desire to maintain an adequate work force during the period of advance notice also applies to the need for handling seasonal peaks in demand or unanticipated production requirements. Once the plant closing has been announced and "permanent" layoffs have been put into effect on this basis, the employer may anticipate difficulties in enlarging the work force for the short period required. That is, there is some expectation that the employees who have been terminated will be reluctant to return. On the other hand, recruiting new employees from the labor market may prove to be a troublesome task.

Actually, the problem has been handled effectively through a variety of methods:

1.  Extension of the period of operation and the re-engagement of permanent employees previously laid off. This action was taken by the manufacturer of electrical equipment and the meatpacker when unexpected increases in orders occurred. Enough former employees returned so that no additional measures were required.

2.  Hiring workers on a temporary basis. This method was used successfully by the soap company and one of the farm equipment firms to handle a seasonal upturn that occurred during the period of advance notice.

3.  Offering overtime to employees remaining on the payrolls. Almost all the companies used this approach successfully.

Obviously, the effectiveness of these devices for meeting unanticipated or peak demands depends upon the state of the labor market in which the plant is located. If full employment exists in the labor market, such expedients may be less successful since employees who have been laid off might be expected to have found new employment, while the attractiveness of temporary employment for unattached job seekers will be considerably diminished. On the other hand, the fact that many of the displaced workers for various reasons do not immediately seek new employment creates a labor pool which the company may conveniently tap. In any case, none of the companies covered by the study experienced serious difficulties in meeting variable labor requirements.

In addition to these expected problems, one unanticipated difficulty arose in the sample of cases studied. Within a chemical plant, in particular, the internal allocation of the work force became a vexing matter as the two-year shutdown was carried out. The plant was composed of three separate chemical operations and, under the existing seniority arrangement, it was possible for a senior employee laid off from one operation to bump a junior employee on another operation. Since considerable differences existed in the skill requirements of the three operations and the sequence of the over-all closing was such that the operation with the lowest skill requirements was shut down first, senior employees could bump into jobs for which they were technically unqualified.

When confronted with these circumstances, the company agreed to train employees who had bumped down into jobs for which they were not qualified. Moreoever, as different workers bumped through the occupational structure, the company had to train several people for the same job. In some cases, the senior man was obviously unable to learn the job in a reasonable period of time, and management persuaded the union to waive the seniority clause. Nonetheless, company officials stated that considerable training costs were incurred because the problem was not fully anticipated.

In conclusion, the first step in the development of programs for advance notice has been recognition of the essential differences between the dislocations arising from a plant shutdown and those associated with conventional layoffs. Because the displacements are permanent and affect the entire work force within a relatively short period of time, the traditional advance notice provisions are usually ineffective. In all cases, recognition of these differences has meant that management and the union have, to some extent, been forced to go outside the regular collective bargaining agreement in establishing programs for advance notice.

Advance notice, by itself, has only limited functions. As analyzed in this study, the concept of advance notice basically encompasses a set of procedural techniques that facilitate the formulation and implementation of substantive measures to handle the problems of labor displacement. In unionized situations, the period of advance notice has been used to reach agreement on common objectives and supplementary provisions for cushioning the impact of the closing on the employees. The details and timing of the elements of an advance notice program have varied with the emphasis placed on different measures such as interplant transfers, retraining or intensive efforts to find jobs (in the labor market) for the displaced employees.

There are wide variations in the duration of advance notice of plant shutdown both within the same company and among different companies. These variations have largely reflected the economic and technical factors that have determined the permissible period of notice. Because the nature of these factors differs from case to case there have been few attempts to prescribe a minimum period of notice either by collective bargaining agreement or management policy. In most situations, any effort to negotiate such a minimum period of notice would implicitly involve the judgment that the duration of advance notice is a cost item subject to joint determination.

Over all, the problems associated with the administration of advance notice programs have not been unduly burdensome in the sample of cases covered by this study. Most of the difficulties relate to the need for maintaining desired levels of performance during the period of advance notice rather than sharp conflicts of interest between the parties. In this respect, labor-market conditions, the possible forfeiture of severance pay, and personal inertia generally have minimized the loss of personnel before the plant closing actually occurs. Even the most ineptly administered notification program studied did not result in lower production or premature attrition of the labor force. In addition, as long as the duration of advance notice exceeded the period specified by the labor agreement, management retained wide discretion in modifying the shutdown schedule without penalty.

Most of the cases investigated here involved large multiplant firms or government agencies. In these situations, management usually enjoys some latitude for innovation without a critical concern over cost and competitive pressures. Doubtless, an important motive for establishing elaborate advance notice programs was to reap a measure of good will among the employees and the wider community. However, such ventures have also contributed to sound labor-management relations by creating a framework for handling severe problems of displacement with a minimum of acrimony. In view of the apparent negligible costs incurred, they offer a pattern of experience that other enterprises, in a variety of economic circumstances, might advantageously emulate.

# 11
# Public Policies for Industry

*Edward J. Blakely*
*Philip Shapira*

I n the two decades immediately after World War II, the United States experienced a long wave of economic growth. The role of the war itself, the stimulus of public infrastructure investment, and the global dominance of American political power were key factors in this growth. There were, of course, millions of Americans in disadvantaged groups and lagging geographical areas who did not benefit from the job expansion and economic growth of the 1950s and 1960s. However, these were seen as residual problems—the direct or indirect result of such factors as the locational disadvantages of rural or inner-city areas, lack of skills, or discrimination in employment. While government programs and redistributional policies might be necessary to tackle these specific problems, there was a broad assumption that the private sector would continue to expand, thus providing sufficient employment in the aggregate. The prevailing view was that macroeconomic policy could be fine-tuned to ensure that the nation experienced consistent overall increases in output and employment.

In the 1970s these assumptions ran into trouble. First, it became clear that the policies to help the disadvantaged gain from overall economic growth were ineffective. Second, ever-deeper economic downturns and rising unemployment cast doubts on the very capacity of the private economy and traditional macroeconomic policies to ensure continued growth.

By the early 1980s America was in its most severe economic crisis since the 1930s. Problems of structural unemployment and inadequate incomes, once thought to be the residual difficulties of a few groups or areas, are today experienced by massive numbers of workers across all regions of the country and virtually every sector of the economy. Measured unemployment, now 8 percent, is more than double the 1969 rate of 3.5 percent; real earnings in 1982 dollars were down to 93.3 percent of 1977 levels; and poverty, especially among female-headed households, has noticeably increased. The average annual growth

Edward J. Blakely and Philip Shapira, "Industrial Restructuring: Public Policies for Investment in Advanced Industrial Society," *The Annals*, American Academy of Political and Social Science 475 (Sept. 1984), pp. 96–109.

rate in real gross national product (GNP) was 2.7 percent between 1974 and 1981, compared with 4.1 percent from 1960 to 1973. Basic industry is being restructured, with dramatic increases in plant shutdowns not just in the so-called rustbelt of the Northeast and Midwest, but in the sunbelt too. And there is increasing concern about whether or not American industries can match the technology and commitment to productive investment of companies based overseas, especially in Europe and Japan.

## The Determinants of Change

Without doubt, the incidence of plant closures, job loss, and poverty has been sharply aggravated by the recessionary macroeconomic policies that have been applied to the American economy since 1980. As and when aggregate demand in the economy increases, some of the workers who have lost their jobs in recent years may be recalled. In early 1984 such recalls to employment did begin. However, beneath these short-term ups and downs in the economic cycle, a more fundamental process of industrial restructuring is occurring. The U.S. economy is suffering from a long-term slowdown in output and productivity growth; the manufacturing sector is being reorganized and deindustrialized; the labor market has become increasingly polarized; and new technology threatens further job loss both in goods production and service-performing activities.

While many factors have influenced these developments, we believe the central driving force is a systematic business strategy to maintain short-term profits and work-place control. This strategy has not only hurt workers and their communities, but has also adversely affected the nation's long-term economic stability and performance. American business has become increasingly concentrated and centralized, which has exacerbated unproductive administrative and supervisory costs, encouraged hierarchical and inflexible production methods, promoted shortsighted management decisions, and decreased workers' job satisfaction and motivation. Government has aided this process through tax laws encouraging "paper entrepreneurship"—the making of profits not from production but from buying and selling subsidiaries, asset stripping, write-offs, and real estate speculation. In many cases, these centralized corporations have set high-profit hurdles for subsidiaries and plants, then either closed those units when the high target rates were not attained or used them as cash cows to engage in wasteful mergers and acquisitions elsewhere. Corporations have used advances in technology and communications to reorganize or relocate jobs to take advantage of lower-wage, less-organized labor, irrespective of the human and social costs incurred.

The result of these corporate strategies has been extremely rapid capital mobility and the virtual deindustrialization of a number of key manufacturing sectors and regions within the United States. Production now occurs on a global scale, with overseas units tied into production and consumption in America through transnational corporations. Meanwhile, within the United States there have been substantial interregional shifts in the location of production, between the frostbelt and sunbelt and between metropolitan and rural areas. These capital shifts have particularly affected the nation's core industrial areas.

Parallel with rapid capital mobility has come increased polarization, segmentation, and inequity in labor markets throughout the country. A large proportion of new employment created in recent years has been in low-wage, low-skill jobs, especially in service industries and in certain sectors of manufacturing, including electronics. Against this, industries that have been heavily hit by plant shutdowns and restructuring have often provided reasonably well-paying, stable, and accessible jobs, as in steel, automobiles, and shipbuilding. The emerging labor market is thus increasingly polarized and segmented, with increases in low-paid, low-skill work, increases in highly credentialed professional and technical jobs, and a dearth of middle-range blue-collar occupations. Further losses of traditional manufacturing jobs can be expected through automation; one estimate suggests that robots will replace up to 50,000 auto workers in the next six years. New office technologies are also likely to affect adversely many clerical and financial jobs.

## The Inadequacy of Current Policy Approaches

As a result of industrial restructuring, large numbers of workers will never return to their former jobs. Millions of other workers will find that the jobs they are able to obtain have been quite radically changed, with adverse implications for working conditions, wages, job satisfaction, and career mobility. Many people will be unable to find any stable work at all, especially in the nation's older industrial areas. Those experiencing economic dislocation or restructuring will invariably face heavy personal costs, including income loss, family stress, adverse physical and mental health, and diminished career and life opportunities. Substantial costs will also be imposed on government, through lost taxes; increased unemployment; welfare, medical, and jobs program expenditures; and higher rates of social pathology.

Of course, the picture is not entirely dismal. Even the regions that currently are most depressed will see some new jobs, while additional financial, service, and technology-based jobs will increase throughout much of the nation. Nonetheless, the present and foreseeable restructuring of employment is deep enough and the likely economic, social, and political costs high enough to push job stabilization and creation to the top of the American political agenda. Not since the Great Depression has there been so much concern about the lack of good jobs, and a growing policy debate has developed over the causes of the problem and the role government can or should play.

Unfortunately, while there is intense interest in the restructuring problem, we believe that much of the current policy debate is misconstrued. The debate tends to be polarized between "industrial policy" advocates who want increased government intervention to deal with industrial transition and "less-government" proponents who believe the problem, to the extent there is one, is manageable with a decreased public role. However, both sets of approaches share a common bias: they both are primarily concerned with restoring the conditions for capital accumulation; the needs of labor and community are subordinated. In our view, this is inadequate. Policy responses to restructuring should start first with the needs of the nation's workers and communities—the human and social resources that have to be at the center of any truly effective and equitable economic development strategy.

*Industrial Policy Approach*

Over the last few years, the idea that the United States needs an industrial policy has gained significant support, particularly among liberals, many academics, and parts of the business community. These groups accept that the United States has considerable industrial and employment problems, but they believe that traditional Keynesian demand-management policies are now insufficient. Instead, they advocate interventionist national industrial and investment policies to deal with industrial transition and encourage high-growth, sunrise companies in advanced technology fields like electronics or biotechnology. Interestingly, there is support within the business community itself for government action to speed industrial restructuring. *Business Week,* for example, has proposed a new era of tripartite, business-labor-government cooperation to develop industrial policies that will preserve resources for production—in part, by moderating wage demands—and encourage investment in new industries through training and tax incentives.

Although the proponents of this form of industrial policy endorse public action of one kind or another to promote growing sectors, they are split on the position government should take toward older, so-called sunset industries like basic steel and automobile production. One view is that public policy errs in trying to maintain old-line sectors. Reich argues that

> preservation strategies often seem preferable to massive unemployment and the destruction of communities. The irony is that these very strategies retard future economic growth by encumbering the movement of resources towards more productive uses.[1]

Indeed, Reich proposes to accelerate the decline of uncompetitive industries and transfer capital and human resources to growing, high-technology sectors.

Public employment and training programs would then help workers and communities adjust to the new technology employment.

A contrasting industrial policy view, however, advocates specific intervention in mature sectors. For instance, investment banker Felix Rohatyn proposes a new federal Reconstruction Finance Corporation, modeled on the old RFC of the 1930s, to help older industries and regions. Learning the lessons of previous bail-outs such as Chrysler and Lockheed—seen as successful, but too ad hoc and political—the new RFC would be a permanent and systematic mechanism for dealing with industrial restructuring. The new RFC would invest in troubled firms and sectors, but it would return firms to the regular private capital markets after those companies had been restructured. Rohatyn makes clear that RFC investments would almost always involve concessions by labor, including wage cuts, employment reorganization, and work-rule revisions.

The Democratic Congressional Subcommittee on Economic Stabilization has adopted much of Rohatyn's approach in its proposal for a new Bank for Industrial Competitiveness (BIC) to aid older industries, as well as high technology. The subcommittee expects the BIC "to bring to the process of revitalization finance the kind of tough-minded negotiation and conditionality which the International Monetary Fund is intended to apply in international lending."[2] Meanwhile the AFL-CIO Industrial Union Department has called for a national reindustrialization board which, while not explicitly demanding labor concessions, still would embody the idea that corporate planning by top business, government, and labor leaders is the best way to deal with economic restructuring.

Unfortunately, both the sunrise and the sunset industrial policy approaches have troubling problems. On the one hand, the jobs generated in growing high technology and service sectors are not comparable with the jobs being lost in traditional industry. These growing sectors offer a far greater proportion of low-wage jobs with restricted opportunities for upward mobility, and, without major policy changes, are unlikely to develop in the places where most of the jobless are. It is also doubtful whether, in absolute terms, there will be enough high-tech jobs to replace the huge number of basic industrial jobs that are disappearing, even if blue-collar workers had access to these new jobs. Moreover, as demonstrated by the loss of nearly 3,000 jobs when Atari moved its production facilities from California to Hong Kong in 1983, high technology jobs are not immune from corporate displacement.

On the other hand, industrial policy proposals to restructure basic industry through a new RFC or BIC have several defects, too. Public capital will be used to restore private profits, mainly in large firms. But it is not clear, particularly if investment is made in labor-saving, capital-intensive technology, that many new jobs would be created. Indeed, the Chrysler experience, often quoted as a success, demonstrates very well that jobs and communities are not necessarily the primary concern of industrial policy advocates. After federal intervention to save the company, the work force was drastically reduced, with minority workers bearing a disproportionate share of the job loss. Fifteen plants were closed, twelve of which were in Detroit, despite loans and grants from the city and the state of Michigan; wages and benefits were substantially cut, job-displacing new technologies were rapidly introduced, and the value of Chrysler stock went from a few dollars to over thirty dollars.

Finally, the current proposals for federal industrial restructuring are basically corporatist and undemocratic. Decisions made by an elite group of bankers, industrialists, government officials, and labor leaders are designed to force workers to accept concessions by skirting the political process. This represents a restructuring of capital to aid capital rather than help workers and communities. The intention of this is simply to restore profitability; the nature of production is not changed, private profits will still guide what is to be made, and the pressing needs for better public goods and services within most American communities will not be satisfied.

### No Problem/Less Government Perspective

A second response to America's changing economy has been the argument that because industrial restructuring is not a major problem in the United States, specific industrial intervention is unnecessary. Indeed, advocates of this view often state that the most helpful policy would be generally to decrease government's role in the economy. These arguments are popular with conservative politicians and business leaders, but support has also come from the unexpected quarter of the generally liberal Brookings Institution, where Robert Lawrence argues that deindustrialization is a myth. In his view, the overall increase in manufacturing employment during the 1970s shows that there has been no deindustrialization in absolute terms and that the aggregate levels of investment and productivity in America are satisfactory. Lawrence believes the real problem is simply a lack of demand and an overvalued dollar, which different macroeconomic policies could correct.

The Republican administration's analysis of the nation's industrial and employment problems adds the argument that misdirected government intervention has resulted in overregulation, higher taxes on both businesses and individuals, and reduced incentives to invest and produce. In this traditional conservative perspective, public interference with industrial change—for example, to stop plant closures—is naturally seen as short-sighted, costly, and doomed to failure. It is believed that efforts to stem capital mobility will impair private property rights, endanger new job creation, and ultimately cause greater job loss. An industrial policy thus is not only unnecessary, but will probably worsen the situation through costly subsidies and protection for declining industries, while government is unlikely to be successful in picking high-technology winners.

However, there are major deficiencies in these no problem/less government arguments. First, there really is a significant restructuring problem. While aggregate manufacturing employment rose by only about 130,000 jobs between 1973 and 1980, this incorporates growing numbers of administrative and supervisory employees. The number of production workers actually fell during this period by some 620,000, a 4.2 percent drop. In many individual sectors, like steel, automobiles, and electrical equipment, the percentage drop in production workers was far greater. The decline of these important basic sectors is not compensated for by the growth of high-technology or service jobs. Besides the wage disparities between growing and declining sectors, there are substantial geographical inequities in employment change. Furthermore, since 1980 both restructuring and the introduction of job-displacing new technology have intensified, which suggests that the employment scenario for the rest of this decade will present even more difficulties than in the 1970s.

Second, there is little evidence that American business suffers from too much government. The Western European economies and Japan have been comparatively successful, although they have more explicit public intervention and administrative guidance than the United States. Indeed, the complex arrangement of federal subsidies, tax supports, and research assistance to both military and civilian industries in the United States is a de facto, although not very good, form of public industrial intervention. Thus the problem in the United States is not that there is too much government, but that government is not doing the right things.

Third, despite the Reagan administration's argument that corporate investment has been hurt by overtaxation, the burden of taxation carried by business—measured as a proportion of all federal and state tax receipts—has, in fact, been decreasing over the last two decades. During the same time period, the level of fixed investment in business has been stable: it was 11 percent of GNP in the 1970s, compared with around 10 percent during the fast-growth years of the early 1960s. Thus, while the level of business investment in the United States has been below its main international competitors, this cannot be attributed to high corporate taxation.

Finally, we need to note that although the aggregate level of investment has not been greatly affected by diminishing business taxation, there have been important changes in the composition of investment. In recent years, while investment in communications and high-technology equipment has grown rapidly, spending on new factories and heavy machinery has declined. At the same time, investment in commercial structures—office buildings, retail stores, fast-food restaurants—has grown. Hence, there is a real problem in where investment is being placed in the economy, a point Lawrence overlooks as he expresses satisfaction with overall rates of investment spending.

In short, further tax concessions are unlikely to have major impacts on employment generation. Few business investment decisions of long-term consequence result from short-term tax advantages or similar incentives. Instead, new tax breaks tend to foster new rounds of unproductive paper speculation in tax shelters, lease-back arrangements, spin-offs, and acquisitions.

At the same time, it is also apparent that while increasing aggregate demand and devaluing the dollar are both important, such macroeconomic policies are no longer sufficient to ensure full employment. The reality of the advanced industrial economy is that job displacement will not abate. Investments in new technology will continue to dislocate workers, and large corporations will continue to relocate and reorganize jobs. Moreover, international competition will continue to present problems for the United States since cheap labor and/or deep export subsidies in other countries will continue to outweigh changes in the value of the dollar.

## An Alternative Policy Approach

If current approaches to reindustrialization and reinvestment are flawed, what are the alternatives? We believe there is a need for a different approach, one that explicitly emphasizes the needs and interests of the nation's workers and communities. There is now an ever-widening gap between the global power and mobility of corporate capital and the need for stability and community in the United States. We need to narrow this gap, not by uprooting or subordinating workers and communities, but by implementing an alternative economic strategy that will create a new context for reinvestment in the United States, curb wasteful and nonproductive corporate strategies, provide good jobs and raise standards of living more equally shared, produce goods for social needs as well as profits, increase worker and community involvement in decisions affecting them, and take a longer-term view about the kind of society we wish to foster.

Developing and implementing the policies necessary to achieve these ends presents a tremendous challenge, as much political as economic. Considerations of space make it impossible to discuss adequately all the complexities involved here. But, to stimulate debate, the balance of this article outlines a number of specific measures that should form the basis of an alternative policy approach to industrial restructuring. Three overall themes guide our proposals:

1. National policies are needed to increase social control over economic change and investment. Macroeconomic policy, whether monetarist or Keynesian, is no longer sufficient in and of itself. An alternative economic framework is necessary to directly promote investment, increase industrial democracy and corporate accountability, and provide for full employment. This framework needs to be built on social equity and human development, not just private profitability.

2.  Communities need a greater role in their local economies to promote stability and quality of life. The positive values of community in American life should not be sacrificed to rapid mobility of capital, nor to corporate—and government—decision making that ignores the community costs of investment and disinvestment.

3.  Workers should have increased control and certainty over their livelihoods. Individuals and their families should not have to bear the tremendous costs they are now facing in terms of lost income, personal stress, family disruption, lack of health protection, loss of pension rights, and lack of stability in employment or comprehensive planning for alternative employment.

### New National Framework for Reinvestment

The long-term future of industry and industrial communities in the United States needs to be based on a new framework for national industrial investment. An essential preliminary step is to change the federal tax code to eliminate the loopholes that provide incentives to close plants, move production overseas, and participate in wasteful conglomerate acquisitions. Similarly, new powers to review the industrial implications of mergers and takeovers should be introduced. U.S. Steel's decision to buy Marathon Oil, rather than reinvest in steel, is one example of the need for such measures.

However, while increased powers to regulate the direction the economy takes are certainly necessary, the central element of our alternative framework involves not top-down planning, but a new thrust to give workers and communities the capacity to know about, participate in, and bargain over investment and disinvestment decisions that will affect them. There is already a strong right-to-know tradition in the United States, through legislation on environmental hazards, the Community Reinvestment Act in banking, and freedom of information in government. We believe that this tradition should be extended fundamentally into the work place.

To establish work-place rights to know, major corporations should be required to provide ongoing information to their work forces on their corporate strategies, product and regional investment plans, overseas subsidiaries, and future employment plans. Corporations would be required to provide this information not only to increase accountability to those they employ, but also to allow real consultation and bargaining over investment strategy. As a matter of national policy, major firms would be required to develop mutually agreed procedures to consult with and involve their work force during planning and decision making. These procedures would vary according to circumstances, particularly if the firm is unionized, but could include joint labor-management boards, plant-level councils, or worker-directors.

As part of this program of increased information provision, advance notice and an economic displacement statement should be required of all firms considering major shutdowns. The displacement statement, which broadly follows the precedent of already existing environmental impact reports, would describe the rationale for closure, alternatives considered, the earnings and investment record at that facility, and the socioeconomic impact of closure. Implementation of this approach would provide American workers and communities with protections that exist in European economies. Workers and communities could use the notice period and displacement statement to investigate a closure, negotiate to keep it open, or explore the viability of community or worker ownership. To encourage compliance, firms failing to provide notice or the displacement report would lose any tax benefits associated with the shutdown and would be required to repay all public subsidies and tax allowances enjoyed while the plant was open.

Our belief in the need for workers, through their organizations, to be involved directly in decisions affecting them also guides our approach to new public reinvestment instruments. Although we have criticized the proposals of current industrial policy advocates, we do accept the need for mechanisms to promote, consistently and on a long-term basis, reinvestment in industrial plants and enterprises. But we differ on two key points: first, there has to be worker involvement; and second, narrowly defined private profitability is an insufficient criterion for national investment decision making.

In our framework, a national industrial reinvestment board or corporation would be established to invest in large firms and basic sectors, accompanied by state-level industrial development boards to make investments in medium-sized regional enterprises. Where either national or state investments are to be made, development agreements would be entered into, with the participating firms specifying investment and employment plans. These agreements would be negotiated among the corporation, the work force, and the reinvestment agency. Additionally, national and state reinvestment agencies would be empowered to develop permanent equity positions in the firms and sectors they involve themselves in, rather than returning them to the stock market. In this way, social control over key elements of the economy, such as steel or rail transportation, would be strengthened. Reinvestment agencies would be empowered to make public balance-sheet calculations of overall costs, including lost tax revenues, unemployment costs, and community disruption, when making investments.

In growing sectors, particularly high technology, our approach also differs somewhat from many current industrial policy proposals. While growth of high technology may mean massive individual returns for a few entrepreneurs and venture capitalists, too often the result is a poor job, a lost job, or a new weapon system; however, other outcomes are possible. We believe that the truly innovative role for public policy with regard to high technology is to generate new, more socially useful directions for the industry. Initiatives along these lines are already occurring in Europe, inspired by the pioneering alternative corporate plan developed by Britain's Lucas Aerospace workers. In the United States, we would propose that federal and state reinvestment programs, combined with research and educational resources, be applied to develop new, advanced technological products that will directly improve working and living conditions and create employment. Human-centered machine tools, new transportation, health, and energy technologies, and broad-based community communications systems are among the many possibilities.

Finally, policies need to be developed for the orderly growth of international trade and finance. From an international perspective, the current economic policies of the United States, particularly high interest rates and an enforced recessionary regime, have adversely affected many other countries and intensified international trade friction. In addition, U.S. corporations have relocated operations to exploit cheap labor in Third World countries, while their lack of basic investment and innovation in home production has placed many American-manufactured goods at an international disadvantage. Narrow protectionist policies within the United States are quite inadequate to deal with these issues, and they unfairly put the blame on foreign countries and workers. At the same time, adherence to outmoded conceptions of free trade in an era of multinational domination is equally unsatisfactory.

There is, instead, a need for policies that can promote balanced trade—increased exports and imports—on a bilateral basis and provide genuine economic development assistance, rather than support continued Third World exploitation. One possible measure is to require U.S. companies to create one domestic job for every job they create through overseas investment. Another step would be to end U.S. support for authoritarian regimes that ban trade union organization.

## Increased Role for Communities in Investment and Employment

A proposal we have already discussed—the requirement for firms to provide advance notice of plant shutdowns and prepare an economic displacement statement—would significantly assist local communities in averting or responding to disinvestment decisions. In addition, use of federally subsidized tax incentives and revenue bonds to finance the movement of jobs between different communities should be ended. Instead, communities should be encouraged to pursue local development agreements to improve conditions of employment equality, wages and benefits, and stability where firms receive public financial assistance. A step in this direction has already been taken by the city of Vacaville, California, which now requires firms receiving redevelopment funding to file local affirmative action plans, give up to one year's notice of closure, and agree to bargain with workers and the city over disinvestment decisions.

A further means of increasing community involvement in employment is through a job-generating program of investment in public goods production and services, including transportation, energy, housing, health, and development. While policies of direct intervention to reindustrialize basic industry and promote new growth sectors will retain and create jobs, the continued development of new technology is likely to diminish the aggregate job-generating capacity of manufacturing industry in the United States. At the same time there are continuing needs in America's communities for improvements in the quality of basic public goods and services. But we are not advocating a temporary job creation or countercyclical public works programs. Rather, we envisage a permanent and sustained investment in public goods and services, targeted to those areas hardest hit by industrial restructuring. To some extent, this massive investment in public goods and services would be self-sustaining, in view of the alternative—continued lost tax revenues and umemployment costs. Additional funding, however, could be obtained from a shifting of resources wasted through military expenditure into more useful and peaceful products.

### Policies for Individual Protection and Development

Because economic development needs to start from and be built around people, investment policies for human resource development are particularly important. First, we need to give immediate protection and relief to workers affected by restructuring. Advance notice and severance pay must be introduced as basic human rights. In addition, new national policies need to be implemented to protect health benefits and pensions and to provide improved unemployment and social service benefits. Individuals and their families should not be crippled through job loss.

Second, workers should have vastly increased access to educational resources through a more comprehensive employment and training system. We propose the establishment of a human resources trust fund, financed by employer contributions and general tax revenues. This trust fund would be permanent, replacing the temporary and ineffective Job Training Partnership Act. Through the trust fund, workers on the job could build up credits via years of service that would allow paid time off or leave to utilize increased opportunities for maintaining and improving their education and skills. This measure would also have the beneficial effect of reducing aggregate work hours, essential in the age of new technology. Workers who become permanently displaced would be eligible for stipends to support themselves and their families while undergoing retraining. It is important to note that retraining would occur in conjunction with alternative investment strategies, not in place of them as in conservative/liberal proposals, so that there would be jobs within the community for which workers could be retrained.

## The Challenge of Economic Alternatives

We have not attempted to examine all the details and implications of our specific policy proposals, nor have we tried to be exhaustive. Many further elements are needed to form a comprehensive alternative economic strategy. But we have tried to show that there is a strong need for increased attention to the issue of industrial restructuring, that current policy proposals from both liberals and conservatives are inadequate, and that innovative and bold alternative policies are now necessary to meet the needs of America's people and its communities. Corporate reindustrialization is not the answer, nor is high technology, while a policy of decreased attention would compound the waste and inequity induced by current economic and industrial restructuring. Instead, human resources and social needs have to be the building blocks for future industrial policies. Workers and communities are resources too valuable and important to be discarded.

There is little that should alarm in the approach we have advanced in this article. America has a long and rich history of deep and far-reaching public interventions that have fundamentally changed the nation's economic and social course. The Louisiana Purchase is one example; the development of the West, the promotion of the railroads, and the foundation of the land-grant universities are others. Public investment in infrastructure, community services, or new research programs is well established, and precedents exist for public development banking and extensive rights to information and participation. We even have an implicit, although not beneficial, set of national industrial planning policies through the current system of corporate tax breaks, import quotas, and support for military production.

Thus, few of the mechanisms we propose are new. What is different, however, is the type of problem to be addressed and the new role for government, workers, and communities. Two roads are open to us at this juncture of history. One road leads to a society increasingly subjected to rapid capital mobility, wasted human resources, widening social inequalities, destructive technologies, unmet community needs, and heightened social tension. This is an avenue already paved and open.

The second road leads in an alternative direction: a reemphasis on the value of labor and community, full employment and more equitable working and living conditions, controls over corporate investment decisions, production for social needs as well as profits, the beneficial application of new ideas and techniques, and more democratic work environments. We strongly believe in the second direction, although much work needs to be done to build the path. The fundamental reorientation of policies needed to achieve this alternative strategy is unlikely to occur easily or quickly, but it is the direction the nation, its workers, and its communities need resolutely to pursue.

# Notes

1. Robert B. Reich, "Take the Money and Run," *New Republic* (Nov. 15, 1982), pp. 28–30.

2. U.S., House of Representatives, Committee on Banking, Finance, and Urban Affairs, Subcommittee on Economic Stabilization. *Forging an Industrial Competitiveness Strategy* (Washington, D.C.: Government Printing Office, 1983).

# 12
# Consequences of Relocation Restrictions

*Richard B. McKenzie*

Contrary to the beliefs of the proponents of restrictions on business closings, disinvestment, and movements, such restrictions can, in an otherwise free economy, retard economic development in all regions of the country. People trade with one another in part because of the comparative advantages they have in production. A person who is relatively more efficient in the production of a good will give up less in producing that good than someone else. In other words, a person with a comparative advantage in the production of a good can produce that good more cheaply than others. If people produce those goods in which they have a comparative advantage and trade them for other goods they want, production costs will be lower than if all people try to be self-sufficient. Since costs are reduced by specialization and exchanges, more output than otherwise can be produced with the resources available to the community.

When evaluating the economic consequences of closing and relocation rules, we must keep two points in mind. First, people in different parts of the country have comparative advantages in different goods, and trade occurs between people in different parts of the country when they find that they can gain from trade. People receive a better deal (as they define the term) by trading with others than by trying to produce all the goods they want by themselves. They benefit by the exchanges they make, and they benefit because specialization results in reduced production costs and expanded output.

Second, the comparative advantages of people in different regions continually change because the conditions of production—that is, the availability of resources, technology, and consumer preferences for work and goods—continually changes. Changes in comparative advantage spell changes in relative costs. What was once relatively less costly to produce in the North can, because of, for example, a change in production technology, become less costly

Richard B. McKenzie, *Fugitive Industry: The Economics and Politics of Deindustrialization* (San Francisco: Pacific Institute for Public Policy Research, 1984), pp. 87–89, 90–104, 107–108.

to produce in the South. By moving from the North to the South, a firm can lower its production costs.

Regional shifts in comparative advantage may occur because a strategic resource used in making product A becomes relatively more scarce in one region than in another, and therefore the cost of producing good A in the region rises above the cost of producing it in another region. Additionally, increases in workers' education level or the discovery of more abundant supplies of a given resource may cause a region to initiate production of good B. Again, a shift in production within the region from A to B can keep the costs of production below what they would otherwise be. The industry that expands as a result of beneficial cost changes begins to impinge on the resources available to industries that once dominated the region. The costs of producing A in that region go up, and, as a consequence, the producer of A moves someplace else. Finally, the preferences of people in a region can change. For example, an increase in the demand for services within a particular region may induce resources like labor to move from the manufacturing sector into the service sector, causing the wages and production costs in the manufacturing sector to rise. One result may be that some manufacturing plants close or relocate, releasing their labor to a growing economic sector.

Describing the causes of changes in regional economic structures is difficult under the best of circumstances because costs are based on subjective evaluations of goods, and those evaluations cannot be *directly* observed in the market process. Over recent decades, however, the comparative advantages enjoyed by many sections of the North have changed for several reasons. First, the demand for services in the North has increased more rapidly than in other parts of the country. Second, environmental legislation has placed more severe restrictions on production in the congested northern region than in many other parts of the country and has increased the relative cost of production in the North.

If the comparative advantage of a region changes, for whatever reason, a restriction on business migration will keep resources tied up in that comparatively inefficient sector. The sector that should be expanding given its comparative advantage will be restricted from doing so because resources will not be released as quickly. Firms will be forced to retain employees and other resources that could be moved to the expanding sector.

In summary, governmental rules that impede the movement of manufacturing industry by actual relocations or by a process of "deaths and births" of firms out of one region will retard not only the development of the manufacturing industry in the South but the development of, for example, the expanding service sector in the North. Furthermore, restrictions on business closings and mobility will cause production costs to be higher than they would otherwise be, and, to that extent, will reduce national production and income.

# Job Movement

Many see closing restrictions as a means of keeping northern firms from moving south or west. Relocations account for a small fraction of all northern employment losses. Nonetheless, even if existing manufacturing firms in the North are restricted from moving to the South by legislation, the movement of manufacturing jobs to the South will be impeded but not stopped. Firms are willing to incur the costs of relocation because moving will enable the firms to protect their competitive positions or to gain competitive advantages over their rivals because the cost of production is lower in the new location than in the old. If firms are unable to move to new, more profitable southern locations, profitable opportunities are left to be exploited by others. Restrictions on business mobility will cause other firms to spring up in the new southern locations and existing southern firms to expand more than they otherwise would. Because their costs of production are lower, emerging and expanding firms in new locations will be able to undersell firms in old locations. Firms in the old locations will eventually be forced to contract their operations or go out of business. Because of births of new firms and deaths of old ones, the employment structure of regional economies will shift in the long run in spite of the relocation statutes. Such laws will only slow the adjustment process and, because new firms are required to form, increase the cost of adjustment.

Under the 1976 version of the National Employment Priorities Act, the Secretary of Labor, with the advice of the National Employment Relocation Advisory Council, is empowered to keep a firm from relocating if the move is without "adequate justification." Presumably this means that if a firm is deemed to be making a "fair rate of return on its investment," then it can be prevented from moving. We have noted, however, that an expanded output in new locations will reduce the profitability in old ones, and therefore even firms in old locations that are not at first able to provide "adequate justification" for a move will eventually be able to present the needed justification. What was once a move for so-called economic greed will soon become a move for economic necessity. Eventually, the government will find itself subsidizing these firms that it wants to see remain in their old locations.

# Power of Unions

Unions are interested in improving wages and benefits for their members. They know, however, that their ability to do so is restricted by the threat of relocation. If businesses are prevented from moving, the immediate threat of job loss is removed and, as a consequence, unions can be expected to increase their demands on employers. The result will be even higher production costs in the old locations; the benefits of relocation rules can be reaped in the short-run income of union members.

## Worker Wages

Wages in many nonunionized labor markets will tend to fall relative to what they would have been because of restrictions on business mobility. The supply of labor in any market is dependent on a number of factors: skill and education required, location, climate, risk of injury, and social prestige. Since such factors affect the supply of labor in individual markets, they influence the wage rate. Steeplejacks tend to make more than janitors because the skills required are more difficult to acquire, which restricts the number of people who are able to do the job, and because there are risks associated with working on metal beams in high places. Ph.D. accountants teaching in universities tend to make less money than their counterparts in private industry, but they receive the nonpecuniary benefits of flexibility in scheduling their time and tenured employment, which has been described as a one-way, lifetime contract.

As noted above, the supply of labor in any labor market is also affected by any expectation that the firms will close down altogether or move elsewhere. Generally, the greater the threat that an industry's firms will move, the higher the wage employees with given skills will demand before accepting employment in that industry. Restrictions on business closings propose to reduce the threat many workers now face in losing their jobs. In the short run the effect will be to increase the relative attractiveness of employment and, hence, the supply of labor in markets from which firms might otherwise move. To the extent that those markets are competitive, the wage rates will fall. If firms are forced to give severance pay to employees they leave behind, the same thing will happen: in competitive labor markets, the relative attractiveness of employment and the supply of labor will increase, pushing the wage rates down. On the other hand, closing requirements increase the relative cost of labor, depress the demand for labor, and cause a substitution on the marginal of capital for labor.

Of course, markets are not perfectly competitive, and they do not always fully adjust to changes in the economic environment. The general tendency, however, is clear. Wages will tend to be depressed by labor and capital supply responses. Wages may be observed rising in spite of closing laws on the books. The point of this section is that they will not rise by as much as they otherwise would because more people will want to work in those areas where the threat of employment loss in the near term is reduced. Workers who were once willing to accept the risk of losing their jobs in order to receive the higher wage will no longer have that opportunity; their real income will be reduced.

In the long run, new labor market equilibrium will be achieved as regional employment structures change to what they would have been in the absence of the relocation rules. On balance, however, the country will be poorer: closing rules cause resources to be inefficiently allocated among regions, and fewer goods and services will be produced. The per capita income of the country, in other words, will be restricted; if it grows, it will grow by less than otherwise because of the relocation rules.

Before we move on to another topic, a few comments need to be added regarding relative labor costs, a particularly troublesome part of the North-South, Frostbelt-Sunbelt controversy. An added inducement to this alleged southward flight of capital, it is often argued, is the "wage attraction" of the South. It is more illuminating, however, to assess the impact of "wage push" in the North. From the wage-attraction perspective, it may appear that low-paid workers in the South are "stealing" business from and causing economic harm to the North. The wage-push perspective, however, suggests that wages in the North are higher and on the rise for such classical economic reasons as competition for workers from the developing service sector in the North. In other words, manufacturers are forced to pay higher wages or risk losing their labor force to more rapidly expanding sectors of the economy. Firms that move south are pushed south, having been outbid for labor resources in the North. From this perspective, industrial movements to the South are a consequence of gains made by many workers in the North—and the "runaway plant phenomenon" is a positive force in the dynamic and growth economy, South and North.

## Wealth Transfers

Restrictions on closing will cause wealth transfers and losses. When people invest in business, they are actually buying a bundle of legal rights whose market value is equal to the present value of the future income that can be received from those rights. Many people have invested in businesses either directly through the purchase of plant and equipment or indirectly through the purchase of stocks and bonds on the assumption that they were purchasing not only the right to operate the business but also the right to move the business to any location they perceive to be more profitable. Because they thought they had the right to move their business, they paid more for the business or the stock than they otherwise might have.

Restrictions on closing propose to take the right of closure away from business owners. Such restrictions thus in effect reduce the income stream of the business and the wealth of business owners. Under the National Employment Priorities Act, as well as under any one of the many proposed state laws, businesses cannot move without having government and labor consent, so the demand for the assets of a firm that would otherwise choose to move will be reduced.

When employees secure legally recognized rights to prevent a plant from closing, which can lead to continued company losses, the employees effectively secure a strong legal position to negotiate their plant's closing. To avoid losses, the company should be willing to "buy off" the workers, a payment that adds to severance pay, giving them some portion of their capital in exchange for an agreement from labor that labor will not oppose (either through specially established labor commissions or the courts) the plant's closing. In this way and to this extent, closing laws transfer ownership rights from investors to workers. Accordingly, the value of the owners' portfolio of rights is reduced because the portfolio then contains fewer rights. The implied transfer of wealth from investors to workers will in the short run show up in reduced prices for the stocks of companies that are prevented from closing. The wealth gained by the workers, however, will not equal the wealth lost by firms prevented from closing. A part of that wealth transfer given up by the firms prevented from closing will go to owners of companies in other places, such as the South, that are able to emerge or expand and to survive more profitably because other businesses have found their movements impeded. A portion of the wealth lost by northern firms with restricted mobility will simply be a deadweight loss because of the subsequent decrease in production efficiency. A part of the wealth transfers will be picked up by government agencies that will be expanded for the purpose of increasing government supervision over business decisions. In the long run the wealth transfer will be partially reversed as wages drop to account for the instituted closing costs.

Although transferring wealth from one group to another may be a socially desirable goal, closing rules seem like a particularly clumsy way of achieving it, especially since the rules amount to a negative-sum game. However, when those closing rights were effectively bought by investors and could be purchased by labor by an appropriate reduction in other employment, one must ask if imposed closing restrictions are not a form of legalized theft. Furthermore, the resulting shifts in rights or wealth are rather haphazard, with unexpected and unintended consequences. Many people who will lose wealth in the transfer process are, of course, high income investors, but many losers are likely to be on low incomes from retirement plans tied to the market prices of various stocks. If the value of these stocks goes down, then the prospective income from retirement programs will also fall. In short, it is doubtful that any analysis has been made by proponents of restrictions to determine who will be harmed by the proposed legislation and by how much each person will be hurt.

## Competitive Governments

Restrictions on business closings will increase the short-run monopoly power of state and local governments. The framers of the Constitution attempted to place restraints on government by sharply defining the role of government and by introducing market principles into its organization. The founding fathers believed that the power of government could be constrained in part by competition among governments. A government, operating in a world of many governments, would face market constraints similar to those faced by private companies. If a company raises its price or lowers the quality of its product, it can expect to lose customers to competing firms that hold prices down or maintain quality or both. Similarly, a government that raises taxes or reduces the quantity and/or quality of its public services can expect to lose people and its tax base to other (competing) governments. Although the degree of potential competition among governments may be less than it is among private firms, competition can still occur. To the extent that it does, the power of state and local government to raise taxes and lower the quality of services provided is restricted.

Restrictions on business mobility will hinder the short-run response of business to tax increases imposed by local governments. To the extent that this occurs, closing restrictions temporarily increase the power of governments to raise taxes and reduce the quality of the services they provide. Consequently, the enactment of relocation laws is likely to lead to higher taxes and lower-quality services in many jurisdictions—but only in the short run.

State and local governments are deluding themselves if they think they can reap hefty, long-run monopoly profits from restrictions. Restrictions on closings may suppress competition among governments during the short period of time that capital cannot be moved or depreciated away, but restrictions do not eliminate forever the tendency of governments to compete. Eventually governments will catch on to the fact thay they can offset, partially or totally, the costs of closing restrictions by eliminating state and local taxes on business or by expanding services and subsidies provided to business by state and local governments. Once a few governments start competing in these ways, then most must enter the competitive game. The result will be that the costs of the closing restrictions can in large measure be passed on to the state and local citizens in the form of higher taxes. Again, someone must pay the costs of the closing restrictions. Closing restrictions are a means of forcing people who do not benefit from them to pay a part of the tab. To the extent that that happens, we should expect a misallocation of resources in the economy.

## Corporate Social Responsibility

Backers of restrictive legislation fervently contend that firms have a social responsibility to their workers and to the communities in which they exist. They point to the social disruption caused by plant closings: the loss of tax base, the idle workers and plants, the impairment of community services because of lower tax revenues, and the higher taxes imposed on others because of higher unemployment and social welfare expenditures. Dayton, a medium-sized manufacturing city in western Ohio, is to these proponents an excellent example of what plant closings can mean. In early 1980, three companies, including Dayton Tire Company, a subsidiary of Firestone, announced their intentions to close. Eighteen hundred jobs were lost at the Dayton Tire plant alone. Workers and town officials interviewed for television recounted the personal hardship the closings had imposed on them, their families, and the communities; several denounced the companies for giving them little notice of the closings and for being socially irresponsible.

A program on plant closings was broadcast on *The McNeil/Lehrer Report,* on June 18, 1980. During the same week, the "Today" program on NBC television ran a series of programs on plant closings, focusing on the closing of Ford Motor Company's plant in Mawah, New Jersey. Not all of the workers interviewed appraised the closings in the same way; a few saw the closings as an opportunity to take another job.

No one seriously contends that firms do not have a responsibility to their communities, but it is hard to accept the assumption that entrepreneurs and their management teams are any less socially responsible than their workers. If nothing else, the development of good community relationships is sensible from a profit motive. Rather than enlightening listeners and readers, claims that firms have a social responsibility only camouflage basic issues. Where does a firm's social responsibility end; who is at fault in plant closings; and what are the alternatives open to workers and communities for dealing with the problems encountered when a plant shuts down? Admittedly, plant closings create hardships for some people. The important question, however, is whether the remedy of closing restrictions is more damaging than the disease. Furthermore, should a firm's social responsibility remain a moral obligation or be made a legal one? For any society that wishes to retain the remnants of individual freedom, a sharp but important distinction must be recognized between voluntary acceptance of social responsibility and forced compliance with a government edict.

Through wages and an array of taxes, from property to sales to income, businesses contribute to the welfare of the community. It is not at all clear, incidentally, that businesses use more community resources than they pay for. In the intense competition for plants, many communities effectively "pay" plants to locate in their areas through below-cost sewage and water facilities and interest rates. Whether or not the competition that now exists among communities is socially beneficial is a question that needs careful attention.

Through personal saving, workers can secure their own individual futures against job displacement. As noted, wages tend to reflect the risk of plant closings. Proposed restrictive legislation will, if its proponents are correct, reduce the risks of job displacement for some (but by no means all or even most) workers. The reduced risk will lead to a reduction in their wages. These workers will, in effect, be forced to buy a social insurance policy that, because of its national coverage, will not always be suitable to many of their individual needs, but may be suitable for the limited number of people orchestrating the legislative drive.

Proponents of the plant closing legislation seem to imagine that in the absence of government restriction almost all workers will be exploited and, as a consequence, be unable to prepare for their futures. They also seem to imagine that the risk of job displacement will somehow disappear with government restrictions and that the costs of the restrictions, which are either overlooked or presumed to be trivial, will be borne fully by the "firm." For a firm like Dayton Tire, however, employing 1,800 workers and paying the average wage in Dayton in 1979, the costs of the two-year notice, plus the one-year severance pay (at 85 percent of the previous year's pay), plus the fringe benefits, plus the community payments, are anything but trivial; they can easily exceed $110 million! If the proposed National Employment Priorities Act or the Ohio closing restrictions had recently been enacted and Firestone had been prevented from closing the Dayton plant (along with four others scattered around the country), the company might have had to incur more than a half a trillion dollars in production costs and losses of upwards of $60 million over the next three years, for which it would have been unprepared. Very likely, the financial solvency of the entire company and the jobs of hundreds of other Firestone workers, which were precarious at the time, would have been placed in even greater jeopardy.

To operate in a financially sound manner under such a law over the long run, a company must prepare for the eventual expenditure associated with closing: it can establish its own contingency fund or buy insurance against the risk that it must assume. Either way, the cost will be recovered either from wages that would otherwise have been paid or from higher prices charged consumers, in which case the purchasing power of the workers' incomes is reduced. Although there is no question that owners of companies will be hurt by the legislation, the point that needs emphasis is that most workers will not escape paying for the benefits received under the restrictions.

Instead of restricting business rights, communities could set aside funds from their taxes, and these funds could be used to alleviate social problems created by plant closings. In the absence of national or state legislation, community contingency funds could be established to meet local needs and to account for the trade-offs that people in the different communities were willing and able to make. If current tax collections were insufficient to meet community needs, then tax rates could be raised. Of course, such an increase would discourage firms from either establishing or expanding their operations. But the proposed restrictions have the same effect. They are a subtle form of business taxation that, like all taxes, would deter investment and thereby further erode growth in productivity and wages. Contrary perhaps to the good intentions of its advocates, the new restrictive legislation increases the social cost associated with business operations.

Plant closing restrictions would probably increase not decrease the number of business failures because, as noted, they raise the cost of doing business and of staying in business. Indeed, closing restrictions can force closing decisions. Firms must continually evaluate their future prospects, asking whether they can make it through the next six months, year, or two years. Once they announce their decision to close, their coffin can be sealed. Their employees can be expected to leave and their suppliers, buyers, and creditors can be expected to deal more cautiously with them. Because of the notification requirements, many firms will announce their intentions to close at a time when they might otherwise try to ride out what might be temporary adverse market conditions. In this respect, "plant closing laws" are laws that close plants.

## Mobility: A Key Economic Liberty

Proponents of restrictions insist that because firms draw on the resources of the communities, they have an obligation to compensate the community for the benefits the firms have received over the years. Proponents are particularly concerned when companies use the profits made in one place to expand elsewhere. Does not the company owe the community a fair share of any future expansions? Messrs. Bluestone and Harrison describe with some eloquence how northern firms are disinvesting themselves of their plants in the North by earning a profit and then expanding their operations in the South and West. A principal problem with such a line of argument is that it is perfectly applicable to employees: workers also draw on a community's services and the resources of their plant. When they decide to resign their employment and move elsewhere, do they not owe a social debt to their community, and should they not compensate their employers, as restrictive legislation proposes that firms repay their employees and communities? Through wages received and purchases made on household goods, employees send their incomes out of the community. To be consistent, should not proponents propose that the public interest dictates that employees spend a fair share of their incomes in the community? Should not employees (and their unions) be told how much of their incomes must be invested in their companies?

These questions are not intended to make the case for restrictions on employee earnings and expenditures. Rather, the point is that we allow individuals the freedom to do what they wish with their incomes and to move when and where they please for very good reasons. First, a worker's income is only one half of a *quid pro quo*, a contractual agreement between the employer and employee that is freely struck and presumably mutually beneficial. Second, freedom gives workers the opportunity to seek out the lowest-priced and highest-quality good compatible with their preferences; that very same freedom forces sellers to compete for the purchases of the workers and provides workers with the security of having a choice of places to work and buy goods.

Third, but foremost, there is the firm belief—call it faith—that people are indeed created with certain inalienable rights. Individuals know, within tolerable limits, what is best for them in their individual circumstances, and they are the ones best qualified to say what they should do and where they should live and how and where they should invest their resources, labor, *and* financial capital. The right of entrepreneurs to use their capital assets is basic to a truly free society; the centralization of authority to determine where and under what circumstances firms should invest leads to the concentration of economic power in the hands of the people who run government. Private rights to move, to invest, to buy, to sell are social devices for the dispersion of economic power.

There are those who believe that the case made against this restrictive legislation is obviously an apology for the corporate giant. Not so. Embedded in the proposed federal legislation are provisions that effectively institutionalize the Chrysler bailout of 1979 and 1980. The government is given broad discretionary authority to provide unspecified forms of aid to companies that go bankrupt. This type of bill could effectively swing the doors of the federal treasury wide open to any firm sufficiently large and with sufficient political muscle to enlist the attention and sympathies of the Secretary of Labor. The bill destroys, in part, the incentive firms now have to watch their costs and avoid bankruptcy. Because votes are what count in politics, under the proposed law the incentive firms have to avoid losses diminishes as the size of the firm (meaning number of employees) grows.

Large rather than small companies will be most likely to secure access to the discretionary authority of government. Chrysler was bailed out in 1979 not because it was the only firm that went broke that year (there were hundreds of thousands of others), but because it was large and had, through its employees, stockholders, and suppliers, the necessary political clout. Of course, the taxes of many smaller businesses will help finance these subsidies for their larger competitors. The marginal firms will be forced to close because of the extra tax burden. We can only imagine what such a bill portends.

Businesses, like their workers, do not always like to compete or to worry about someone else outproducing or underpricing them. Many businesses actively seek government bailouts, subsidies, and buyouts. For these reasons alone, we should be cautious in interpreting the consequences of restrictions, no matter what their stated intentions. Restrictions on plant exits from are also restrictions on entry into competitive markets. They keep competition out of local markets by restricting the ability of businesses to move from other areas of the country, and they discourage the emergence of new businesses, meaning both new investment and new competitors. Thus, such laws can protect some established wealth by protecting established businesses with market-proven products from potential competitors with goods untried in the market. Northerners can be assured that the southern textile industry, for example, does not look kindly upon northern businesses expanding into their low-wage markets.

## Visible and Invisible Effects of Restriction

Supporters of the restrictive legislation frequently point to the emotional and physical difficulties of those who suffer job displacement. These problems can be serious; there is no debate on that point. However, the political attractiveness of restrictive legislation can be appraised by the visibility of the harm done by plant closings and the invisibility of the harm done by restrictions on closings. The hardship associated with closings is easily observed. The media can take pictures of idle plants and interview unemployed workers; researchers can identify and study the psychological effects of job displacement. On the other hand, restrictions on plant closings are also restrictions on plant openings. They reduce the competitive drive of business, deter investment, and reduce the growth in productive employment, in generally retarding the efficiency of the economy. However, it is impossible for the media to photograph plants not opened because of the restrictions on plant closings, or to interview workers not able to find employment (who, as a consequence of unemployment, develop hypertension, peptic ulcers, and severe depression) because of the inability or lack of incentive of firms to open or expand plants.

Proponents contend that they support both the little person, the low-income, uneducated worker, who may otherwise be exploited by the system, as well as the relatively highly paid, skilled worker. The fact of the matter is that the proposed protective legislation will work to the detriment of some of the lower-income, uneducated workers. The legislation imposes a severe penalty on entrepreneurs who seek to establish production facilities whose chance of success is slim. Plants that would otherwise be built will, with this law, not be constructed, and therefore the law would work to the detriment of workers in low-income neighborhoods in the inner cities because that is where the chance of success is often lowest. Furthermore, if the law were enacted, it would freeze in place for a period of time many of the production facilities of the country. Relatively depressed areas like Dayton, Ohio, would lose one of their best opportunities for recovering from the recent loss of jobs: the recruitment of plants from other parts of the country.

Television coverage often fails to consider the widespread economic growth occurring over time in a particular area. Dayton Tire shut down in 1980; manufacturing employment in the Dayton area was down slightly from what it was in 1970. These are the facts we hear repeatedly. What we do not hear is that total employment in Dayton and in Ohio rose during the 1970s by 10 and 16 percent, respectively; that the average weekly wage in Dayton is 50 percent higher than in Greenville, South Carolina; and that earnings during the 1970s, after adjusting for inflation, rose modestly but several times faster than the earnings in the rest of the country. These are the positive results brought about, to a significant extent, by the ability of firms to adjust by closings and openings to changing economic circumstances.

It is difficult to measure the value of goods that are never produced because of the greater (government imposed) cost of capital. Nonetheless, if restrictive legislation is passed, goods will go unproduced, and many of the goods produced will be goods consumers do not want. Firestone closed Dayton Tire because it produced bias-ply tires. Only months prior to the announced closing of Dayton Tire, the tire market had turned down dramatically; domestic car sales had plummeted because of higher automobile prices (brought on partially by safety and environmental regulations), higher fuel prices, and the shift in consumer tastes to smaller imported cars. In addition, consumers revealed through their purchases that they wanted safer, more fuel-efficient, and more reliable radial tires. If Firestone had been required to keep the Dayton plant open, along with five others scheduled for closing, Firestone would have been forced to produce tires that consumers did not buy and consumers would have been forced to purchase bias tires that they did not want.

There are costs associated with tying up an entire economy for even short periods of time, and people will bear those costs. The residents of Gary, Indiana, may reason that closing restrictions will keep factories in place and that therefore their employment opportunities will be improved. What they may fail to see is that restrictions on plant closings across the state or nation will retain firms in other cities and prevent those firms from moving to Gary, denying Gary in the process the opportunity to make use of its comparative advantages in production and to establish a strong, viable employment base for the future. Proponents of restrictions argue that free marketeers ignore the human costs of unfettered markets, but such a position requires that its advocates, for no other reason than consistency of argument, take account of the human costs of a fettered economy, especially one with restraints as extensive as those proposed.

If a plant closing law is ever passed, its victims will be largely invisible. Disenchanted consumers and unemployed workers may very well not realize they have been victimized, and if they do realize it, they will probably be unable to determine who is at fault. Therein lies the political appeal of restrictive legislation; Congressman Ford and others can champion this cause of the political left without ever confronting those harmed by it.

## Concluding Comments

Advocates of restrictions chide supporters of a market economy for their opposition to restrictions, claiming their arguments are based on models of a perfect economy that are never duplicated in the real world. Real world markets, so the argument goes, are fraught with imperfections in the form of restrictions on the mobility of labor and the ability of prices and wages to adjust to new market conditions, such as changes in income and tastes. Such a criticism is totally misdirected. Economists who understand their models have never seen them as "descriptive reality," but as means—methods or theoretical tools—designed to accomplish one objective, to make correct predictions concerning the impact of public policy changes. The usefulness of such theories must accordingly be judged in terms of that objective.

Descriptively speaking, the case for a free market is strengthened by a recognition of its imperfections. People, including state government officials, know little of what others want and how those wants can best be satisfied. We need some social means by which information on people's preferences is revealed to others. The free play of individual actions, resulting in millions of daily exchanges organized through business firms is what we call "the market." However, the market must be understood as more than the sum of what individuals do, for through the forces of competition the market generates critically important information on individuals' preferences and relative costs of production that cannot effectively be generated in any other way. Competition forces people to reveal what they are willing to accept for the labor and other resources they have to offer. One of the more important bits of information passed around by the market is where the most cost-effective production locations are. If firms are denied the right to move to less costly areas of production, those firms can be certain that other firms will emerge to take advantage of the unexploited cost savings and to outproduce and underprice firms held in place by plant closing laws. Inevitably, plant closing laws will be transformed into firm shut-down laws.

Many of the arguments of the proponents of restrictions amount to the simple proposition that not all people want to compete, which is to a degree understandable. That is basically what is implied when advocates of restrictions deride cheap labor in the South, as if highly paid workers in the North, for example, have an inherent right to demand whatever they want in the form of wages and fringe benefits, no matter what other people are willing to do, and to deny by way of government restrictions workers in the South higher paying jobs.

Competition is a hard taskmaster, but it is also the glory of the free market system. It is the social process that insures that, on balance, the incomes earned in the North or South or West have significant purchasing power in terms of prices and qualities of goods and services. We must question whether restrictions on plant closings actually improve on the imperfect competitiveness of the market system or simply make the system more imperfect. Opponents of restrictions suggest the latter is the case.

# 13
# Management's Social Responsibilities

*Archie B. Carroll*

*Item*: General Electric recently announced plans to close its electric iron plant in Ontario, California and shift its operations to Singapore. One employee of 37 years was heard to exclaim: "It's a dirty rotten shame. So many years of work and then all at once they're locking the gates on you."

*Item*: Ford Motor Company revealed its intention to shut down its Sheffield plant in Alabama. The Sheffield plant is the world's largest aluminum casting plant. In what shapes up as a local disaster, the Chamber of Commerce estimates the closure will result in a loss of 1,100 jobs directly and another 2,200 jobs indirectly.

*Item*: Paragon Gears, Inc., a marine-engine transmission factory in Massachusetts, is scheduled to close down by summer. Union efforts to keep the plant open included a parade, posters in downtown stores, a school essay contest, and a visit to the company's headquarters in Wisconsin by an employee delegation. Despite these efforts, the company plans to close down the plant.

Throughout the United States, union, employee, and community anger over plant and business closings is reaching unprecedented heights. New forms of resistance are being experimented with daily. The resistance ranges from community protests and employee marches on headquarters of national firms to union demands that plants be kept open with negotiated concessions on wages, work rules, and benefits.

## Magnitude of the Problem

Although no one keeps official records on the number of plant closings and jobs lost in the U.S., David Birch of MIT has done pioneering work on data collected by Dun & Bradstreet and has presented an analysis that estimates the numbers of jobs lost due to business closings. His findings, which Barry Bluestone and Bennett Harrison, authors of *The Deindustrialization of America*

Archie B. Carroll, "When Business Closes Down: Social Responsibilities and Management Actions," *California Management Review* 26, no. 2 (winter 1984), pp. 125–139.

(1982), cite as the best available, conclude that 32 to 38 million jobs were lost during the decade of the 1970s due to plant and business closings. Bluestone and Harrison judge that the figure is closer to 38 million and that even this may undercount the total jobs lost for the decade because of the need to estimate the losses in the major wave of retrenchment and closings in the nation's most basic industries—autos, steel, and tires—during the last few years of the decade.

One of the most interesting findings to come out of the analyses of job losses is that business closings are not confined to the old industrial Frostbelt and thus must be considered a national phenomenon. The greatest surprise comes in the South, where the overall rate of economic growth has been greater than anywhere else. In spite of its legendary "good business climate," almost 7 million jobs were lost as a result of shutdowns and another 4 million due to cutbacks in existing operations between 1969–1976 alone.

## Reasons for Business Closedowns

There is no single reason why so many business and plant shutdowns have occurred over the past decade, although the recession of the early 1980s has been a recent catalyst. Some companies are simply in declining industries, others have outdated facilities or outmoded technology and see the shutdown as their only alternative. A number of companies seek nonunion regions of the country while others seek access to new markets. Automation, rising energy costs, inadequate capital investment, availability of transportation, foreign competition, depressed demand for products, costly regulation, labor-management conflict, poor long-term planning, and changes in corporation strategy are other reasons often cited.

Throughout the country companies are closing or selling operations as a reflection of the fundamental redeployment of corporate assets that is taking place. This trend has been referred to by one consultant as deconglomeration—abandoning low-return assets to maximize long-term return on remaining assets. A good example of this type of strategic shift can be seen in the case of Bethlehem Steel Corporation, the nation's second largest steel producer. Bethlehem has taken steps to shut down two west coast steel plants and to rid itself of its ship-repair business. A company executive indicated that this was a part of a strategic decision that had been on the books but was accelerated due to the state of the economy. BTK Industries, Inc., of Dallas provides another illustration as it closed down seven different subsidiaries and divisions due to corporate repositioning of where it wants to be over the longer term. A final example is seen in the case of National Steel Corporation's 1982 decision to close its giant Weirton steel mill in West Virginia. The company decided to close down or sell out as part of a strategy to lessen its dependence on steel.

## Legislative Activity on Business Closings

The degree of legislative activity on the issue of business closings is an index of its viability as a legitimate social issue which management needs to address. To date no federal legislation has been passed, although several bills have been introduced and have failed. The legislative proposals have typically been structured such that prenotification and severance pay would be required of employers before closing down. One bill would have provided for continuation of employee benefits for displaced employees. The bills presented thus far would have required anywhere from six months to two years advance notice depending upon the degree to which the firm could have reasonably predicted its intention to close or move.

In 1982, at least 12 states were considering legislation regulating employers who close, relocate, or reduce their operations. Such legislation was considered in California, Illinois, Indiana, Michigan, Minnesota, Missouri, New York, Ohio, Pennsylvania, Rhode Island, Vermont, and Wisconsin.

The legislation typically would have required an employer of more than a specified number (for example, 100) to give prior notice of six to twelve months of its intent. Groups to receive such notice usually included affected employees, labor organizations, and local and state government officials. The measures often required the company to file an economic impact statement which would show impact on budgets in affected communities. The impact statement would have been required to show, for example, the projected effect on the community's tax base and projected losses to local businesses and stores.

A main thrust of many of the bills had been to provide displaced employees with increased benefits—severance pay, pension benefits, continued health insurance benefits, job retraining, and job relocation assistance. In Vermont, for example, the proposal was for employers to deposit 15 percent of the affected annual payroll in a special Community Assistance Fund. The proceeds would have provided severance pay to employees. In Ohio, the proposed measure would have required the employer to make a discharge payment equal to one week's pay for each year of employment. Bills in Indiana, Michigan, and Rhode Island would have required the employer to maintain medical benefits six to twelve months beyond termination. Wisconsin and Maine have passed laws requiring advance notice. Legislation died in at least seven states in 1981.

Laws such as these have met considerable resistance by business leaders because of their cost and other problems. In Connecticut, such legislation was abandoned after business leaders convinced legislators that such measures would not only scare away new business but would contribute to a climate of hostility.

The city of Philadelphia, over its mayor's veto, became the first *city* to pass a law requiring firms to notify employees of intention to close "all or substantially all" of its operations. Sixty day notice is required before a plant closes or the worker can sue to get paid for the period of time short of 60 days.

Plant shutdown and relocation problems are not unique to the United States. In at least three other countries—Great Britain, West Germany, and Sweden—advance notice and other concessions must be given prior to shutdowns. In Great Britain, companies are required to give severance pay, advance notice, and time off to seek alternative employment and must consult with the union and consider their proposed alternatives to dismissal. In West Germany, where employee works councils are required, companies must give timely notice to the works council before the closedown decision is finalized. The decision to close down must be negotiated with the works council and management must consider alternatives to the dismissal of employees. The agreement then must be put in writing in the form of what is called a "social plan." The situation in Sweden is much like that of West Germany. There, handicapped and older workers are given preferential treatment relative to order of layoff.

On a number of fronts, both domestic and foreign, legislative activity on the issue of business closings suggests it to be a serious problem that demands businesses' attention. State legislation has a way of becoming the forerunner to federal legislation and initiatives in Europe often precede action in the U.S. Consequently, this could become yet another realm in which business may be faced with unwanted regulation if it does not fashion responsive actions that address some of the problems created.

## The Union's Role

As with so many other issues, unions have attempted to fight business closings. In those cases where they have failed to stop closings, they have attempted to make the conditions of closing subject to contract agreement and/or collective bargaining. It seems only natural that the unions would be in on the early organized efforts to combat business closings since so many of the industries and communities affected are dominated by unionization. Also, union leaders think that the widespread shutdowns that are occurring could fatally weaken organized labor. Union members who lose their jobs might take nonunion jobs elsewhere or possibly turn sour on the union movement. Plant closings could also weaken the union's ability to organize or bargain elsewhere in the same community.

Court rulings have appeared to be somewhat mixed thus far on management's rights to close a plant. In one case a federal district judge ordered a company to keep its plant open beyond the planned shutdown date. His argument was that a property right might have been established from the lengthy, long-established relationship between the company and the town. On the other hand, the Supreme Court has ruled that it does not constitute an unfair labor practice for an employer to close his entire business even if the closing is due solely to antiunion animus. The difficulty begins when one attempts to assess whether a change in business operations is of such a magnitude that the courts will consider it a complete closing. More recently the Supreme Court has ruled that employers need not bargain over the initial decision to close part of an operation.

Most union activity has focused on getting companies to bargain in contract negotiations over the plant closing issue. The law is not clear regarding the employer's duty to bargain on the decision to close one of its plants, but unions have continued to press this issue. Beginning in 1979 major unions such as the United Auto Workers and the United Rubber Workers have made demands for earlier notification of plant closings and negotiation over the terms of closing. Some of the terms of closings discussed included advance notice, improved transfer rights, retraining and relocation benefits.

Union activism has grown because labor leaders foresee continuing business closures in the future and because of their frustration at the lack of assistance they have received from state legislatures and the courts. Although a number of state legislatures have considered bills intended to ameliorate the impact of plant closings, only Maine and Wisconsin have actually adopted legislation. If the rate of business closures continues, we might reasonably expect to see more states and perhaps even the federal government assume a more active role in attempting to solve the problem.

## What Should Business Do?

By all appearances the business closing issue is evolving much like many of the other social issues business has had to address during the last twenty years. The effect of the recession of the early 1980s has been to put the issue more dramatically on the front page. Although it is somewhat premature to assess where public opinion on the issue will eventually come to rest, a strong case could be made for a positive or proactive response given the magnitude of the problem, the level of legislative activity, and the activism of labor unions. In addition to the desire to circumvent more government regulation and a continued hostile relationship with labor, responsive corporate action could be justified on the grounds of long-run, enlightened self interest, preserving business as a viable institution in society, preventing further social problems, and creating a favorable public image for the corporation.

Though the right to close a business down has long been regarded as a management prerogative, the business shutdowns since 1970—especially their dramatic effects—call to attention the question of what rights and responsibilities business has vis-à-vis employees and communities.

The literature of business social responsibility and policy has documented corporate concern with the detrimental impact of its actions—indeed, businesses' social response patterns over the past fifteen years have borne this out. No less a business advocate than Peter Drucker has suggested what business owes regarding the social impacts of management decisions:

> Because one is responsible for one's impacts, one minimizes them. The fewer impacts an institution has outside of its own specific purpose and mission, the better does it conduct itself, the more responsibly does it act, and the more acceptable a citizen, neighbor and contributor it is.[1]

The question is raised, therefore, whether businesses' responsibilities in the realm of plant closings and their impacts on employees and communities is any different than the host of responsibilities that have already been assumed in areas such as employment discrimination, employee privacy and safety, honesty in advertising, product safety, and concern for the environment. In fact, at least from the perspective of the employees affected, might not the employee role in plant and business closings be considered just an extension of the numerous employee rights issues that many corporate social policy experts think may dominate employer-employee relations throughout the 1980s?

A number of executives have spoken to this issue and several have indicated that there is an obligation to employees and the community when a business opens up or decides to close. As D. Kenneth Patten, president of the Real Estate Board of New York, illustrates:

> A corporation has a responsibility not only to its employees but to the community involved. It's a simple question of corporate citizenship. Just as an individual must conduct himself in a way relating to the community, so must a corporation. As a matter of fact, a corporation has an even larger responsibility since it has been afforded even greater advantages than the individual. Just as a golfer must replace divots, a corporation must be prepared at all times to deal with hardship it may create when it moves or closes down.[2]

Others have also argued that there is a moral obligation at stake in the business closing issue. In a rather extensive consideration of plant closings, philosopher John Kavanagh has asserted that companies are not morally free to ignore the impact of a closing on employees and the community. His argument is similar to those that have been given on many other social issues; namely, that business should minimize the negative externalities (unintended side effects) of its actions.

The scope of businesses' responsibility in the closing issue includes actions taken before the decision to close or relocate is made and those taken after the decisions to close or relocate is made.

*Before the Decision to Close Is Made*

Before a company makes the decision to close down it has a responsibility to itself, its employees, and the community to thoroughly and diligently study whether the closing is the only option available. A decision to leave should be preceded by in-depth discussions with community leaders, testing their willingness to cooperate in meeting the difficulties faced, and by a critical and realistic investigation of economic alternatives. This would include a study of long-term productivity and cost estimates, the employee base and the nature of skills available, and the likelihood that the identical problems might recur in a different location.

After a careful study has been made, it may be concluded that finding new ownership for the plant or business is the only feasible alternative. Two basic options exist at this point: find a new owner or explore the possibility of employee ownership.

**New Ownership.** Malcolm Baldrige, chairman of Scoville Manufacturing Co., has argued that the first obligation a company has to its employees and the community is to try to sell the business as a going unit instead of shutting down. This may not always be possible, but it's an avenue that should be explored to its fullest extent. Quite often the most promising new buyers of a firm are residents of the state who have a long-term stake in the community and are willing to make a strong commitment.

For example, when Viner Brothers, a Bangor, Maine, shoe manufacturer, filed for bankruptcy in 1980, its three plants presented an attractive investment opportunity for area shoe companies. Within several weeks, Wolverine, the maker of Hush Puppies, was the new owner. Part of the multimillion dollar sale agreement was that Wolverine hire at least 60 percent of the laid-off workers. About 90 percent of the 900 workers who were laid off were eventually rehired.

**Employee Ownership.** The idea of the company selling the plant or business to the employees as a way of avoiding a close-down appears appealing on first blush. Hundreds of U.S. companies with at least 10 workers each are employee-owned. Most of these arrangements are the result of last-ditch efforts to stay in business. According to the National Center for Employee Ownership, some 50,000 workers have saved their jobs by taking over companies. With inflation, high interest rates, and a recession, buy-outs of marginally productive firms appear likely to increase.

In the last several years, such national firms as General Motors, National Steel, Sperry Rand, and Rath Packing Co. have sold to employees plants that would have been closed otherwise. The experience of many of these firms is not extremely favorable, however. In numerous cases, employees are forced to take significant wage-and-benefit reductions to make the business profitable. In other cases, morale and working conditions have not been deemed acceptable as a result of the new method of ownership and management.

Before Ford Motor Co. closed its Sheffield, Alabama, plant, it offered to sell it to the employees. The only way for this to work, according to Ford officials, would be for the employees to agree to 50 percent wage and benefit reductions. Union officials and employees declined the offer. One individual who had worked at the plant for eighteen years summed up the thinking of many: "If Ford can't make the plant pay, I can't see me making it pay."

In a rather dramatic case, however, negotiators in 1983 worked out an agreement whereby the employees of National Steel's Weirton (West Virginia) mill would purchase the mill. This new company has now become the nation's largest employee-owned enterprise, as well as its eighth largest producer of steel. Experts give the mill a surprisingly good chance of succeeding, though Weirton's workers have had to take about a 32 percent cut in pay. As the mill's union president argued, however, "32 percent less of $25 an hour is a whole lot better than 100 percent of nothing."[3]

For the employee ownership option to have a chance, a long lead time between the announcement and the actual closing is needed to organize employees while they are still on the site. In addition, time is necessary to conduct complete and detailed feasibility studies. These studies need to assess such factors as:

Employee readiness for ownership

Union attitudes

Management/entrepreneurial skills present among employees

Company's products and its markets

Technology

Proposed organization structure

Potential funding sources

The National Center for Employee Ownership has found that most employee-owned firms have a good chance of survival.

### After the Decision to Close Is Made

There are a multitude of actions that business can take once the decision has been made that a close-down or relocation is unavoidable. The overriding concern is that the company seriously attempt to mitigate the social and economic impacts of its actions on employees and the community. Regardless of the circumstances of the move, some basic planning can be accomplished that would help alleviate the disruptions felt by those affected. Possible actions that management can take include:

Conducting a community impact analysis

Providing advance notice to employees/community

Providing transfer, relocation and outplacement benefits

Gradually phasing out the business

Helping community attract replacement industry.

**Community Impact Analysis.** If, as Drucker stated, management is responsible for its impacts on employees and the community, a thorough community impact analysis of a decision to close down or move would be in order. The initial action would be to realistically identify those aspects of the community that would be affected by the company's plans. This would entail asking such questions as:

What groups will be affected?

How will they be affected?

What is the timing of initial and later effects?

What is the magnitude of the effect?

What is the duration of the impact?

To what extent will the impact be diffused in the community?

Once these answers are provided, management is in a better position to modify its plans so that negative impacts can be minimized and favorable impacts, if any, can be maximized.

A Community Impact Analysis was proposed as a part of the Corporate Democracy Act introduced in Congress by Rep. Benjamin Rosenthal (Democrat, New York) in 1980. The act was the outgrowth of some of the work inspired by Ralph Nader and Mark Green. Though the act failed, it provides further evidence of the kinds of considerations that are now being discussed in an effort to increase corporate accountability. It seems much more reasonable that companies conduct such impact analyses upon their own volition than to have government involved in yet another aspect of business.

Since it is inevitable that management is going to be drawn into economic action teams that will be formed in the community, initiatives should convey a spirit of cooperation in facilitating community action. For example, once National Steel decided it had to close its Weirton mill, its involvement in attempting to assess community impact eventually led to management's joining with union members, local business people, and government leaders in raising the money that was needed for the expensive studies and sophisticated investment advice that led to the employee-ownership decision.

**Advance Notice.** Perhaps one of the most often discussed responsibilities in business or plant closing situations is the provision of advance notice to workers and communities. As discussed earlier, political pressures are already mounting for laws that would mandate such pre-notification before firms close down. Two-year advance notice was one of the key provisions in the Corporate Democracy Act of 1980 that failed to pass, but was included in the legislation created in Wisconsin and Maine.

The advantages of advance notice accrue primarily to the affected employees and their communities. Workers are given time to prepare for the shutdown both emotionally and financially. Advance notice makes it easier for employees to find new jobs because research has shown that employees have an improved chance at reemployment while they still carry the "employed" label. Part of advance notice is motivational in that once one joins the ranks of the unemployed there is a tendency to coast until benefits start to be exhausted. Also, the company is in a better position to provide references, retraining, or counseling during the advance notice time period.

The disadvantages of advance notice, particularly that which extends over a longer time period, accrue principally to the organization. Once word leaks out in the community, financial institutions may be reluctant to grant credit, customers become worried about items purchased or promised, and the overall level of business activity can decline rapidly.

One of the major disadvantages of a lengthy prenotification is the task of motivating workers who know they are going to lose their job. Employee morale, pride in work, and productivity declines can be expected. Absenteeism may increase as workers begin to seek other employment. In addition, there is the likelihood of vandalism, pilferage, and neglect of property as employees lose interest.

A Bureau of Labor Statistics survey shows that only about 10 percent of current collective bargaining agreements contain advance notice provisions. The notifications typically run from a week to six months. It is expected that these will increasingly appear in labor contracts, but the question of what is the appropriate period of time will continue to be the subject of much discussion.

**Outplacement Benefits.** Enlightened companies are increasingly recognizing that the provision of various separation or outplacement benefits are in their own and their employees' and communities' long-range best interests. Whether it be to avoid adverse effects on internal employee relations or external public relations, or to be responsive to perceived social responsibilities, everyone is better off if disruptions are minimized in the lives of management, displaced workers, and the community. Outplacement benefits have been used for years as companies have attempted to remove redundant or marginal personnel with minimum disruption and cost to the company and maximum benefit to the individuals involved. Increasingly these same benefits are beginning to be used in plant shutdowns.

The outplacement activities of American Hospital Supply (AHS) provide a useful model of a socially responsive firm. AHS announced its intent to sell its medical manufacturing company of about 275 employees in October 1979. A meeting was held of all employees to communicate the rationale for the decision. Though AHS received numerous inquiries from outside firms, the business did not sell. In preparing for this possible outcome, department heads prepared termination lists specifying those employees essential to the phaseout. In December 1979 another all-employee meeting was held, and it was announced that the business would be gradually phased out. A retention/outplacement program was prepared and explained in detail immediately following the meeting.

Terminated employees with no specific skills or with unique situations (illnesses, etc.) were identified early so that special outplacement support could be provided. The first group of terminated employees included all of the sales department, half of R&D, all of marketing, and various others not crucial to the wind-down.

The outplacement activities resembled a placement office of a college campus. Over twenty-five firms visited to conduct on-site interviews. Résumés were drafted by the employees, reviewed and proofed by the personnel department, and typed on the company's word processing equipment. The volume of outplacement correspondence was so high that additional word processing capability and clerical support had to be acquired. Final survey results showed that one-fourth of the outplaced employees received "similar" and another 65 percent received "superior" compensation packages as compared to their previous positions.

In addition, severance pay and a benefits plan were created for the employees. The benefits plan provided for 100 percent vesting in the company incentive investment, retirement and profit sharing plans, three months' basic benefit coverage (medical, dental, life insurance) beyond the date of termination and various other extensions (maternity, orthodontia) as considered necessary.

The efforts of two British companies to ease the pain of plant closings by providing outplacement benefits offer some creative alternatives of positive social responses. Tate and Lyle, faced with a need to switch from imported cane to domestic beets, had to close three of its sugar refineries and sharply reduce production between 1977 and 1979. The company knew of the devastating impact this would have and decided to assume responsibility for finding alternative jobs for its displaced workers, most of whom were middle-aged with an average of twenty to twenty-five years of experience. Its most successful effort was to act as a merchant banker, prepared to invest in existing viable companies that needed capital to expand and would give first refusal for jobs to former Tate and Lyle workers. In one case, a small electrical engineering firm moved to the economically weakened area and created 150 new jobs in return for financing from Tate and Lyle.

In another case, British Steel Corporation set up BSC, Ltd., a subsidiary whose purpose was to find jobs for approximately 50,000 displaced workers. The subsidiary assists small firms willing to rent space in its old plants with an objective of creating new jobs. The results of the firm's persistence are quite impressive. In 1979, BSC met its target of 3,000 jobs; in 1980, 5,000 more jobs were created; and, in 1981 it hoped to develop 10,000 more jobs. As seen in the cases presented, outplacement benefits can include assisting the displaced workers in finding new jobs, severance pay, insurance coverage, and a host of other separation benefits.

**Gradual Phase Outs.** Another management action that can significantly ameliorate the effects of a business shut down is the gradual phasing out of the business. American Hospital Supply employed a gradual phase out in addition to other benefits offered. The gradual phase out buys time for employees and the community to adjust to the new situation and to solve some of its problems.

A case from the early 1970s gives us a good example of a gradual phase out combined with other benefits to the community being left. The Olin Corporation decided it had to close down its operations in Saltville, Virginia, because, among other reasons, it could not meet new water quality standards being required by the Virginia Water Control Board and the federal EPA. Olin attempted to lessen its impact on the employees and community by taking a number of actions. For its employees it set up a generous severance plan and relocation assistance service. Another major decision it faced was the disposal of its plant, property and equipment after the operation had closed. The company made several attempts to attract replacement industry but had little success. Finally, Olin approached the town of Saltville with an offer to donate the plant, 3,500 acres of property with mineral rights, and all remaining tangible property in the town over a period of several years. The company also gave the town $30,000 to hire experts to mark all the equipment with its probable market value, $150,000 to compensate the town for lost tax revenue over the next three years, and $450,000 to be used for planning, developing and rehabilitating purposes over the next four years.

The closedown of Saltville represented such a massive economic upheaval that, of course, the townspeople were not completely satisfied with what the company did for them. In addition to their large donations, Olin Corporation received sizable tax writeoffs that eased their own financial burden. Nevertheless, what the company did for the community and the people over a period of several years represented a real attempt to help, and this fact cannot be overlooked.

**Attract Replacement Industry.** Another significant management action that can be taken in close down situations is to assist the local community in attracting industry to replace the closed business. The principal responsibility for attracting new industry will fall on the community, but the management of the closing firm can provide cooperation and assistance. The closing company can help by providing inside information on building and equipment characteristics and capabilities, transportation options based upon their past experience, and contacts with other firms in its industry that may be seeking facilities. The strategy of helping the community to attract replacement industry has the overwhelming advantage of rapidly replacing large numbers of lost jobs. Also, attracted businesses tend to be smaller than those that closed, so this strategy enables the community to diversify its economic base while gaining the jobs.

A number of factors go into helping a business decide the extent to which it should assist displaced employees and communities. Some of these factors include the general size of the negative impact it is creating, the extent of commitment its employees and the communities displayed over the years to the firm, how large an employer the firm was relative to the total economic base, the length of time the firm was in the community, the length of time employees had worked in the firm, the economic options available to it to assist, and, its overall sense of corporate responsibility or corporate social policy. Any one or several of these factors, along with other issues, assume a major role in dictating management and the firm's response.

## What Does the Future Hold?

Whether you consider the theory of corporate social responsiveness or the actual views of important executives in business, the case is clear that organizations must in the future be responsive to social problems.

Among others, the views of Preston and Post and of Sethi provide representative justification for the sensitivity of the corporate world to the business closings issue now and in the future. The former posit that the responsibility of the corporation for social concerns arises from its own operations and their necessary consequences and impacts. Furthermore, they assert that the crisis concerning the status of the corporation is in part traceable to specific business behavior or lack of behavior, and that the firm's basic legitimacy depends on its ability to meet the performance expectations of its stakeholders.

The socially responsive business, as described by Sethi, is one that is "willing to account for its actions" and take the lead in developing new postures. Furthermore, the socially responsive business "evaluates the side effects of its actions and eliminates them prior to the action being taken."[4] As it applies to the case of business closings, the responsive firm attempts to anticipate actions that might have deleterious effects on stakeholder groups and engage in actions that will minimize the negative impacts that are perceived and experienced.

A very recent example of a firm that has taken all this to heart and might serve as a model for socially responsive plant closings in the future is that of Brown & Williamson Tobacco Corporation that in 1982 shut down a fifty-one-year-old facility in Kentucky. The company won praise from its employees and union because of how it handled the closing. Though the labor pact called for eighteen months' notice of closing, the firm told the workers three years in advance. Several hundred of the workers were relocated in a new plant, and all others got six months' separation pay and health-policy coverage. They also received vocational training for new jobs and financial counseling. The transferred workers kept seniority rights for vacation, holiday, and retirement benefits, rights often lost in closings. The union said, "There's no such thing as a good plant closing, but they gave us a fair deal." Both sides thought the effort succeeded because of the advance notice, transfer and severance benefits, job training, and gradual phase out—all ideas, as we previously noted, that are important management actions.

From the perspective of business executives who have thought through the responsibility business should assume, several major justifications for present and future management action emerge. Among these are the desire to avoid more state and federal legislation that will dictate responsibilities in business closings, to avoid a continued hostile relationship with labor and labor organizations, to preserve business as a viable institution in society, to prevent further social problems, and to create a favorable public image for the corporation. Implicit in each of these is a central perspective which is "enlightened self-interest."

To date, research has not been conducted as to the specific degree to which executives perceive a responsibility and need for management action in business closing cases. However, inferences can be made from their views on balancing constituency interests and their actions, such as in the case of Brown and Williamson. A number of executives support the strategy of positive response to legitimate constituent interests, as cited in George Steiner's latest book *The New CEO* (1983). According to Steiner, all of the CEOs interviewed, without exception, accepted a responsibility for responding as best they could to the social (as distinct from the economic and technical) forces exerted by individuals and groups on the corporation.

If this is so, why have we not seen more positive responses in business closings? There are a variety of answers. One would be that executives still regard as foremost their economic performance and survival. To the extent that this is the issue, of course, this is quite legitimate. Another answer might be that sufficient exposure and analysis as to what is really at stake in business closings has not taken place. Charles R. Dahl, past president and CEO of Crown Zellerbach stated that

> corporations do try to be aware of social trends and embryonic issues. But until there is sufficient evolution of the thought process among the various constituencies to clearly define the issues, the corporation is not in a position to make an appropriate response.[5]

At this point we are just beginning to define the parties interested in the issue, the impacts that business closings are having on these interest groups, and the possible actions management can take to be more responsive. From past experience, one cannot avoid arguing that if further federal and state government regulation and social protests are to be avoided or minimized, positive and constructive steps must be taken by business to be responsive to the employee and community needs that arise when closedowns are imminent. What is needed is for business to engage in the same kind of sensitive, thoughtful, and deliberate decision-making process when it is planning to leave a community as when it made the initial decision to enter the community. If this is done, and if the interests of constituency groups are carefully weighed before decisions are made, it is possible that yet another social issue can be managed without culminating in new laws or a knotty regulatory apparatus—with the attendant costs, frustrations, and divisiveness that comes from its administration.

The firm that is sensitive to the changing social environment, that practices public issues management as a way of organizational life, and has integrated corporate social policy into its overall strategic management, will likely see the business closing issue as one in which anticipation and prevention will be a natural and logical social response. The proactive adaptation will reap benefits that will be felt for many years to come in the realm of businesses' relationships with society and the groups with which it experiences its most vital dependencies.

## Notes

1. Peter F. Drucker, *Management: Tasks, Responsibilities, Practices* (New York: Harper & Row, 1974), pp. 327–328.

2. "A Firm's Obligations: To Employees, Community," *Atlanta Journal* (Sept. 19, 1977), p. 4–C.

3. "A Steel Town's Fight for Life," *Newsweek* (March 28, 1983), p. 49.

4. S. Prakash Sethi, "Dimensions of Corporate Social Performance: An Analytical Framework," *California Management Review* (spring 1975), pp. 58–64.

5. George A. Steiner, *The New CEO* (New York: Macmillan, 1983), pp. 38–39.

# 14
# Management and Public Policy

*Dale Yoder*
*Paul D. Staudohar*

**M**anagers who are concerned about plant closures will find the recent plant closings in California to be valuable lessons. The shutdown of two large automobile manufacturing facilities illustrates poignantly the traumatic impact that closures have on companies, workers, and communities. The one firm, General Motors (GM), created an environment of uncertainty and anxiety, which caused inordinate suffering among workers and sharp public criticism. The other firm, Ford, was able to generate a smooth shutdown transition because its policies elicited understanding and cooperation from its employees and the community. Thus, in the one closure, serious doubt was raised about the competence of management because of the crises it allowed to occur; while in the other, the management team came out of the experience looking good.

To some readers, the moral of this comparison will be simple: society must provide legislation to regulate a plant closure—that is, legislation that warns and insulates the local community from a plant closure's impact. This is the public policy side of the issue, and its adherents will find support for this approach in our case studies. But there is yet another side of the issue: the management policy side. This policy should be one in which management extends its responsibilities beyond merely giving employees advance notification of a shutdown and protection from economic insecurity. In fact, such a policy should point to multidimensional plans that give specific attention to several facets of employment relationships. Fundamental to all these plans is a management policy that accepts responsibility for employees' interests, welfare, and problems. (One must keep in mind that if management policy does not take the initiative, the role of public policy becomes that much more important.)

Dale Yoder and Paul D. Staudohar, "Management and Public Policy in Plant Closure," *Sloan Management Review* 26, no. 4 (summer 1985), pp. 45–57.

## Who Is Affected by Plant Closures?

The issues of plant closure center on the appropriate exercise of management and public policy. There are criteria for judging these policies, such as adequate notice to employees, income maintenance, job-training opportunities, and job placement. The criteria are viewed in terms of their impact on companies, workers, and communities.

### Companies

The dynamics of changing technology and consumer demand require that companies continually revise their production practices if they are to survive and remain profitable. Plant closings are perhaps the most vivid reminder of this restructuring of American industry. Management's responsibilities do not begin with the decision to close down a plant. Ongoing concern must be communicated to employees, and their feedback must be recognized as an important source of trust and mutual respect. This may not only help avoid a shutdown in the first place, but it may also reduce the adverse impact if one were to occur. In this respect, companies have policy options: they can either walk away from a closed plant without showing any regard for workers and the community, or they can plan against obsolescence and try to lessen the consequences of a plant shutdown. Companies are judged by which course is taken and the ensuing results.

### Workers

The impact on workers from a plant closure is fourfold: (1) their jobs are lost; (2) the old job experiences are not readily transferable to new jobs in growing industries; (3) their new jobs typically provide pay and benefits that are significantly reduced from previous levels; and (4) they and their families experience high levels of anxiety and stress. During the period of transition to new jobs, displaced workers usually need help in such areas as retraining, counseling, job search, and income maintenance. Thus, management and public policies are judged by their effectiveness in cushioning adversity.

### Communities

There are many diverse community groups that are affected by plant closures. For example, local merchants, from grocery stores to dry cleaners to bowling alleys, derive incomes from workers' expenditures in local manufacturing industries. Citizens, such as school children whose welfare is supported from tax revenues and job seekers who compete with laid-off workers in the labor market, may feel the rippling effects of a closure. Television, radio stations, and newspapers report and editorialize on the impact. Government officials, such as mayors, city managers, and other public employees in the community, are left with the task of providing services to ease the burden of laid-off workers. Although volunteer groups may lend a hand during plant closures, the extent to which a community is able to weather the storm of a shutdown depends on the magnitude of the closure and the ability of management and public policies to adjust to and resolve the problems.

The plant closings in California are also important in that they are part of a larger mosaic, the decline in U.S. manufacturing. Economic recession, combined with ongoing foreign competition and technological change, has caused widespread unemployment from extended layoffs and plant closings. Joblessness has been especially severe in the traditional smokestack industries manufacturing automobiles, steel, and rubber. Despite the fact that many of these industries are now returning to economic health, thousands of jobs have been permanently lost. In addition, foreign competition remains intense and new production facilities are increasingly relying on robots and other labor-saving equipment. It is clear that a deep shift is occurring in the labor force and that the problem of plant closings will not disappear.

## The GM Shutdown

In 1963, production began in the state-of-the-art GM–Fremont facility located in Northern California between Oakland and San Jose, with outstanding railroad and port facilities nearby. At its peak output in the late 1970s, nearly 7,000 employees worked on three shifts. However, a steadily weakening demand for American-built cars and trucks, caused especially by Japanese imports, became the most important reason for the plant's closing. By the early 1980s, Californians were buying approximately one-half their automobiles from abroad, at more than twice the rate of the rest of the nation. Enough California-assembled cars could not be sold in the state to justify the cost of shipping parts west. Thus, GM decided to concentrate production nearest to its busiest markets. It closed the Fremont plant in March 1982. At the same time, it closed its South Gate assembly plant in the Los Angeles area, which idled 4,200 employees.

Other reasons for the Fremont plant shutdown were the acrimonious labor relations, high rates of absenteeism, and alcohol- and drug-related problems. For many years local union officials and company representatives had frequent disagreements over plant working conditions and enforcement of rules. Grievances filed annually by the union ran into the several hundreds and were among the highest rates per employee in the GM system. Absenteeism at the plant in 1981 was the highest of any GM assembly facility. The company had extensive alcoholism and drug abuse programs for rehabilitation of troubled employees. Yet, it was not uncommon for chemical abuse problems to result in production shortcomings.

The truck production segment of the plant was shut down in October 1981, idling 1,400 employees. These workers had some grounds for optimism that they would be recalled, however, since truck production at the plant had been of high quality, and the company undertook an expensive retooling for production of smaller vehicles. This suggested that the future of the plant was at least somewhat encouraging.

Nevertheless, in February 1982, with just three weeks' notice, the company announced it was closing the entire plant in March for an indefinite period. Although the company indicated that the plant might never reopen, the closing was not generally perceived by the workers to be permanent. The company had reason to keep its options open for resumption of production, but the uncertainty produced in the minds of workers became a hindrance to their seeking and getting alternative employment. What made this problem particularly acute was that the period of indefiniteness dragged on for so long. Moreover, the uncertainty continued in discussions of job rights of these workers in a possible new joint venture with Toyota at the Fremont plant. It was not until April 1983, thirteen months after the closing, that the company announced its final decision to make the closing permanent.

## How GM Dealt with the Shutdown

Cushioning the overall impact of industrial decline on workers is the task of programs providing income and job security. Yet the auto industry traditionally has placed far greater emphasis on income maintenance than on incentives to make companies more competitive, which, in the long run, is a much more effective way to strengthen job security. Recently, there has been some shift in policy toward profit sharing and concession granting from employees to help keep plants open under an employment guarantee program. But until the joint venture with Toyota actually materialized, GM did not seriously entertain these options.

### Income Security

Income security at the GM–Fremont plant concentrated on state unemployment insurance, combined with supplemental unemployment benefits (SUBs) under the company's contract with the United Automobile Workers. SUB payments are made to augment unemployment insurance up to a certain proportion of wages, with funds created by employer contributions. For the auto industry, it provides 95 percent of former base take-home pay (exclusive of overtime). Although SUB benefits allow an extra protection not generally available to employees outside the auto, steel, and rubber industries, these benefits can run out during prolonged periods of unemployment, as is illustrated by the GM–Fremont experience.

Unemployed hourly employees at GM–Fremont received 95 percent of their base take-home pay for varying periods depending on seniority. Salaried employees received 75 percent of their net pay for six months and 60 percent for another six months. After a year of unemployment, benefits for most of the GM–Fremont labor force had been reduced substantially. About a year after the initial layoffs, which were caused by the closing of truck production, additional funds were paid to workers under the Trade Adjustment Act (TAA) of 1974 because the workers lost their jobs as a result of foreign competition. Most eligible workers received about $8,000 from this source, with benefits depending on the duration of the layoff. However, it was later discovered that payment of company SUB funds could be reduced by TAA payments because employees could not receive double benefits for the same layoff period. This caused some workers to have their company payments reduced by about $100 a week. Employees 55 years old with more than ten years of seniority received additional funds ($15 per month for every year worked, up to thirty years) for early retirement when the company announced the permanent closing of the plant in April 1983.

## Job Security

In another program, GM sought to create job security by transferring workers to other facilities. Early in 1982, the company provided fifty jobs for Fremont workers at the Kansas City assembly plant as a result of an expanded shift. About a year later, the company sent letters to 2,000 Fremont workers who had at least ten years' seniority and invited them to meet to discuss jobs at the Oklahoma City assembly plant. Under the Guaranteed Income Stream provision of the GM-UAW contract, employees who relocated would receive a secured income of 50 to 75 percent of their average annual wage until retirement. However, if they failed to accept a job at another GM plant, they would lose the benefits. Some 270 former Fremont employees went to work in Oklahoma City; approximately 400 others took jobs at GM plants in Kansas and Missouri.

**Discouraging Factors.** One factor limiting the success of getting Fremont workers to accept jobs in other geographic locations in the first layoff year was the generous SUB program. Only when these funds ran out did employees become more amenable to moving, especially senior workers who would otherwise have to forfeit their guaranteed income until retirement. However, other factors diminished the attractiveness of relocating. One was the desirability of the San Francisco Bay Area as a place to live: workers had sunk their roots there. Another was the workers' hope that the Fremont plant would reopen, either under its previous status or in a new combined GM–Toyota enterprise.

**Relocation Efforts.** The relocation efforts at GM–Fremont stemmed from separate agreements among the UAW, GM, and the Ford Motor Company to give transfer options and benefits to high seniority employees. Nationally, however, the program has not worked out well for the total 3,600 GM and Ford transferees. Moving allowances were not very generous: a married blue-collar worker was given $2,000 to move. In addition, many employees were forced to leave their families behind, which ultimately strained relationships. Although corporate seniority remained intact for retirement and other purposes, transferees were typically treated as new hires in the new locations; therefore, they got the more difficult jobs and unattractive night shift assignments.

*Training and External Job Placement*

Refugees from the declining auto industry have little chance of getting good paying jobs without extensive preparatory job training. Apart from relocating to another company plant, workers find it extremely difficult to get another job, and when they do it is almost always at less pay. In addition, employers are reluctant to hire people with limited skills who may not even stay on the job because their former employer may reopen.

Table 14–1 shows certain employment outcomes for displaced GM–Fremont workers as of the end of 1983. Nearly all of the laid-off workers registered with the Employment Development Department (EDD), which is the state agency responsible for payment of unemployment benefits to workers in California. This agency also assists persons in job search. As indicated in table 14–1, the most common way for GM workers to find work was through EDD assistance. As discussed above, a sizable number of workers obtained new jobs by transferring to other GM facilities. Many also became reemployed through their own initiative. Approximately 40 percent of the GM–Fremont workers remained unemployed at the end of 1983.

Auto production is one of the nation's highest productivity industries because of its capital-intensive nature. Wage rates for relatively unskilled work are seldom duplicated elsewhere. The average hourly wage rate at GM–Fremont at the time of the closing was approximately $11.50, plus benefits of about

**Table 14–1**
**Job Placement of GM–Fremont Workers**

|  | From November 1, 1981, to December 31, 1983 | From November 1, 1982, to December 31, 1983 |
|---|---|---|
| Registered with EDD | 5,836 | 2,829 |
| Entered employment:[a] | 2,694 (46%) | 1,927 |
| Placed by EDD | 1,082 (19%) | 645 |
| Rehired by GM | 827 (14%) | 827 |
| Obtained by own effort | 785 (13%) | 455 |
| Retired | 69 ( 1%) | 58 |

*Source:* Data provided by the Employment Development Department, State of California, Fremont office.

[a]Because GM and the EDD lost contact with some workers, the number that entered employment is understated somewhat.

$8.50. Those workers who got new jobs outside the auto industry did so at an hourly rate of about $7.00, with substantially reduced benefits. This pay reduction is consistent with a study of auto workers who were found to be working in jobs that pay on the average 43 percent less than their former positions. Salaried auto workers who are laid off have an easier time finding a job because they are better educated and have management skills that are more readily transferable to other industries. Apart from self-placement workshops for salaried GM–Fremont employees, which focused on résumé writing, job search, and interviewing techniques, most of the training efforts were aimed at hourly employees.

Hourly workers in many cases need remedial training in basic subjects of reading, writing, and mathematics before they can even begin more advanced vocational training. Such courses were widely available to GM–Fremont employees through local adult schools. However, overall efforts toward retraining and job placement were handicapped because well-organized and funded programs were late in being fully established. This is largely a result of the magnitude of laid-off workers at the plant and the feeling that it might be reopened. Funds, training facilities, and personnel came from a variety of sources as programs unfolded and became operational. The primary initial sources of support were GM and the EDD, with assistance coming from the UAW and federal and local government agencies.

At the national level GM established special retraining funds by contributing five cents for each hour worked by UAW members. This resulted in an annual fund of $40 million nationally, available for teaching new skills, courses at community colleges and technical schools, job counseling, and job placement. In late 1981, the EDD established a counseling center that was staffed by four counselors and volunteer UAW members. The center helped eighty-four laid-off auto workers find jobs and provided stress and financial counseling during the first six weeks of operation. However, the center was not sufficiently staffed or funded to handle the large numbers needing assistance.

Shortly after the total shutdown of the plant in March 1982, a company-state-federal retraining and job development program was established. Tests were administered for determining hand-eye coordination and verbal and quantitative skills. By July nearly 500 GM–Fremont employees had found jobs through the EDD. However, the economic recession made good jobs scarce. Despite the availability of job-finding services many jobless workers simply did not seek assistance.

By mid-1983 the former auto workers began to get more serious about training opportunities as income maintenance programs waned. Employers who hired graduates of the training programs in such areas as robotics and cooking received reimbursement for up to half the employee's wages during the period of company in-house training. With the end of the economic recession, a greater number of jobs were available, which made placement easier. Training and job placement programs were ended on December 31, 1983.

Table 14–2 shows the number of GM–Fremont workers who were referred to, participated in, and completed certain training programs. About 42 percent

**Table 14-2**
**Training Program Activity for GM-Fremont Workers**

|  | From November 1, 1981, to December 31, 1983 | From November 1, 1982, to December 31, 1983 |
|---|---|---|
| Registered with EDD | 5,836 | 2,829 |
| Referred to training | 2,045 (42%) | 1,892 |
| Entered training: | | |
| Classroom training | 888 (15%) | 871 |
| On-the-job training | 202 ( 3%) | 187 |
| Remedial training | 266 ( 4%) | 266 |
| Completed training: | | |
| Classroom training | 624 (70%)[a] | 624 |
| On-the-job training | 163 (81%) | 163 |
| Remedial training | 237 (89%) | 237 |

*Source:* Data provided by the Employment Development Department, State of California, Fremont office.

[a]Percentage of participants who completed the training.

of the persons who registered with the EDD were referred to training. The table indicates that for those persons who entered classroom, on-the-job, and remedial training, the completion rates were good. However, the proportion of persons who took these forms of training was not very high. Again, this is in large part due to the feeling by many employees that they would be returning to work in the auto industry.

### GM-Toyota Joint Venture

In February 1983, a year from the time the plant closed, GM signed a "memorandum of understanding" with Toyota Motor Company for the creation of a $300 million joint venture, known as New United Motor Manufacturing Inc., to produce 200,000 small cars at the Fremont facility. This agreement is important because it justified the hopes of many former GM workers that the plant would reopen. However, it caused apprehension over whether they would be rehired in the new venture. The companies also agreed to build a new stamping plant adjacent to the Fremont site. Secretive talks had been under way for several months between the automakers. Shortly after the announcement a confrontation between Toyota and the UAW emerged over the union's priority hiring rights for the 3,000 new jobs expected to be created. Toyota's chief executive indicated that laid-off UAW workers could not be given priority in hiring. Moreover, the GM chairman said that since the plant was new, it would "start from scratch without any legal obligations."[1]

Technically, these executives were correct in their assessment. The union's master contract with GM contained no successor clause, so Toyota had no legal obligation to honor the GM-UAW contract. Toyota undoubtedly wished to operate the new facility nonunion, like the other Japanese plants in Marysville, Ohio (Honda), and Smyrna, Tennessee (Nissan). However, the UAW carried the big stick of a strike at other GM plants in the U.S. if the union were not recognized in the new venture.

As a conciliatory move the two automakers hired W.J. Usery, former secretary of labor, to smooth over the emerging conflict. Usery subsequently announced that former workers at GM-Fremont will be the "primary source" for recruitment and that the UAW will most likely represent workers when the plant reopens. However, this was less than a complete victory for the union, which wanted an exclusive opportunity for GM-Fremont workers in the new jobs, based on prior seniority.

Although the workers hired will probably be paid the same hourly rates as other GM employees, Toyota is expected to press for relaxation of seniority pay, which could mean a significant real wage cut. Also, it is anticipated that Toyota will want to operate under new work rules that are far less restrictive than in UAW-organized plants. In other words, Toyota wants the freedom to assign workers to a greater variety of job tasks.

In September 1983 the UAW reached a hiring agreement with the companies, in which preference will be given to former workers provided their skills and abilities are equal to less-experienced applicants. Thus, the joint venture is not obligated to hire workers back on a strict seniority basis. Seniority, however, will be taken into account in considering experience. A formal contract is expected to be negotiated by June 1985, which will resolve other work issues.

## The Ford Shutdown

Ford's San Jose assembly plant, located in the city of Milpitas, was closed for essentially the same reasons as the GM-Fremont plant: lack of demand for new American-built cars in California and the high cost of shipping component parts west for assembly. Unlike GM-Fremont, however, the quality of production at Ford-San Jose was the highest in all the company's plants for automobiles (Escort and Lynx), and among the best on trucks. Ford did not experience. the acrimonious labor relations that occurred with the UAW local at GM. Instead, the Ford plant had a history of cooperation between labor and management. For example, in 1980 an intensive "employee involvement plan" was begun with full union cooperation. This program, which lasted up to the time of the plant closing, provided a type of quality circle. Each Thursday at 9:00 A.M., the assembly line was stopped for a half hour. About 100 groups, with twenty to twenty-five employees each, were formed to make suggestions for improvement of quality control methods, safety and health, and other aspects of production. Managers and union representatives interacted together with employees in these groups. Although participation in the groups was optional, nearly 100 percent of the employees were involved.

Situated about ten miles down the freeway from GM-Fremont, the Ford plant was shut down on May 20, 1983. This was about fourteen months after the GM closing. Ford management looked carefully at the GM experience and was able to benefit from its example.

Several characteristics of the Ford program stand out, reflecting a more sensible approach than GM in its planning for obsolescence. One is that employees were told six months in advance that the plant would close. This gave the workers enough time to adjust psychologically to the shutdown. Second, elaborate programs for testing, stress counseling, training, and job placement were fully operational when the plant closed. These services were mostly provided at the plant itself, and employees felt comfortable going there. Third, Ford drew upon numerous specialists from within the company to assist the unemployed. Ford's operation was smaller, with only about 2,300 employees, yet the company's generous initial commitment of financial and professional resources is in stark contrast to GM's. Lastly, the Ford plant closing was truly a joint effort between the company and union. All important decisions were reached through agreement between UAW and Ford representatives. Because the plant had a history of cooperation on programs to improve quality, morale, and operation, the parties were able to draw on mutual respect and trust in planning for the shutdown.

Table 14–3 shows that 63 percent of the Ford employees obtained jobs after only fifteen months, which is significantly higher than the 46 percent of GM workers who were employed after over two years (see table 14–1). Much of this difference appears attributable to the fact that the GM–Fremont workers delayed getting jobs because of the expectation that they would be hired back in the GM–Toyota venture. However, it is clear that by having its programs in place at the time of the shutdown, Ford–San Jose was able to help its displaced employees get training and jobs far sooner than was possible at GM–Fremont. Also, the Ford programs were operating during a period of economic upturn, which made jobs somewhat easier to find.

Table 14–4 indicates that about 70 percent of the Ford employees who registered with the EDD were referred to training. This points out a key

## Table 14–3
## Job Placement of Ford–San Jose Workers

|  | From May 20, 1983, to September 1, 1984[a] |
|---|---|
| Laid-off workers | 2,300 |
| Entered employment | 1,460 (63%) |
| Eligible for retirement | 371 (16%) |
| Retired | 138 ( 6%) |
| In vocational training | 118 ( 5%) |
| Remaining to be served[b] | 308 (13%) |

*Source:* Data provided by the Ford Motor Company and the Employment Development Department, State of California, San Jose office.

[a]The Ford Training Center was closed in September 1984.

[b]The company lost contact with these employees.

Percentages do not add to 100 because of the category overlap.

**Table 14-4**
**Training Program Activity for**
**Ford-San Jose Workers**

|  | From May 20, 1983, to September 1, 1984 |
| --- | --- |
| Registered with EDD | 2,111 |
| Referred to training | 1,427 (70%) |
| Entered training: | |
| Classroom training | 761 (36%) |
| Remedial training[a] | 770 (36%) |

*Source:* Data provided by the Employment Development Department, State of California, San Jose office.

[a]341 persons went from remedial training to vocational training.

difference between the policies of Ford and GM. As shown in table 14-2, only 42 percent of the GM employees were referred to training. Rather than emphasizing direct placement of workers in the job market, as GM did, Ford decided that the biggest need was in retraining. It therefore stressed "employability plans" for individuals, which first determined workers' aptitudes and interests through testing. Test results were then compared to existing opportunities in the labor market. Ford strongly urged employees to undertake training to qualify for these opportunities.

Accordingly, the proportion of employees who registered with the EDD and who undertook its training programs was far higher at Ford than it was at GM. Thirty-six percent of the Ford employees entered classroom training and remedial training, compared to only 15 percent (classroom) and 4 percent (remedial) at GM (see tables 14-2 and 14-4). Also, it is not surprising that training completion rates were higher at Ford than they were at GM. For example, 99 percent of the Ford employees who entered remedial training completed it, compared to 89 percent at GM. Again, these results reflect the heavy emphasis on retraining by Ford.

To cite one vital bottom-line statistic from Ford, there have been no suicides reported among its former employees. In contrast, eight GM-Fremont employees took their lives by suicide. Numerous other deaths were reported from heart attacks and cirrhosis of the liver among GM–Fremont employees. Child abuse incidents reported to local police increased 240 percent in the first four months after the plant closed. Stressful circumstances, from lack of adequate notification of the closing and disheartening uncertainty over reopening, pushed nerves to the limit.

## Differences between GM's Management Policy and Ford's

Probably the most apparent shortcomings in the GM-Fremont plant closing were the scant three-week notification to employees and the thirteen-month delay before the decision was made by the company to close the plant permanently. This left workers with little time to plan for alternative careers and for government support agencies to gear up for training and job-finding programs. The indefiniteness as to whether the plant would reopen left many workers twisting in the wind. Moreover, other companies were reluctant to hire the idled auto workers because they felt that the workers might quit and return to GM if given the chance.

At the same time, the GM–Fremont problems were in some ways unavoidable. The resources of the company eventually did help alleviate problems for many workers. Also, the EDD, community colleges, numerous volunteer workers, and other agencies in California did their best to make the most out of a bad situation. That the company truly believed the plant might reopen is evidenced by its talks with Toyota, which began shortly after the closing. More important, these talks came to fruition, and the plant will indeed reopen and provide jobs for many former employees. There will not be enough jobs open in the GM–Toyota operation to hire back all the nearly 6,000 laid-off employees, but those who do get the new jobs will be paid well and will have an opportunity to use their skills to their best advantage. Also, the size of the company enabled it to offer numerous jobs in other auto assembly plants around the country. The reluctance of workers to leave their homes in the Bay Area made this option less attractive, but at least the opportunity to relocate existed and many workers took advantage of it.

The closings present interesting contrasts in the operation of management policy. When the big layoffs started at GM–Fremont, the company seemed to feel little sense of obligation to its employees. It was as if GM was going to let the state and local government income maintenance and training authorities take care of the problems. This exercise of policy by GM did not sit well with the Bay Area community. The print and broadcast media had substantial coverage of the plight of the laid-off employees. The company was chastised by charges that it walked away from its problems. Thus, when a large segment of the citizenry concluded that GM lacked a social conscience, the company shifted its policy toward greater commitment of its resources.

Meanwhile, Ford was on the sidelines during the early episode at GM, and thus it saw a chance to get favorable publicity by showing up GM as a poor example of social responsibility. From the start Ford's programs were better organized, executed, and achieved better results. Ford became a media darling and sought to deserve the plaudits. It exercised a beneficient policy voluntarily, absent public pressure.

## Effects of Management Policy

How did the management policy at GM and Ford affect the three actors—companies, workers, and communities?

*Companies*

Through our analysis of GM and Ford, we conclude that the companies had good reasons for closing the plants. Assembling cars and trucks in California was no longer economically feasible in light of the diminished market and long distances that component parts had to be shipped. However, managers at GM failed to recognize the necessity of accepting responsibility for the interests and welfare of their employees. This is in large part explained by the poor labor relations that existed for a long time prior to the closing.

In contrast, Ford–San Jose laid a foundation for mutual trust and respect through employee involvement, which allowed for participation between labor and management in designing programs for training and job search. Proof of the importance of this foundation is the success that Ford had in having programs well established and staffed by the time the plant closed. That these programs were far more successful than those at GM is not surprising because an environment ensuring good results had been created beforehand through Ford's use of a sound management policy.

*Workers*

The exercise of management policy also shows up clearly in the impact on the workers. Nothing could be done to avoid closure, but application of the criterion of a cushion against adversity yields disparate results. GM's programs were poorly established at the outset, causing a heavy burden to be placed on the limited job placement services available in the community. The resulting anxiety among workers took a severe toll in physical and psychological problems that led to an alarming number of suicides. Ford workers, on the other hand, were able to make the transition to new jobs much sooner and at significantly higher placement rates. The company's emphasis on training enabled them to have a better chance at a good job rather than simply being recycled into a low-paying dead-end job. This help and this hope reduced stress on Ford employees and their families.

*Communities*

What of the community impact? Judging from the reaction of the print and broadcast media, the adversity was substantially greater in Fremont than it was in Milpitas. However, it is difficult to make a precise assessment because the Ford programs were operating during a period of economic upturn, while GM's programs were operating during a recession. Also, GM's operation was far larger than Ford's, and thus it was bound to have a greater spillover effect on the community. The impact on Fremont is further clouded by the GM–Toyota venture, which is already having a rebounding effect on the community. Nonetheless, comparisons of data in the tables and the results of operation of other variables strongly indicate that management policy acted to mitigate adverse community impact more in Milpitas/San Jose than it did in Fremont. For example, when the last car rolled off the assembly line at Ford, the company celebrated by donating the car to the city of Milpitas in recognition of the city's support over the years.

## The Legislative Controversy: Should Public Policy Be Changed?

The outcome of the two California auto plant closings raises an important question: Should public policy be changed to require advance notification and added protections to the victims of plant closure? Before addressing this issue, however, it is appropriate to review the current legal status and indicate the principal arguments for and against plant closure legislation. Thirteen states have bills in their legislatures that propose to regulate plant closings, and four states already have such laws: Connecticut, Maine, Massachusetts, and Wisconsin. There have also been various plant closure bills introduced in Congress. Whether to pass such a law has been a controversial subject between business and labor groups.

By examining worker rights in seven countries, it was found that the United States as a whole is the only country that does not guarantee employee advance notification before plant shutdowns. Legislation providing notification protections exists in Germany, France, Japan, Sweden, the Netherlands, and Great Britain. For the 80 percent of the American labor force which is not unionized, there is no protection from plant closings unless it is provided by state law. For unionized employees, about three-fourths of the 400 collective bargaining agreements surveyed by the Bureau of National Affairs required prenotification of employees, but 81 percent of these agreements provided for only one week's notice or less.

The Maine law, initially passed in 1971, currently requires advance notice of plant closings of 60 days. Also, firms that close a plant with over 100 employees are required to provide severance pay to workers who have at least three years' seniority. In Wisconsin, since 1975, firms with 100 or more employees were required to give 60 days' notice; however, in 1984 the law was changed to replace mandatory notification with voluntary guidelines. The 1983 Connecticut law requires partial continuation of health benefits for certain eligible workers, but contains no notification provision. The Massachusetts law, passed in 1984, is voluntary in that it encourages businesses to give either 90 days' notice or severance pay to employees. In addition, this law guarantees health insurance benefits for 90 days, either from the company or from a state fund, and provides some counseling and training services. These state laws are similar to those in Western Europe on timing of prenotification. In Britain, for example, a 60- to 90-day notice is required for plant closings.

*Pros and Cons of Changing Public Policy*

Most proposals for changes in public policy center on the areas of (1) advance notification; (2) provision of income maintenance, such as severance benefits and payment of health and welfare benefits for a certain period of time after the closing; (3) opportunities for training to acquire skills that are transferable to the labor market; and (4) requirement of efforts by the company to aid employees in finding other jobs, usually in cooperation with local and state agencies.

Supporters of legislation point to the need to insulate employees from the impact of plant closures, especially since victims of shutdowns possess few skills that are in demand by other employers and are particularly vulnerable to prolonged unemployment. Without income, training, and job-search assistance, it is difficult to be optimistic about one's future career. Psychological stress, depression, and even suicide may result. Supporters note the especially adverse impact on small communities that suffer declining tax revenues from major plant closings, necessitating cutbacks in public services.

Opposers of legislation, on the other hand, point to unemployment insurance and job-search facilities that are already available at the state level. They argue that income maintenance programs are sufficiently generous to allow time for workers to find new jobs. Employment services such as computerized job banks, which are available in nearly all large metropolitan areas, are said to be an adequate source for persons who are serious about finding job opportunities. In free labor markets, individuals choosing blue-collar manufacturing careers receive high pay for low skills (however, they also accept a tradeoff by assuming a greater risk of unemployment and lower transferability of skills). Moreover, opponents contend that the problem is one of management rather than of public policy. They feel that firms, through their own initiative, should be free to determine what services they will provide in the event of a closure. Many employees are represented by unions that can seek to negotiate protections on their behalf.

Part of the controversy over public policy on plant closures concerns financial cost. Although it is possible to extend substantial protection to employees, it is questionable whether some firms would be able to afford the expense of underwriting the programs. For example, it can be argued that extensive training opportunities should be mandated for permanently displaced workers with limited skills. However, when economic recessions occur, corporate budgets are strained. Many firms are under continuous pressure from foreign competition and are struggling to retain an edge in domestic and world markets. They argue that less, not more, regulation of their activities is needed. Legislative proponents counter that because recessions hit goods-producing industries like the automobile industry particularly hard, permanently displaced employees have an even greater need for additional training opportunities, and these firms have a moral obligation to provide such training.

*How Adequate Are Existing Public Policy Programs?*

Workers affected by plant shutdowns are concentrated in certain industries. Among the most important of these industries today are automobiles, steel, rubber, and copper. Both current and past experiences indicate that the cycle of permanent decline is concentrated in regional areas: the manufacture of shoes and textiles in New England, lumber and iron ore in Minnesota, meatpacking in the Middle West, and coal in Appalachia. In the future, other industries and geographic areas will be the victims of the continual restructuring process. As attention is drawn to the affected companies, workers, and communities, political pressures will mount for developing solutions to moderate the intensity of these impacts. We look to public policy for at least a part of the remedy. As shown from the General Motors and Ford experiences in California, government programs were used primarily for income maintenance and job placement to augment the operation of management policy. Because impacts on workers were particularly devastating at GM–Fremont, questions are raised about public policy. Do existing public policy programs provide sufficient help? If not, how can they be strengthened?

Ideally, management policy could be exercised to anticipate and resolve the problems in conjunction with existing public policy. Ideally, both policies would operate separately with adequate financial resources, as the cost of such programs is high. However, that four states have already passed plant closure laws and that many other states are considering passing them suggests that the needs are not being adequately filled by either of these policies. Moreover, some companies try simply to get out and avoid any obligations.

## What Are the Public Policy Choices?

There are essentially four public policy choices: (1) no additional law, (2) a voluntary law that sets standards but allows discretion in compliance, (3) a mandatory law that requires full compliance by covered firms, and (4) a law with mixed voluntary and mandatory features.

While it is not suggested that decisions on legislation be based solely on the two cases presented in this article, the cases indicate what companies do on their own to ameliorate the impact of a plant closing. Certainly in the case of Ford, we see how management has taken the initiative. It is for legislators to decide how to best shape public policy so it can be supportive of voluntary actions such as Ford's, while, at the same time, fill in the gaps resulting from a more traditional handling of a closure as exemplified by General Motors.

Although the monetary outlays necessary for strict compliance with protective requirements can be supplied easily by industrial giants like GM and Ford, many smaller firms cannot afford to do so. Strong mandatory features would thus be expected to result in low overall compliance rates. On the other hand, a shift to a voluntary law would probably be better than no plant closure law at all. Simply having a law on the books should generate more protective action by firms for the benefit of employees and the community. It also has more political appeal. Unions advocate mandatory plant closure laws, while business interest groups favor maximum flexibility. Elected officials can bridge this gap by compromise legislation that leaves compliance up to the firms. The firms can then make decisions on compliance based on their financial resources, sense of responsibility toward employees, and desire for a favorable public image, and at the same time, set standards that allow for protections that the unions want. In this way, companies get the kind of public goodwill and respect that they deserve. Uncaring abandonment in the face of a law, even if it is a voluntary law, would not successfully serve a company's future marketing campaign of its products.

We feel that advance notification has particular merit. There is evidence that advance notification alone is relatively unimportant to firms while it is of great benefit to displaced workers. A study of thirty-two plant closings in the 1960s determined that advance notice rarely led to higher quit rates or lower productivity. Absenteeism rates at Ford dropped after announcement of the shutdown. Even if notification were to impose costs on an employer they would not be significant and would be trifling compared to the costs on employees from lack of sufficient opportunity to revise career plans plus accompanying stress on them and their families. Research on advance notification in Maine indicates that it substantially lowers local unemployment rates.

An additional consideration for proposed legislation is that employers who close their operations should be required to provide financial support for retraining, if funding does not exist under current public programs. This requirement could be made conditional in that unemployed workers must undertake retraining in order to receive legally provided severance pay. The rationale for this suggestion is to ensure that appropriate training is available to the permanently displaced and to encourage them to take advantage of it.

## Conclusion

It is not in the nation's interest to place artificial barriers on imports of foreign-made products via tariffs and import restrictions that invite retaliation and higher prices for our consumers. Nor is it prudent to interfere with the capital investment decisions of corporate managers, even though the process of techno-logical change replaces people with machines or renders plants obsolete. What does seem vital is to provide at least a minimal legal safeguard to people who are permanently displaced from their jobs, through a three-month advance notifi-cation requirement. This measure, apart from its merit on humanitarian grounds, provides a buffer against unemployment and allows employees more time to adjust to reentry into the job market. Other changes in public policy—such as severance pay, health insurance, and training opportunities—have merit, but their political feasibility would be enhanced by making compliance voluntary.

## Note

1. H. Weinstein, "GM Joins Toyota in Saying UAW to Get No Preference at Fremont Plant," *Los Angeles Times* (Feb. 18, 1983), pt. IV, p. 1.

# 15
# Union Responses to Plant Closure

*Anne T. Lawrence*

U nions have been hard hit by the recent wave of plant closures. Between 1979 and 1982, American union membership declined from 22.6 to 19.8 million, a drop of over 12 percent; during the same period, the rate of unionization dropped from 21.5 to 17.9 percent of the civilian labor force, the lowest recorded in the postwar period. Of course, there are many reasons for the decline in union membership and coverage, including the growth of managerial opposition; fewer union resources devoted to organizing; and the changing industrial, occupational, and regional distribution of the labor force.[1] However, recent plant closures in industries in which unions have traditionally been strong, such as automobiles, steel, rubber, and meatpacking, have accelerated labor's losses. In California, for example, slightly over one-half the decline in union membership in manufacturing between 1979 and 1983 can be attributed directly to the closure or relocation of plants.[2] Plant closures are not the only cause of organized labor's decline, but their acceleration in the late 1970s and early 1980s has clearly heightened this trend.

Organizational decline on a scale unparalleled since the Great Depression has led unions to depart from business as usual. In recent years, innovation has come not only from employers, but from unions as well, in response to economic dislocations occurring in American industry. Some unions have stepped up efforts in collective bargaining negotiations to obtain job security provisions requiring, for example, advance notice of closure, severance pay, continued health insurance benefits, transfer rights, and retraining for displaced workers. Others have offered concessions in wages, benefits, or work rules in exchange for an explicit or, more often, implicit promise of greater job security for their members. Some unions have supported legislation that would require advance notice or mandatory severance benefits, impose import controls, or extend unemployment benefits. Finally, some have adopted strategies of membership

Anne T. Lawrence, "Union Responses to Plant Closure," an original essay based on an unpublished doctoral dissertation completed in December 1985 in the Department of Sociology, University of California, Berkeley. The author is grateful to Harold L. Wilensky, George Strauss, Philip Shapira, and Amy Glasmeier for critical readings of an earlier draft of this paper.

mobilization and direct action, ranging from demonstrations, strikes, and consumer boycotts to joint labor-management programs at the shop floor level designed to improve work productivity and competitiveness.

This chapter presents a typology of unions' strategic responses to plant closures during the recent period of deindustrialization in the United States. Drawing on published evidence and my own study of union performance and collective bargaining practice in California manufacturing, I will examine the prevalence of various union responses to plant closure and assess the effectiveness of labor's efforts to retard membership loss and prevent plant shutdowns. My central argument is that although unions have engaged in a wide range of innovative responses to economic dislocation, organized labor has for the most part been unable to prevent plant closures or mass layoffs. In conclusion, I will speculate briefly on some of the reasons unions have found it so difficult to respond effectively to recent deindustrialization.

## Union Responses to Plant Closure: A Typology

Table 15-1 presents a typology of union responses to plant closure. In attempting to prevent mass layoffs and plant closures, unions have adopted three main strategies: collective bargaining, political action, and direct action. These three strategies form the vertical axis of the typology. In collective bargaining, unions attempt to influence employer decision making by negotiating a legally enforceable contract that limits managerial rights to terminate or relocate production or that makes it more expensive to do so. Second, unions may seek these same goals indirectly, by supporting government policies and laws that either expand union or public control over firms' location decisions or favor investment in sectors in which unions are strong. Finally, unions may also attempt to prevent or postpone plant closures by mobilizing their members and other organizations to save jobs, using a variety of direct action methods.[3]

Each of these three strategies may, in turn, be classified by the underlying philosophical approach: adversarial or cooperative. Some union strategies are adversarial, in that they presume unions and employers have fundamentally opposing interests in a plant closure situation. Union efforts to restrict the mobility of capital by requiring negotiations over the decision to relocate production, for example, are typically opposed by employers who view this as an unwarranted incursion on managerial rights. On the other hand, other strategies are cooperative, in that they presume a congruence of interest between labor and management. The use of concession bargaining as a job-saving strategy, for example, is based on the premise that, in the long run at least, job security of workers is best protected by improving the firm's overall competitiveness. Similarly, in the pursuit of tariffs on imports or the abatement of

**Table 15–1**
**Union Responses to Plant Closure**

| | Philosophical Orientation | |
| --- | --- | --- |
| *Method* | *Adversarial* | *Cooperative* |
| Collective bargaining | Negotiation of job security provisions, such as: | Negotiation of concessions, such as: |
| | Restrictions on management rights to terminate or relocate production | Cuts or freezes in wages |
| | | Cuts or freezes in benefits |
| | Advance notification of closure and/or union rights to negotiate over decision | Modification of work rules favorable to the union |
| | | Two-tier compensation systems |
| | Severance pay; supplemental unemployment benefits; extended health benefits; early retirement | |
| | Interplant transfer | |
| | Retraining and job placement assistance | |
| Political action | Support for plant closure laws that require, for example: Advance notification of closure | Support for protectionist policies such as: Import controls |
| | Severance pay | Local content requirements |
| | Extension of health benefits | Support for abatement of environmental or other governmental regulations |
| | Payments to affected community for economic development | |
| | Right of workers or community to purchase plant at market value | |
| Direct action | Strike | Employee stock ownership plan (ESOP) |
| | Consumer boycott | Joint shopfloor programs to improve productivity and quality |
| | Formation of alliances with community or religious organizations to protest plant closures | Alternate use committees to promote product conversion |

environmental controls, both unions and employers may seek to protect their industrial sector from external competition or government regulation.[4]

## Collective Bargaining

The most common union response to the threat of plant closure has been to influence the employer directly through collective bargaining. In recent years, two important trends have emerged in this area. First, unions have shown

204 · Deindustrialization and Plant Closure

renewed interest in noneconomic areas of the contract, especially those related to job security. Second, many unions have negotiated concessions in wages, benefits, and work rules in an attempt to improve the profitability of the businesses in which their members are employed, hoping thereby to improve the long-term chances of survival of these companies. Both trends reflect, in large part, declining employment and plant closures in union jurisdictions.

In recent years, many unions have responded to the threat of plant closure by negotiating new or strengthening existing job security provisions in their collective bargaining agreements. Labor has sought to restrict management rights to terminate or relocate production, as well as to extend various benefits and opportunities to dislocated workers. For example, the United Food and Commercial Workers in 1981 negotiated an 18-month ban on plant closures in the meatpacking industry. At Ford Motor Company, the Autoworkers won an agreement that the firm would not close any plants because of "outsourcing" (the purchase of parts from other manufacturers). Many union contracts now require the employer to notify the union as much as six months in advance and to negotiate over the effects of a closure. Other contracts provide specific benefits to workers dislocated as a result of plant closure. Many agreements require severance payments, early retirement, and extended medical benefits for displaced workers. Others provide for transfer to other company facilities or retraining at the employer's expense.[5] Under the terms of innovative contract provisions negotiated with the Autoworkers in 1984, for example, both Ford and General Motors participate in a Job Security Program that permits certain displaced workers to be paid their previous wages while assigned to training or replacing other workers undergoing training.

The prevalence of such job security provisions in union agreements has expanded over the past two decades. A periodic survey of collective bargaining provisions conducted by the Bureau of National Affairs suggests that plant closure provisions rose from negligible levels in the early 1960s to over 20 percent of manufacturing pacts by the early 1980s. Using somewhat less restrictive definitions, the Bureau of Labor Statistics estimated in 1981 that 23 percent of manufacturing contracts provided for advance notification of closure or union participation in the decision to close; 37 percent provided for interplant transfer. Although only a minority of contracts provided for job security, these levels were nevertheless the highest ever recorded by the bureau. Union representatives have stated repeatedly that job security has become their highest bargaining priority in the 1980s.[6]

Union proponents of an adversarial bargaining strategy argue that job security provisions function to prevent closures and layoffs both directly and indirectly. For its part, advance notice gives unions an opportunity to offer concessions, mobilize opposition to the closure, or recommend alternatives to shutdown, thus boosting chances the plant will survive. Severance pay, retraining programs, and other economic benefits and programs for dislocated workers may discourage closures by increasing their cost to employers.

However, there is little empirical support for these claims. In a study of the performance of thirty-one California manufacturing unions during the years 1979–1983, I found that the presence of job security provisions in union contracts did not improve union members' job security. In fact, unions with many job security provisions actually experienced greater erosion of membership and displacement due to plant closure, relative to industry employment, than did unions with less extensive coverage.[7] These findings suggest either that employers prefer to close plants where unions have attempted to restrict management rights or that poorly performing unions are highly motivated to win job security provisions—most likely, both.

Although job security provisions have not been effective in preventing layoff or plant closure, other evidence indicates they have clearly benefited dislocated workers. In a sample of 108 closures of unionized plants, I found that increased advance notice improved the likelihood of the union obtaining severance benefits at the time of closure.[8] Other studies indicate that advance notice reduces aggregate unemployment in the month following closure, probably by permitting displaced workers to begin an early job search.[9] Severance pay, the extension of medical benefits, and retraining programs have been shown to ameliorate some of the adverse effects of unemployment and facilitate workers' transition to new jobs and occupations.[10] However, there is little empirical support for the argument that job security provisions actually prevent plant closure or layoff.

The second major approach unions have taken in the collective bargaining arena has been a cooperative one: to negotiate concessions. In accepting "givebacks," unions hope both to improve the firm's competitiveness, thereby indirectly strengthening their members' job security, and possibly also to gain tradeoffs in the form of strengthened job security provisions. Researchers have identified four major types of concession: freezes or cuts in wages, freezes or cuts in benefits, modifications of previously negotiated work rules favorable to the union, and two-tier compensation systems that lower the wages or benefits of new hires while maintaining or increasing those of current employees. Although concessions are not unprecedented, the early 1980s witnessed a dramatic shift in the direction of wage and benefit negotiations, reflecting both weakened bargaining power of unions and their increased concern with the financial health of their members' employers. Between 1979 and 1983, the proportion of major U.S. agreements containing a wage freeze or cut in the first year of the contract rose from zero to 28 percent. Median first year wage adjustments dropped during this period from 8.4 to 4.7 percent. In 1985, 11 percent of nonconstruction contracts included two-tier provisions, up from 5 percent in 1983.

Union proponents of concessions make two main arguments. First, concessions may be necessary to restore the competitiveness of many American firms in an era of increasing import penetration and domestic deregulation. Without wage restraints and relaxation of traditional work rules, many businesses would fail, boosting the number of workers displaced by plant shutdowns. Second, concessions have been beneficial for labor in that they have represented an historic opportunity for unions to gain in noneconomic areas of the contract. In this view, *quid pro quos* negotiated by unions include gainsharing programs, increased participation in managerial decision making, and provisions to increase the job security of members.[11]

Research on the results of the concession strategy is scanty, but available evidence casts doubt on its effectiveness. Undoubtedly, concessions have saved some jobs. Case studies point to a number of instances where thoughtfully negotiated concessions prevented, or at least postponed, layoffs, especially where the union won specific employer commitments to invest in the facility or to preserve employment levels in exchange for givebacks. However, overall concessions have had little effect on union performance. My study found that California manufacturing unions that negotiated concessions covering large proportions of their members fared no better or worse, relative to underlying industry conditions, than those that negotiated few or none. Moreover, concessions produced few *quid pro quos*. In major California manufacturing contracts negotiated between 1981 and 1983, most new job security language was not exchanged for union givebacks but instead appeared in nonconcessionary pacts.[12]

### Political Action

A second strategy unions have adopted is political action: the attempt to influence investment decisions and employment levels through law or regulation. Most commonly, this means support for so-called "plant closure laws." At the federal level, several bills have been introduced since 1974 that address the issue of economic dislocation. The National Employment Priorities Act of 1974 and its successors sought to provide federal assistance to communities and workers affected by plant closures and to require that employers provide advance notification, severance pay, and continued health insurance coverage. Despite vigorous lobbying efforts by labor, none of these bills passed either house of Congress, and few were even reported from committee. With these failures, many unions shifted their political attention, leading to a dramatic increase in the number of state and local plant closure bills in the late 1970s. In 1982, probably the peak year for such efforts, twenty-one state legislatures considered 60 plant closure proposals.[13]

Typically, such legislation has generated considerable controversy. Unions view these statutes as a key element in their effort to restrict shutdowns and actively support them.[14] In fact, one study shows that the extent of unionization in a state is the best single predictor that plant closure legislation will be introduced.[15] Organized labor's enthusiasm has been more than matched, however, by the vigor of employer opposition. Plant closure legislation has been fought strenuously by employer groups, which see it as punitive and antibusiness.[16] Public debate over the issue typically has been highly acrimonious, with employers referring to proposed legislation as "industrial hostage" bills and unions retorting that they seek only to curb "lawless runaways" and "corporate outlaws." In large part because of vigorous employer opposition, little of this legislation has passed, and the laws that have been enacted have been weakly enforced.[17]

Although comprehensive plant closure legislation has repeatedly failed to pass both national and state legislatures, organized labor's strong backing has probably prompted several employer groups to adopt voluntary guidelines. For example, those adopted in 1982 by the California Chamber of Commerce recommend as much advance notice "as practical," cooperation with employee organizations, and the establishment of "employee assistance centers" to provide outplacement counseling and retraining. Similar voluntary guidelines were adopted by the California Association of Manufacturers in 1983 and more recently by the Business Roundtable, a national organization of corporate chief executive officers. These initiatives from the business community probably represent an effort to deflect more extreme legislative proposals backed by organized labor.

Although unions and employers have gone head to head over plant closure bills, in several other areas the two parties have formulated collaborative legislative agendas to save jobs and slow plant closures. Some unions and employers have joined forces to seek legislation aimed at protecting their industry from foreign competition or abating environmental regulations perceived as onerous. For example, the Glass, Pottery, Plastics and Allied Workers and employer associations in the glass bottle industry have a joint lobbying program in many states to prevent passage of laws that would reduce demand by requiring deposits on glass beverage containers. The union is motivated by a desire to avoid plant closures, which have occurred in the wake of deposit laws in several states. The Steelworkers union on several occasions has supported industry efforts to weaken or postpone implementation of emissions standards. Especially in industries hard hit by foreign imports, such as garments, steel, automobiles, and some food products, unions and employers cooperate fully in seeking import restrictions to protect domestically produced goods.

American manufacturers do not always support protective legislation. In many cases, U.S.-based firms that manufacture some of their goods domestically also maintain overseas operations and are themselves importers. In the garment industry, for example, the International Ladies' Garment Workers Union has found that its protectionist programs are backed only by the exclusively domestic segment of the industry; firms that are mainly or partially based overseas tend to oppose such initiatives. These cases are in the minority, however; cooperative protection is the rule.

The effect of protectionist policies on domestic employment is difficult to measure. Some studies suggest that such policies save jobs; for example, voluntary quotas on Japanese auto imports are estimated to have prevented the layoff of tens of thousands of American autoworkers and encouraged job-creating Japanese investment in the United States.[18] However, these gains must be weighed against the long-term effects on employment of protectionism: reduced incentives for American industries to become fully competitive, higher costs to consumers, and possible retaliation by foreign producers. The consequences of protection remain a topic of controversy both among academics and within the ranks of organized labor.

## Direct Action

Unions have also sought to prevent or postpone plant closures and layoffs through mobilizing their members and organizational allies for direct action. The methods and tactics used here are diverse, ranging from traditional adversarial methods, such as the strike or boycott, to shop floor cooperation aimed at improving the productivity of marginal plants.

In general, unions have been reluctant to use adversarial direct action to prevent plant closures. Few unions choose to strike when faced with a shutdown because the union's leverage is severely diminished when the employer plans to discontinue production. In multiplant companies in which the same union represents several branches or subsidiaries, unions have the option of conducting a national strike—or one at another plant—to protest a threatened closure, but this tactic has apparently not been used. The only reported successful strike to prevent a closure occurred at an automobile parts factory in Windsor, Canada, where workers conducted a sit-down inside the plant, supported by a large demonstration outside, to protest management's decision to relocate production. After more than a week, the employer, apparently fearing damage to valuable dies and equipment, reopened the plant and signed a three-year agreement.[19]

Boycotts of company products in order to pressure an employer to reverse a closure decision are also rare. Several unions have considered but rejected them, doubting their ability to mount an effective national or regional boycott in response to a local closure. The only reported instance where a boycott may have been effective in preventing a closure occurred at the General Motors assembly plant in Van Nuys, California. In 1982, the company notified Autoworkers Local 645 that the plant, at that time the last remaining auto assembly plant on the west coast, was on the danger list. The local, which has a predominantly minority membership, formed a "Committee to Keep GM Van Nuys

Open" and joined with Hispanic and black churches, community organizations, and other unions in threatening a boycott of GM cars in Los Angeles if the plant were closed. The plant remains open, although the company denies it has succumbed to union and community pressure.[20]

One of the central difficulties unions face in attempting to use muscle, of course, is their reduced leverage when the employer no longer requires their members' labor. In response to this problem, unions in many areas have formed grassroots coalitions with other unions, religious organizations, and community groups to protest plant closures. According to a recent survey, by late 1983 such jobs coalitions had formed in at least eight states.[21] Borrowing tactics from the civil rights and antiwar movements, these coalitions have conducted demonstrations, rallies, petition campaigns, public hearings, and press conferences to dramatize the plight of unemployed workers and pressure companies to reverse closure decisions. The most developed labor-community coalition to protest plant shutdowns emerged in Youngstown, Ohio, and surrounding communities during the period 1977–1981 in response to a series of steel mill closures in the Mahoning Valley.[22]

The success of this strategy has been mixed. An evaluation of the experience of Pennsylvania's Delaware Valley Coalition for Jobs found that the group had been unable to avert any closures, although in several instances their intervention improved severance payments.[23] In California, the Oakland Plant Closures Project, a local affiliate of the California Coalition Against Plant Shutdowns (CalCAPS), assisted workers at a closing garment factory to set up a workers' cooperative, but has been unsuccessful in preventing the shutdown of any major industrial plants. As in Pennsylvania, however, in several instances the group's action probably resulted in improved severance packages for displaced workers.[24] Probably the most effective use of the coalition strategy occurred at Morse Cutting Tool in New Bedford, Massachusetts. In this case, the United Electrical Workers local at the plant mounted a vigorous campaign in alliance with other unions, churches, and local politicians. The union succeeded in winning a pledge from Morse's owner, Gulf and Western, to reinvest in the plant, which was apparently slated for closure or sale.[25]

Unions have also worked with employers cooperatively by mobilizing their members at the firm or shop level to participate in activities to halt plant closures.

One approach has been for workers to buy their plants under the terms of a negotiated employee stock ownership plan (ESOP).[26] Employers usually do not resist ESOPs because they are typically organized only after management has decided to withdraw. Unions historically have been cool to the idea of worker ownership, fearing a diminished role for the union, a weakening of pattern bargaining, and a blurring of the traditional distinction between labor and management. Survey evidence suggests, however, that union leaders' attitudes toward worker ownership have become markedly more favorable in the early

1980s, largely in response to the recent wave of plant closures.[27] This strategic approach remains confined to a small minority of cases, however. According to the most recent estimates, there are currently somewhere between 700 and 900 majority worker-owned firms in the United States; less than 5 percent of these were organized in response to a threatened shutdown.[28] Worker ownership, although on the rise, is still very much on the margins of organized labor's response to plant closures.

Another approach has been for unions to join with employers in establishing joint employee involvement and quality-of-worklife programs as a strategy for improving job security as well as productivity. Typically, such programs involve informal problem-solving by unionists and managers at the shop level, outside traditional collective bargaining and grievance-handling procedures. Whereas managers are usually motivated to initiate such programs to improve productivity and quality, unions may hope to improve their members' job security. However, there is little evidence that employee involvement programs have had this effect. Recent studies indicate that the effects of quality-of-worklife programs on the economic performance of the firm and hence its likelihood of survival are marginal at best.[29] Most unionists who have participated in such programs discount them as avenues of influence over plant location decisions.[30]

Developing proposals for alternative products is a final direct action strategy unions have pursued when faced with declining demand or obsolete technology. This approach has probably been furthest developed by the United Electrical Workers (UE), which has established local "alternate use committees" to promote product reconversion. For example, when General Electric announced plans to close its steam turbine factory in South Carolina, the UE local proposed instead that management convert the plant to produce environmental protection and alternative electric-generating equipment.[31] Although such initiatives often founder on management's unwillingness to accept organized labor as a partner in product planning and investment decisions, in a few cases cooperative relationships have emerged. Since 1982, for example, the Autoworkers and Machinists have been involved in a joint effort with management at McDonnell Douglas in Burbank, California, to develop proposals for alternative products, such as electric rail cars, wheelchairs, and commuter aircraft. Currently, the company is highly dependent on defense contracts, and its production levels and employment have been volatile. Diversifying the company's product mix may serve both management's objective of stabilizing production and labor's objective of promoting job security.[32]

## Conclusion

Neither adversarial nor cooperative strategies that organized labor has adopted in response to the threat of plant closure have been notably effective in stemming the tide of deindustrialization. Despite innovative efforts in collective bargaining, political action, and direct action, unions have been largely unable to save their members' jobs in the face of high rates of capital mobility.

In my view, an effective strategy for preventing plant closures would require unions to involve themselves squarely in what McKersie has termed the "entrepreneurial decisions of business"—that is, decisions on such issues as investment policy, product mix, plant location, and staffing levels.[33] Although unions have shown increasing interest in such involvement in recent years, largely in response to threats to their members' job security, the legal, institutional, and political settings in which union strategy is enacted in the United States make control over investment difficult for organized labor to achieve. In conclusion, I will briefly highlight some of the structural constraints that make successful union prevention of plant closure so elusive.

The legal framework of collective bargaining limits labor's ability to use contract negotiations as a mechanism for extending its influence over a firm's investment decisions. Under American labor law, only wages, hours, and working conditions are "mandatory" subjects of bargaining; employers have no obligation to negotiate over investment decisions. Historically, these limitations on the scope of collective bargaining have been supported by both managers and unions, although for different reasons. American employers have viewed investment decisions as a managerial prerogative and have resisted union incursions into this area. For their part, American unions have been reluctant to participate in entrepreneurial decisions, fearing that their involvement would blur the traditional distinction between labor and management, weaken their advocacy of worker interests, and open them to blame if business turned bad.

The traditional reluctance of both parties to become involved in a mutual discussion of investment decisions is often reinforced by the institutional structure of bargaining. Contract negotiations in the United States are highly decentralized relative to those in other industrial democracies, reflecting the decentralization of both labor and employer associations, the greater size of the country, and the absence of government involvement. Bargaining over investment decisions is clearly less feasible at the local or plant level than it would be at peak levels of the corporate hierarchy, where most strategic planning occurs. The exception illustrates the rule: some of the most innovative provisions unions have won in exchange for recent concessions have occurred in nationally negotiated contracts, such as those in the automobile and meatpacking industries. In many industries, however, the fragmentation of bargaining mitigates against its use as a forum for discussing overall business policy.

The absence of a labor or social democratic party and the political weakness of organized labor in the United States also limit unions' ability to restrict plant closure. In several European nations, lengthy uninterrupted rule by social democratic parties with close ties to organized labor has facilitated the passage of legislation providing for advance notification and other restrictions on management rights. By contrast, in the United States the multiconstituent Democratic Party, traditionally supported by organized labor, has been less responsive to union involvement in policy formation, particularly on issues where the business community is in opposition. There is evidence, further, that labor's clout within the Democratic Party has declined in recent years. Despite the persistent popular belief that "Big Labor" wields substantial political power, unions have failed to achieve most of their legislative goals over the past decade, including labor law reform, common situs picketing, and federal and most state-level plant closure legislation, even during Democratic administrations. On current evidence, unions are not likely to be able to garner sufficient political muscle to place any significant legislative limits on capital mobility in the near term.

Yet unions have made some small gains in responding to plant closures. A few unions, sometimes in exchange for concessions, have obtained limited contractual guarantees of job security; a few have mobilized union and community pressure to reverse or postpone a closure decision; and a few have been able to effect conversions to worker ownership or alternative products. These cases hold out promise. For the most part, however, union efforts to control capital mobility have proved unsuccessful. The narrow economic focus and decentralization of traditional collective bargaining, coupled with organized labor's present political weakness, have limited unions' ability to win a real voice in determining managerial investment policy. Without such a voice, labor's most effective strategy to maintain membership or grow during a period of deindustrialization is to expand organizing efforts in growing economic sectors, such as services and high technology industries. As long as unions are unable to bargain for or legislate real influence over investment decisions, labor's losses are more likely to be reversed by recruitment of new members than by attempts to save jobs in declining sectors of the economy.

## Notes

1. Richard B. Freeman and James L. Medoff, *What Do Unions Do* (New York: Basic Books, 1984); Henry S. Farber, "The Extent of Unionization in the United States," in *Challenges and Choices Facing American Labor*, Thomas A. Kochan, ed. (Cambridge: MIT Press, 1985); William T. Dickens and Jonathan S. Leonard, "Accounting for the Decline in Union Membership, 1950–1980," *Industrial and Labor Relations Review* 38, no. 3 (April 1985).

2. Anne T. Lawrence, *Plant Closing and Technological Change Provisions in California Collective Bargaining Agreements* (San Francisco: Division of Labor Statistics and Research, California Department of Industrial Relations, 1985), pp. 54–57.

3. The method of response shown in table 15–1 is similar in some respects to one proposed by Craft (1984). However, Craft's purpose is somewhat broader than mine; he attempts to classify all responses to plant closures, not just those adopted by unions. Craft does not share my classification of approaches by philosophical orientation.

4. The typology presented in table 15–1 addresses only union strategies for preventing or postponing plant closures; it excludes strategies for alleviating their adverse

effects on union members or for compensating organizationally for membership loss (for example, through campaigns to organize new members). Thus, union strategies for assisting displaced workers will not be considered here, except insofar as severance benefits represent a disincentive to closure.

5. Anne T. Lawrence and Paul Chown, *Plant Closings and Technological Change: A Guide for Union Negotiators* (Berkeley: Center for Labor Research and Education, University of California, 1983); Anne T. Lawrence, "Organizations in Crisis: Labor Union Responses to Plant Closures in California Manufacturing 1979–1983," Ph.D. dissertation, University of California, Berkeley, 1985).

6. Union concern with obtaining job security provisions in their collective bargaining agreements has been heightened by several recent developments in labor law. Under the National Labor Relations Act, employers are required to bargain in good faith with respect to wages, hours, and other terms and conditions of employment. These issues are considered "mandatory" subjects of bargaining; all others are considered "permissive." Employer failure to bargain over mandatory subjects is an unfair labor practice.

In 1981, the Supreme Court ruled in the *First National Maintenance* case that an economically motivated decision to close part of a business is *not* a mandatory subject of bargaining, in effect stating that an employer has no legal obligation to bargain with the union over a closure decision. The effects of a closure on employees, however, are still considered mandatory subjects. In a more recent development, the National Labor Relations Board ruled in 1983 in the *Milwaukee Spring* case that in the absence of a contract provision explicitly prohibiting such a move, the employer can relocate a facility in mid-contract in order to reduce labor costs. This decision was upheld by a federal court in 1985.

These rulings have increased organized labor's concern with obtaining contract provisions that obligate the employer to bargain over the closure decision and/or restrict management rights to relocate production during the term of the contract, because the courts have made clear that these protections are not automatically provided by labor law. For a discussion of recent legal developments affecting union propensity to bargain in these areas, see Sharon Simon, "Plant Closings and the Law of Collective Bargaining," in *Labor and Reindustrialization: Workers and Corporate Change,* Donald Kennedy, ed. (State College: Department of Labor Studies, Pennsylvania State University, 1984); Patricia A. Greenfield, "Plant Closing Obligations under the National Labor Relations Act," in *Plant Closing Legislation,* Antone Aboud, ed. (Ithaca: New York State School of Industrial and Labor Relations, 1984), pp. 13–32; Barbara Rhine, "Business Closings and Their Effects on Employees: The Need for New Priorities," *Labor Law Journal* 35, no. 5 (May 1984), pp. 268–280.

7. Anne T. Lawrence, *Plant Closing and Technological Change Provisions in California Collective Bargaining Agreements,* pp. 164–169.

8. *Ibid.,* pp. 169–172.

9. Nancy R. Folbre et al., "Plant Closings and Their Regulation in Maine, 1971–1982," *Industrial and Labor Relations Review* 37, no. 2 (Jan. 1984), pp. 185–196.

10. Robert B. McKersie and William S. McKersie, *Plant Closings: What Can Be Learned from Best Practice* (Washington, D.C.: U.S. Department of Labor, 1982); Melva Meacham, "Coping with Long-Term Unemployment: The Role of the Trade Union Movement," in *Proceedings of the 36th Annual Meeting of the Industrial Relations Research Association* (Madison, Wisc., 1984), pp. 138–144.

11. Everett M. Kassalow, "Concession Bargaining—Something Old, But Also Something Quite New," *Proceedings of the 35th Annual Meeting of the Industrial Relations Research Association* (Madison, Wisc., 1983); Peter Cappelli, "Union Gains Under Concession Bargaining," *Proceedings of the 36th Annual Meeting of the Industrial Relations Research Association* (Madison, Wisc., 1984).

12. Anne T. Lawrence, *Plant Closing and Technological Change Provisions in California Collective Bargaining Agreements,* 1985, pp. 149–164.

13. Daniel A. Littman and Myung-Hoon Lee, "Plant Closings and Worker Dislocation," *Economic Review* (fall 1983), pp. 2–18.

14. Although plant closing legislation has generally received wide support from organized labor, a few unions have supported national legislation but opposed (or remained neutral toward) state legislation, on the grounds that unevenness of coverage might prompt business relocations across state borders, resulting in a shift of jobs from areas where unions are strong to those where they are weak.

15. Richard B. McKenzie and Bruce Yandle, "State Plant Closing Laws: Their Union Support," *Journal of Labor Research* 3 (winter 1982), pp. 101–110.

16. One exception has been reported in the literature. Representatives of Levi Strauss and Company testified at congressional hearings in support of national plant closing legislation. Their argument: that national legislation would prevent the "chaos" of separate state laws (Bennett Harrison, "Plant Closures: Efforts to Cushion the Blow," *Monthly Labor Review* [June 1984], p. 42). This view, however, is apparently an exception in the business community.

17. U.S. Congress, Office of Technology Assessment, *Technology and Structural Unemployment: Reemploying Displaced Adults,* OTA-ITE-250 (Washington, D.C.: U.S. Government Printing Office, Feb. 1986); Folbre et al., "Plant Closings and Their Regulation in Maine," pp. 190–191.

18. U.S. Congress, Office of Technology Assessment, *Technology and Structural Unemployment.*

19. Staughton Lynd, *The Fight against Shutdowns: Youngstown's Steel Mill Closings* (San Pedro: Calif.: Singlejack Books, 1982), pp. 224–226; Edward M. Feigen and Mona R. Hochberg, "Union Responses to Plant Disinvestment and Shutdowns," master's thesis, Department of Urban and Environmental Policy, Tufts University, 1984), p. 234.

20. Eric Mann, "What GM Owes the Freeway Capital," *Los Angeles Times* (April 17, 1983); "The Van Nuys Campaign: Workers and Community Take on G.M.," *The Nation* (Feb. 11, 1984).

21. Arthur Hochner, "Shutdowns and the New Job Coalitions: The Philadelphia Experience," *Labor Research Review* 5 (summer 1984), pp. 15–27.

22. Staughton Lynd, *The Fight against Shutdowns;* Terry F. Buss and F. Stevens Redburn, *Shutdown at Youngstown: Public Policy for Mass Unemployment* (Albany: State University of New York Press, 1983).

23. Arthur Hochner, "Shutdowns and the New Job Coalitions."

24. Alison Givens, "Fighting Shutdowns in Sunny California," *Labor Research Review* 5 (summer 1984), pp. 15–27; Gilda Haas, *Plant Closures: Myths, Realities, and Responses* (Boston, Mass.: South End Press, 1985).

25. Bureau of National Affairs, *Unions Today: New Tactics to Tackle Tough Times* (Washington, D.C., 1985).

26. ESOPs, now the most common institutional form of worker ownership, have existed in the United States since 1974, when they were granted tax-advantaged status under the national Employee Retirement Income Security Act (ERISA). Other forms include worker cooperatives and joint-stock companies or partnerships (Joseph R. Blasi, "The Sociology of Worker Ownership and Participation," in *Proceedings of the 37th Annual Meeting of the Industrial Relations Research Association* [Madison, Wisc., 1985]).

27. Joyce Rothschild-Whitt, "Who Will Benefit from ESOPs?" *Labor Research Review* 1, no. 6 (spring 1985), pp. 71–80; William Foote Whyte and Joseph R. Blasi, "Employee Ownership and the Future of Unions," *Annals of the American Academy of Political and Social Science* 473 (May 1984), pp. 128–140.

28. Joseph R. Blasi, "The Sociology of Worker Ownership and Participation, p. 362; Julia Parzen et al. *Buyout: A Guide for Workers Facing Plant Closings* (Sacramento, Calif.: Office of Economic Policy, Planning and Research, Dec. 1982).

29. Harry Katz, et al., "Industrial Relations Performance, Economic Performance, and QWL Programs: An Interplant Analysis," *Industrial and Labor Relations Review* 37, no. 1 (Oct. 1983), pp. 3–17; Thomas A. Kochan, et al., "U.S. Industrial Relations in Transition: A Summary Report," in *Proceedings of the 37th Annual Meeting of the Industrial Relations Research Association* (Madison, Wisc., Dec. 1984), p. 269.

30. Thomas A. Kochan, et al., "Worker Participation and American Unions," in *Challenges and Choices Facing American Labor*, T.A. Kochan, ed. (Cambridge: MIT Press, 1985).

31. Bureau of National Affairs, *Unions Today*, pp. 21–23.

32. Alison Givens, "Fighting Shutdowns in Sunny California."

33. Robert B. McKersie, "Union Involvement in the Entrepreneurial Decisions of Business," in Thomas A. Kochan, ed., *Challenges and Choices Facing American Labor* (Cambridge: MIT Press, 1985).

# 16
# The Employee Ownership Alternative

*William Foote Whyte*

I n the United States, in most cases employee ownership occurs when there is an employee buyout of a plant being shut down." In those words, an American student was misinforming a foreign scholar. Probably this myth has gained wide currency because this type of case involves a highly dramatic struggle that attracts widespread media attention.

In fact, according to Corey Rosen of the National Center for Employee Ownership, of over 6000 cases in which employees hold some share of ownership in companies, only about 1 percent have arisen out of employee buyouts in the face of impending plant shutdowns. Nevertheless, I shall focus on this type of case for several reasons. With my Cornell associates, I have been tracking this type for the past ten years. Such cases illustrate important problems that may be at the cutting edge of contemporary changes in industrial relations. And finally they do add up to a significant number of jobs saved—well over 50,000—and of the approximately sixty cases, we know of only 4 or 5 where the employee-owned firm subsequently went out of business. Considering that we have been tracking these firms through the most severe recession since World War II, this is an impressive record. Examining that record has also alerted us to phenomena that may occur in cases where employee ownership has arisen in other circumstances.

## Worker Expectations and Labor Relations

When workers come to share in ownership, this changes their expectations about labor relations and this, in turn, leads to changes in the nature of labor relations. In the 1970s cases, without exception, it was key members of local management, sometimes in association with community officials, who organized the buyout to stave off a plant shutdown. Workers and local union leaders

William Foote Whyte, "Employee Ownership: Lessons Learned," in *Proceedings of the Thirty-Seventh Annual Meeting of the Industrial Relations Research Association,* Barbara D. Dennis, ed. (Madison, Wisc, 1985), pp. 385–395.

cooperated passively with management, and high level union officials stood aside, not knowing what to do. Neither workers nor their representatives made any demands for worker participation in decision-making. However, in the general spirit of euphoria that accompanied the successful campaign to save jobs through buying the plant, they did have some vague idea that they would be treated with more dignity by management. They also believed that their views would be taken into account in managerial decision-making. When I asked the president of the blue-colar UE local in the Herkimer plant that was being shut down by Sperry-Rand whether he thought that workers or union leaders should "participate in management," he flatly stated that it should be management's responsibility to manage. But when I asked him if workers should have some "input into decision-making," he said that this was obviously what they wanted.

As the spirit of euphoria wore off and workers found that they were treated just as before, disillusionment set in. This was illustrated by the comment of the president of the white-collar union in the same plant, when I asked him, fifteen months after the buyout, what had changed in labor relations. In the period of euphoria, he and the blue-collar union president had spoken enthusiastically about the new spirit of cooperation between union and management. Now he said, "You ask any worker what it means to work in that plant now. They'll tell you, 'I've got a job.' That's all. Nothing else has changed. This place doesn't run any different now than it did when Sperry-Rand owned it."[1]

In the 1970s we find the change from euphoria to disillusionment taking place within the first six to twelve months following the takeover. Where employee sharing in ownership came into effect long ago in quite a different climate of opinion and attitudes, the clash between worker expectations and maintenance of the traditional pattern of labor relations could remain without resolution for many years and then surface to provoke a conflict. Such was the case with American Cast Iron Pipe Company, which was turned over to the employees in 1924 by its entrepreneur-owner John Egan.

> Under an arrangement that John Egan made while still alive, 12 workers elected from the shop floor sit with 5 executives on the board of trustees. After 59 years of rubber stamp complacency, the workers mutinied last year. The workers reread Egan's will and decided that they, not management, should control the company. Then they announced to fellow laborers that Acipco had improperly withheld $24 million in bonus pay.[2]

The outraged chief executive officer then fired the worker-trustees, who got their jobs back under court order.

So far we have seen a general failure of management to recognize that a shift to employee ownership requires a change from traditional styles of labor relations and managerial leadership. This failure is most dramatically illustrated in the case of John Lupien, the maintenance supervisor in the asbestos plant that was being shut down by GAF. Lupien was the popular hero of what became Vermont Asbestos Group. Whenever Board Chairman Lupien was asked by reporters or researchers what plans he had for worker participation in decision-making, he had a stock answer: "If you own stock in General Motors, that doesn't give you the right to run General Motors."

When I first heard this answer, it seemed to me that someone who could see a close analogy between that giant corporation and a tiny company of 150 workers who owned 80 percent of the stock was bound to run into trouble. Indeed he did. Not long after, in response in part to worker dissatisfaction with a major decision he had pushed through the board of directors, Lupien lost control when many workers sold their stock to an independent entrepreneur whose group took over a majority of the board. In a later stage of the conflict, Lupien was discharged.

When workers and management leaders of the Herkimer plant which had become the Mohawk Valley Community Corporation met with us at Cornell within two weeks after the dramatic success of their employee buyout, I asked President Robert May whether he had any plans for worker participation in decision-making. He replied, "No, we haven't got around to that yet." Then, after a pause, as if he sensed there was something lacking in his response, he added, "Maybe we should install a suggestion box."

Management people don't have to be told that workers need some education so that they can perform the role of worker-owners. However, I have yet to encounter a manager in this situation who has volunteered the statement that management people need to be involved in an educational resocialization process. There seems to be a general tendency to take the view that I have heard expressed by several management people: "Workers should act like workers from 9 to 5 and like owners after working hours." In other words, for managers it should be just business as usual.


## Views of Social Justice


Whatever its origin, a shift to employee ownership tends to change worker views of social justice. We see this in the comparisons they look for in deciding whether their pay and benefits are equitable. Under private ownership, they tend to compare themselves with other workers with similar seniority, skills, and job classifications within their own industry. They are well aware that people in top management are paid many times more than they; if you ask them whether such a wide salary range is fair, they might well reply that it is not, but they don't appear to give much attention to such comparisons. When the company shifts to partial or full employee ownership, the comparison of their financial and social rewards with those of higher management becomes much more salient.

I observed this shift first in Peru in the early 1970s. Those who called themselves "the revolutionary government of the armed forces" had established by decree The Industrial Community. This was a program under which private firms were required to provide employees with stock based upon profit-sharing, along with one representative on the board of directors initially, with additional representatives to come as the workers' shares increased. Along with board representation came the right to examine the company's books and even bring in accountants and lawyers to assist in this process.

As we studied the implementation of this drastic change, we found workers and their representatives focusing with increasing vigor upon managerial perks, salaries, and benefits. They wanted to know why top management people were using company cars for their private purposes. They asked whether it was really necessary, when they travelled to another city, for them to stay in the most expensive hotel. They challenged the expenditures management was making on high-priced consultants and wanted to know why management didn't make more use of the skill and experience of employees.

My favorite illustration of this change comes from a factory in Iquitos, a city on the Amazon, close to sea level and to the equator. Worker representatives complained that it was hard to do heavy work under tropical conditions and asked management to air-condition the factory. After studying this proposal, management came back with a report that the cost not only of buying the equipment but also of operating it would be prohibitively expensive and therefore could not be considered. The worker representatives were duly impressed with management's cost figures but came back with this suggestion: "Then why don't we save the company a lot of money by shutting off the air-conditioning in the management offices?"

That Peruvian case involved no major disruption because the workers were not in a position to push their claims. We found quite a different situation in 1984 at Hyatt-Clark Industries. In buying the plant from General Motors to save their jobs, the employees had made substantial pay sacrifices. Under the terms of the ESOP, it would not be until ten years after the buyout (1991) that the workers would be able to vote that stock. However, they did gain three members on a thirteen-person board of directors and have a militant union, UAW Local 736. The union's concern with financial equity manifested itself in the buyout process in a dispute over the allocation of the stock. Management proposed that the stock be distributed in relationship to compensation levels, as is customary in a management-dominated ESOP. Union pressure finally secured an equal distribution of stock, without regard to pay differentials.

The equity question remained very much alive in the process of recruiting people for top management positions. The union leaders went along reluctantly with management's contention that the job market would have to determine salaries offered. However, in the cases of all subsequent promotions and replacements, the union leaders have given careful scrutiny to high level salaries and demanded extensive justifications. At one point, when the chief executive officer announced a plan for management salary increases averaging 6.5 percent, the union leaders accepted the overall figure as reasonable, but demanded specific salary information so as to determine whether those being rewarded with higher than average salaries were really worth the increases. When the chief executive officer declined to provide this information, the union leaders on the board of directors took the case to court. Thereupon, the chairman of the board released the figures.

The most serious confrontation so far occurred in the late spring of 1984. Productivity had increased, and in the preceding quarter the company had earned a substantial profit. The contract negotiated with the shift to employee ownership had led workers to expect that there would be profit-sharing so that they could gain back some of their sacrifices. In fact they were hoping for as much as $600 each. In a board meeting, the chief executive officer announced that the company would invest all the profits in a technological modernization program. Union President James May said that he recognized the need for further investment but argued that dashing worker expectations would have disastrous results. He argued for at least some sharing of profits.

When the majority on the board supported the CEO, the labor relations situation in the company changed abruptly. The union leaders reported that the workers might have accepted this decision reluctantly if it had not been accompanied by a program to refurbish management offices and a decision that management people could work somewhat reduced hours during the summer months, to make up for extra hard work earlier—but of course, without any cut in management salaries.

Following these decisions productivity dropped off so that management had to resort to paying workers overtime in order to keep up with current orders. Instead of operating at a substantial profit, Hyatt-Clark was now losing money, and the union leaders, in preparing to bargain for a new contract, were determined to push for substantial sharing in power by workers through their union.

The parties began bargaining for a new three-year contract in July 1984 against a September 15 deadline, at which time, with no contract signed, General Motors would cancel its purchase orders. Since 85 percent of Hyatt-Clark's production went to GM, failure to meet the deadline would have meant bankruptcy for the company. When agreement was still not in sight in September, GM agreed to a three-month extension but insisted that midnight, December 15, was the final date. Since the 15th fell on Saturday, the parties were able to continue bargaining until they finally settled at 5:00 a.m. on Monday, December 17.

The union secured a comprehensive structure of joint participation committees from the bottom to the top of the company, but President Howard Kurt retained veto power over committee actions. The union accepted management's offer of about a 5 percent pay increase—50 cents per hour for each of the first two years, 55 cents for the third year. The importance of the equity issue is indicated by the fact that management's pay offer had been on the table weeks earlier, but the union leaders bargained up to the deadline in a vain effort to limit office workers and management people to the same cents per hour figures.

## Will Cooperative Labor Relations Save the Jobs?

As researchers and practitioners, we are inclined to overestimate the importance of industrial relations in business success or failure. Obviously, labor relations conflicts can destroy a company, and cooperative labor relations can strengthen it, but other factors may be overriding.

That is one of the principal lessons of our university involvement in research and action with Rath Packing Company and Local 46 of the United Food and Commercial Workers. The case seemed particularly important to us as the first one where the shift to employee ownership in a major company (in 1980) was engineered by the union leaders and where they secured majority control of the board of directors. Since Rath was forced into Chapter 11 in November 1983, there has been a widespread tendency to attribute the company's financial problems to a failure of worker participation and control. Since ESOP attorney Jack Curtis had made a great creative contribution in working out for Local 46 an employee stock ownership trust which achieved and protected continuing worker control of the board of directors, I was startled to hear him say in a 1984 meeting of the National Center for Employee Ownership that the financial deterioration of Rath was due to the unwillingness of workers to make sufficient sacrifices. Let's look at the record. According to Donald Wade, Director of the Blackhawk County Economic Development Committee, which played a major role in securing the refinancing of Rath, the workers made sacrifices in pay and benefits of over $17 million at the time of the buyout. In September 1982, the workers accepted the termination of their pension program, thus leaving all but the most senior employees unprotected. In February 1983, when the banks threatened to cut off Rath's credit unless further substantial cuts were made, 72 percent of the workers voted to accept a $2.50 an hour "wage deferral"—the money coming back to them only in the unlikely event that Rath once again became profitable.

Throughout this period from early 1980 until 1984, hundreds of workers spent endless hours on their own time in meetings before or after shifts to work with management on improving productivity and lowering costs. This cooperative problem-solving program achieved a plant productivity improvement of at least 20 percent. However, Christopher Meek points out that, throughout this period, labor costs in production accounted for only between 15 and 20 percent of total costs. Thus a 20-percent increase in plant productivity yields only a 3 to 4 percent total gain in cutting overall costs, and that was far from enough.

The productivity gains were overwhelmed by problems in marketing, in top level administration, and in financing the company. Rath was short of working capital and had to be financed at interest rates up to 5 percent above prime during a period of historically high interest rates. At the same time, hog prices were extraordinarily high during a recessionary period when the weakness of the consumer market prevented the company from passing on increasing costs to customers.

The union leaders were aware of all of these problems, but their experience and expertise were limited to the production process. Those of us who were providing applied research and technical assistance to union and management had no mandate to work on the overall problems of business administration and marketing, nor did we have the necessary expertise.

Without downgrading the importance of contributions in the industrial relations field, we have to recognize that the saving of Rath Packing Company through employee ownership—or of any company in dire financial straits—requires an integrated program of technical assistance including labor relations but extending also into various fields of business administration.

## Jobs and/or a Social Movement?

If we judge simply from the standpoint of saving jobs, there is no doubt that employee ownership can and does work in many cases. If we are concerned not only with saving jobs but also with building a social movement, there are additional and difficult problems to be resolved. Besides a viable economic base initially for each employee-owned firm, the growth of a social movement requires two additional factors: a system for maintaining employee ownership and a system for linking together otherwise isolated worker cooperatives or employee-owned firms.

## How Employee Ownership Is Lost

For a social movement, the nature and form of employee ownership is of critical importance. In the past most employee-owned firms have found themselves in a Catch-22 situation: the firm ceased to be employee-owned either because it went bankrupt or because it was highly successful, but the original employee-owners had not devised a system for sharing ownership with new workers. This factor accounts for much of the decline in the numbers of the West Coast plywood worker cooperatives, which overall have been able to compete success-fully with private firms in their field. On the face of it, the plywoods have appeared to have an ideal democratic structure: at the outset, only workers owned the company, with each worker owning one share of stock and only one. As such a company becomes highly prosperous, the value of each share of stock increases substantially. In this situation, what we call collective selfishness tends to take over. The original worker-owners decline to dilute their own equity and require new people to come in simply as hired labor. When the original worker-owners approach retirement, they would be glad to sell their shares of stock, but by this time the value of the shares is beyond the reach of nonowning workers.

Whether employee ownership is successful in such cases may depend upon whether we are considering the interests of the original or early worker-owners or whether we are thinking of the long-run maintenance of employee owner-ship. No doubt the employees of the *Kansas City Star* and the *U.S. News and World Report* believe that employee ownership has been a bonanza to them. When the profitable newspaper was sold, employees found that the market value of their shares had multiplied many times. The magazine story is even more spectacular. David Lawrence, founder of the magazine, sold it to the employees when he retired in 1962. A *New York Times* report (June 15, 1984) estimated that the average employee would receive $50,000 and that 20 to 40 long-service and high-ranking executives would become millionaires from the sale.

A less happy tale is found in the case of one of the plywood cooperatives. In 1938 Anacortes Veneer, Inc. was established as a worker cooperative, with each worker putting up $4,000. Publishers Paper Co. bought out the worker-owners for $55,000 per share in 1969. On September 28, 1981, that company closed the plant. On April 16, 1984 the plant reopened again as a worker cooperative, Anacortes Plywood, Inc. The 150 worker-owners bought back their jobs at between $15,000 and $20,000 each. The newspaper reports (*Seattle Post-Intelligencer,* March 29 and June 25, 1984) do not tell us whether the worker-owners today learned anything from the experience of Anacortes Veneer, Inc. to assure the long-run continuation of worker ownership this time around.

## U.S. Future

This continuity problem can be solved through applying the social policy invented by the founders of the Mondragón cooperative complex in the Basque country of Spain. In this phenomenally successful cooperative movement, con-trol is based on labor rather than on capital. Ultimate power is in the hands of workers on a one-worker-one-vote basis. There is no stock. Worker initial contributions are treated as if they were loans to the company, paying interest each year to the members. These capital accounts of members remain with the firm as long as they stay on the job, and the capital accounts are increased through profit sharing.

Students of employee ownership believe that it is possible to design an ESOP so as to come close to the Mondragón formula in securing continuity of employee ownership. However, the design problems involved in reaching this goal with an ESOP are much more complex than the Mondragón formula. Unless those designing the ESOP take care to apply the best information and ideas now available on the continuity problem, in the future as in the past financially successful employee-owned firms will come to be owned and controlled by outside stockholder interests.

Students of worker ownership have come to the conclusion that an isolated worker cooperative, like an island in a sea of private enterprises, is bound to lead a precarious existence. In Mondragón, the cooperatives are closely linked with supporting organizations. Of these, the Caja Laboral Popular (the bank) is the most important since it provides not only financing but also an extraordinary range of technical assistance in starting and maintaining cooperatives.

A growing number of practitioners in worker cooperatives and nonprofit organizations have recognized the need for linking cooperatives together and building support organizations. The linkages in the U.S. will be looser than those of Mondragón, but the surge of interest in employee ownership and the growing involvement of researchers and practitioners in the field give some promise that one day we may be able to refer to a real American worker cooperative movement.

A major factor supporting worker ownership and worker participation in management has been the striking growth of union interest and involvement in employee buyouts and shifts to partial employee ownership. In the mid-1970s, higher level union leaders were either ambivalent or hostile toward employee buyouts. By the mid-1980s some of the major unions had become involved in employee ownership.

The change is particularly striking in the cases of the UAW and the USWA. In the 1981 employee buyout of a General Motors plant to create Hyatt-Clark Industries, leaders of UAW Local 736 received no help from regional or national officials. By 1984 former UAW President Douglas Fraser had assumed a position on the board of Hyatt-Clark, and regional and national officials had provided essential assistance in the employee buyout that created Atlas Chain Company. Furthermore, we see the lessons learned earlier being applied later as James May and James Zarrello, leaders of Local 736, consulted with the prospective worker-owners of Atlas Chain, and Craig Livingston, attorney for Local 736, working with PACE (Philadelphia Area Cooperative Enterprise), put in place one of the most democratic ESOPs yet to be created.

In one of the earliest of the 1970s buyouts, officials from the international office of the Steelworkers abstained from any involvement in the design of the

South Bend Lathe ESOP. When the local union members reluctantly agreed to trade their pension program for stock ownership, USWA international officials sued the company in a vain effort to void that transaction. In a 1984 meeting of the Michigan Center for Employee Ownership, the assistant to the president of USWA, James Smith, stated that it had been a mistake for the international office to refuse to participate in the design of the South Bend Lathe ESOP. He now believed that the future of the U.S. steel industry would depend in part upon the ability of labor and management to negotiate terms for worker-sharing in ownership.

The Machinists, the Airline Pilots, and the Teamsters have taken the leadership in the restructuring of the airline and interstate trucking industries as they have traded pay and benefit concessions for stock ownership, positions on the boards of directors, and (especially in Eastern Airlines) for worker and union involvement in decision-making at all levels of the companies.

United Food and Commercial Workers, through the Rath Local 46, was the first major union to take a leadership role in establishing employee owner-ship. Beginning in 1982, under the leadership of Wendell Young, president, Local 1357 has made the creation and strengthening of worker cooperatives in the Philadelphia area a major strategy for coping with plant and store shut-downs.

Through continuing study of cases and learning from experience, we can predict that union leaders, with their academic and consultant allies, will play increasingly important roles in the spread of employee ownership and in the democratization of management.

## Notes

1. William Foote Whyte, Tove Helland Hammer, Christopher Meek, Reed Nelson, and Robert Stern, *Worker Participation and Ownership: Cooperative Strategies for Strengthening Local Economies* (Ithaca, NY: ILR Press, Cornell University, 1983), p. 88.
    2. *Forbes* (April 23, 1984).

# Part IV
# Perspectives from
# Foreign Countries

A n important difference between the industrial relations system in the United States and Western Europe is the greater emphasis in Europe on codetermination, in which employee group representatives participate in key management decisions. Codetermination gives European employees an opportunity, through their unions, to be consulted on policy matters such as plant closure and employment level, as well as operational decisions. Western European countries also have specific laws that regulate plant closure. Although U.S. law provides income replacement for employees who lose their jobs as a result of economic exigency, only a few states have enacted legislation to relieve the effects of plant closure. (These laws are discussed in Part V.) In this sense, workers in the United States are well behind their Western European counterparts. It is therefore instructive to examine the content of and experience with legislation on plant closure in other countries.

Bennett Harrison, in the opening chapter, reviews plant closure laws in Western Europe, particularly those concerning advance notification. Even though these laws extend far greater protections than those afforded most American workers, attempts are being made to upgrade and extend consultation, including plant closure protections, within the ten-nation European Community. Harrison discusses this movement. Bennett Harrison is a professor of political economy and planning at Massachusetts Institute of Technology.

Next, Philip L. Martin examines the specific Western European laws on plant closure and how they are implemented, as well as similar Canadian legislation. He compares foreign laws with U.S. public policy concerning advance notice of plant closure, unemployment insurance, training and relocation allowances, and the Trade Adjustment Act, which provides income to American workers who lose their jobs as a result of foreign competition. Martin is a professor of economics at the University of California, Davis.

Although several studies have compared United States and European experience with plant closure legislation, the article by Gregory Hooks is particularly useful for its insightful historical perspective and cross cultural comparisons of policy development. Hooks identifies three major policy options—free market, liberal industrial, and social welfare—and describes the use of these policies in the United States, France, and Sweden. He concludes that none of the three approaches can solve the problems of plant closure and that effective industrial policies must combine elements of all three. Gregory Hooks is a Ph.D. candidate in the sociology department at the University of Wisconsin, Madison.

In the next selection, Bennett Harrison, a leading proponent of plant closing legislation, draws conclusions about the future direction of voluntary and mandatory laws in the United States on the basis of his research in Europe and America. Harrison thinks most of the differences between international practices stem from widely divergent attitudes about business responsibilities and individual rights. In Europe, advance notification is considered a basic human right and shows corporate responsibility, whereas the United States emphasizes managerial preprogatives and freedom of movement in a free-market system. Another important difference Harrison identifies is that unions in the United States are in a far weaker position than their European counterparts, who have the support of political parties in several countries.

The concluding reading, by Tadashi A. Hanami, is on plant closure policy in Japan. The Japanese have a well-developed system of consultation between companies and unions, so that decisions to close a plant are usually well planned and fully discussed in advance between labor and management. Also, employers are required to give thirty days advance notice to dismissed workers. Problems of plant closure have not been significant in Japan because of the high economic growth rate, lifetime employment for many workers, and a well-developed practice of consultation. Tadashi A. Hanami is a professor of economics at Sophia University in Tokyo.

# 17
# Closure Notification in Western Europe

*Bennett Harrison*

he idea that workers should have a legally justifiable collective right to "notification and consultation" emerged in Western Europe as a widespread political issue in the wake of the mass labor unrest of the years 1968–1969. One country after another began to pass laws guaranteeing the rights of "workers or their representatives" to information and to consult with management (but only with local management, a point to which I shall return) prior to the undertaking of major business decisions. The works councils that were being created during this period were typically designated by law as the official recipients of information to be periodically transmitted to the workers by management, and as the bodies invested with the right to respond.

With respect to closures per se, prenotification periods were negotiated or legislated in nearly every country. The longest is the Federal Republic of Germany, with a twelve-month notification period introduced as part of the 1976 Co-Determination Act. Other notice requirements range from two months in Belgium to six months in Sweden (for plants with 100 or more employees). In some cases, compliance is formally voluntary, although important public incentives to the firm may be triggered by advance notification. Unfortunately, there is no systematic evaluation literature reporting on the incidence of compliance.

The German arrangements for anticipating economic dislocation are among the most comprehensive in Europe. Under the Co-Determination Act of 1976, most German firms are expected to provide the regional government and the works council within the plant with a year's notice prior to making a final decision about closure or a major layoff. There are no legal penalties if the firm fails to give such notice, and one evaluator concluded that "penalties under the law are minimal and enforcement is lax."[1] However, those firms that do not comply may be liable for picking up some of the retraining costs later. (I can find no hard evidence on how many companies have been required to pay

Bennett Harrison, "The International Movement for Prenotification of Plant Closures," *Industrial Relations* 23, no. 3 (fall 1984), pp. 391–398.

those social costs.) There is also a strong positive incentive to comply, at least for those companies that expect future layoffs in one or another of their plants. If the firm wants to acquire short-term compensation for any of its other employees during a business downturn, the managers must provide prenotification to the affected workers.

When the final closing announcement is actually given to the government labor office and to the works council, local tripartite (labor-management-government) boards are empowered to grant the workers up to two months' notice from that point. Thus, the board may require the plant to remain in operation for as many as sixty additional days. In practice, this means that public agencies may get as long as fourteen months to set up retraining or other programs in anticipation of the closure.

However attractive these arrangements may appear to American advocates of prenotification legislation, there is a great deal of criticism of them among European trade unionists. The results of my own unstructured interviews with Italian, Dutch, Belgian, British, and Swedish trade unionists, conducted in Europe during the spring of 1983, indicate a belief that employers generally *do* provide periodic information. However, when closures or major technological changes are at stake, there is a widespread feeling in the unions that the information which gets transmitted is incomplete, confusing, and frequently comes too late to be of use. This (it is said) is especially true when the local plant is a branch or subsidiary of a multinational corporation headquartered in another country. In such cases, local managers often make the excuse—thought by the unions to be quite valid, in many cases—that their own superiors at headquarters did not provide *them* with "timely" notice of the impending decision.

The trade unions' response in Europe to what their members believe to be the unequal bargaining power of multinational corporations vis-à-vis a nation-based labor movement has been twofold. First, the unions engage in extensive international interunion consultation and coordination. This occurs in three ways: through official transnational bureaucracies at the branch (for instance, International Metalworkers' Federation) or interbranch (for instance, European Trade Union Confederation) levels; through official international meetings of the leadership of the various national unions; and through "unofficial" international meetings of representatives from the base (especially shop stewards). Second, the unions have urged their political representatives at both the national and European Community levels to pursue legal remedies. Specifically, in order to put the force of law behind the principles of prenotification and consultation, and especially to apply them to the employees of multinational corporations, whether based in an EC member country or not, the European labor movement has for a dozen years sought to get the EC to issue regulations that would apply across national borders and with increased force within each member state, as well. The Community of ten nations is as yet the only region in the world which has such power of transnational legislation.

Thus, in 1972, the European Commission took up draft legislation proposing Community-wide regulation of mergers, and requiring that individual members' state subsidies to companies be contingent upon the appointment of workers' representatives to company boards of directors and upon the recognition by companies of transnational works councils. Even with the recent labor struggles of the 1968–1969 period still fresh in the minds of the members of the European Parliament and the Council of Ministers, such a direct attempt by the Left to legislate continental Co-determination was far too radical to succeed.

Between 1975 and 1980, with scaled-down ambitions, advocates managed to push through into European Community-wide law a number of provisions that effectively harmonized differences among the laws of the individual states. Thus, in 1975 the EC Council of Ministers passed a Directive obliging all governments to require that European employers operating anywhere inside the EC give workers' representatives 30 days' advance notification of intended mass layoffs, short of complete closure. In 1977, the Council passed a Directive stipulating that, in the event of a transfer of ownership, workers in any of the affected facilities retain the "rights and advantages" they had acquired prior to the change (in other words, the previous collective bargaining agreement remains in force). A 1980 Council Directive ensured that workers' contractual claims would be given priority in bankruptcy cases. In the United States, there are still no uniform, comprehensive laws which unambiguously regulate advance notice of layoff, successorship, or bankruptcy.

## The Vredeling Directive

In October 1980, the Commission took up the Vredeling Directive, named after Dutch labor leader Henk Vredeling, who was then Commissioner of Directorate-General V for Employment, Social Affairs, and Education. During the preceding few years, a number of voluntary codes of conduct for multinationals had been promulgated by other international bodies, such as the OECD in 1976 and the ILO in 1977. From the point of view of the unions' and labor's various political parties however, such voluntary codes were insufficient. What was logically needed (according to the unions) was a new international *law*.

The original Vredeling document, approved by the European Commission on October 24, 1980, contained three major sets of provisions concerning information disclosure, consultation over proposed structural changes, and the matter of confidentiality (or "secrecy") of corporate information.

### Information Disclosure

Every six months the home office of any corporation doing business anywhere in the ten-nation EC would be required to provide various forms of operating information to the managers of all of its EC subsidiaries, and to disclose other procedures and plans likely to have a "substantial effect on employees' interests." Local managers of plants with 100 or more employees would then be required to forward this information to workers' representatives.

If local managers would not or could not forward the information, workers' representatives were authorized to appeal directly to the home office. This "by-pass" clause became a major point of contention with the multinational corporations, which seek to shore up the authority of their local managers as much as possible, to facilitate direct day-to-day operations, but also to shield the parent firm from local political and legal entanglements. If the home office was located outside the EC altogether (in the United States or in Switzerland, for example), then the parent firm was asked to designate an agent inside the EC. If it did not do so, then responsibility would fall on the parent corporation's single largest plant or subsidiary presently doing business inside the EC. (Opponents of Vredeling referred to this as the "hostage" rule.)

## Consultation

Whenever home office management was seriously considering major decisions about, for example, the closure or transfer of an establishment, a merger or acquisition, or the initiation or termination of a coproduction, sourcing, or licensing arrangement with another firm, the headquarters would be required to notify the managers of all of its local subsidiaries in Europe within forty days. Local management would then be given ten days to inform the workers' representatives in every subsidiary, who in turn would have thirty days to reply. If the workers requested a consultation with management, it had to be granted, toward the objective of resolving any differences. This was immediately interpreted as effectively meaning that the plan to close, merge, and so forth was itself subject to discussion and, presumably, to some sort of negotiation. In the event that the local managers could not or would not meet with the workers' representatives, the representatives had recourse to other alternatives, including direct appeal to the home office and legal action.

## Secrecy

The original draft of Vredeling had very little to say about the question of confidential proprietary information. There was no granting of any right by management to withhold any information from workers' representatives, on grounds of secrecy or otherwise. Employees' representatives were warned to maintain discretion, especially with third parties, such as technical consultants and American members of the various international trade union secretariats. Member states were instructed to erect their own judicial review procedures in the event of disputes.

*Debate*

The draft directive was debated during 1981 and through the winter and spring of 1982 in the Economic and Social Committee and in three different commit-tees of the Parliament. The European employers' association, UNICE, led the criticism from within Europe, while the European Trade Union Confederation became Vredeling's principal defender (by this time, Henk Vredeling had retired and was replaced as Social Affairs Commissioner by a former British Labour Party Member of Parliament, Ivor Richard). Examination of news-paper stories from the time, and interviews with many of the principals, strongly suggest that the managers of many European based corporations might have been prepared to live with the new Directive.

What no one seems to have anticipated was the well-organized lobbying effort conducted by the U.S. multinationals. It is by now commmonplace to observe that—in the words of the American Productivity Center—"American corporations . . . continue to have, by European standards, an extraordinary freedom to lay off workers when orders shrink."[2] Indeed, European business executives openly express surprise when some U.S. corporations threaten to close down a facility unless wage or other concessions are made by the workers. Even so, few in Europe expected what occurred during the fall of 1982. Led by the U.S. Council for International Business, the U.S. Chamber of Commerce, the American Chamber of Commerce in Brussels, the National Association of Manufacturers, and the National Foreign Trade Council, the "American offen-sive" (as the European newspapers called it) was "unprecedented."[3] There were numerous thinly veiled threats of a capital strike by American companies, should Vredeling pass—a tactic which even businessmen in Europe publicly criticized.

On December 14, 1982, the Parliament voted to approve Vredeling, along with dozens of amendments which effectively gutted the by-pass and hostage provisions of the original version (the Economic and Social Committee had favorably reported out the measure months before). Even though the Parlia-mentary amendments were only advisory, it was well understood by the legisla-tion's new advocate-manager, Ivor Richard, that it would be politically im-possible to proceed without somehow taking these criticisms into account. Together with his principal aide, George Trevelyn, EC staff, and outside con-sultants, Richard spent the winter of 1982–1983 redrafting Vredeling, and going "on tour" to explain it to business and political groups both in Europe and in the United States. He resubmitted a substantially revised draft Directive to the 14 commissioners of the EC in the late spring of 1983. On June 15, they approved the revision and again sent it on to the Parliament for discussion.

## Revised Vredeling Directive

The new version of the proposed Directive contains major changes, most although not all of which substantially weaken the main intention of the original draft. For one thing, the size criterion for coverage is no longer on a plant-by-plant basis, a procedure which was said to discriminate against the largest operations by systematically excluding smaller companies. The new language covers any corporation with a total of at least 1,000 workers among all of its EC facilities. Headquarters of multinational corporations are required to convey information to their European subsidiaries annually instead of every six months, and the list of data items to be disclosed has been reduced.

If the local subsidiary's managers fail to transmit the requisite information to the workers' representatives, appeal is still to the corporate headquarters, provided that the headquarters is located within a member state. If not—if the corporation is based in for instance Sweden or the U.S.—then the workers' appeal must be directed to the corporation's EC agent, if any. If there is no delegated agent, or if the quality of the transmitted information fails to meet the legislated standard, then employee representatives can go to court. They are no longer authorized, as before, to "directly approach" either the non-EC corporate headquarters or the corporation's largest EC facility. This effectively means that, in any lawsuit, the unions would have to name the managers of the local subsidiary as defendants. Thus, the controversial by-pass and hostage aspects of Vredeling have been removed, and the non-EC multinational corporate headquarters' legal shelter is protected.

Concerning consultation, the list of items on which advance notification and the opportunity for subsequent consultation must be provided has been reworded to include "modifications resulting from the introduction of new technologies" and "measures relating to workers' health and to industrial safety." On the other hand, the by-pass and hostage provisions are eliminated here, as well.

Finally, management is for the first time permitted to withhold information which it considers "secret." However, Richard refused to bow to business pressure to leave it at that. In a November 1982 speech before Parliament, he said: "The directive could be fatally weakened if the decision [on confidentiality] was left entirely to management." Instead, he called for the creation of a special EC tribunal which would "review, ex post facto, disputed cases and establish over time a body of case law."[4]

## What Next for Vredeling?

Even in its watered-down form, the Vredeling Directive continues to be strongly opposed by multinational corporate spokespersons, especially those based in America and Great Britain. Such influential organizations as the U.S. Council for International Business base their opposition on the conviction that Vredeling constitutes a step toward the eventual internationalization of collective bargaining. After another round of Parliamentary debates during the winter of 1983, the Council of Ministers took up discussion of Vredeling II in early

1984. No immediate decision is expected. Even if the directive does eventually pass into EC law, the growing political weakness of the Common Market raises serious questions about how well Vredeling II could be enforced.

Whatever happens in the next year or two, the issue is unlikely to fade. In a speech to the European Parliament, of which he is now a member, former German trade union confederation (DGB) president Heinz Vetter called Vredeling "a first attempt in international social history to submit multinational companies to supranational legal discipline," implying that it would not be the last.[5] Or, as Ivor Richard has put it: "If the multinationals succeed in killing Vredeling, there will be a Son of Vredeling and there will be a Grandson of Vredeling. Sooner or later this area of increased consultation is going to have to be worked out."[6]

## Notes

1. Morris L. Sweet, *Industrial Location Policy for Economic Revitalization: National and International Perspectives* (New York: Praeger, 1981), p. 139.

2. Ibid., p. 145.

3. Bureau of National Affairs, *Labor Relations in an Economic Recession: Job Losses and Concession Bargaining* (Washington, D.C., July 16, 1982), p. 650.

4. Ibid., p. 652.

5. Wim Albers, et al., "Hush, Don't Tell the Workers: The Vredeling Case," *Agenor*, no. 90 (May-June 1983), p. 18.

6. "Special Report: Vredeling," *Personnel Administration*, no. 35 (Sept. 1983), p. 57.

# 18
# Displacement Policies in Europe and Canada

*Philip L. Martin*

P ermanent dismissal usually follows the same pattern of advance notice, unemployment insurance, and both training and relocation allowances. Loss compensation, as in TAA and the railroad industry in the United States, is not the basis for worker protection programs in Europe. Because labor market policies are more comprehensive, displacements are anticipated and handled by agencies created to deal with dismissals.

The normal procedure for a company wanting to close all or part of a facility is to provide two to six months advance notice of its plans. In Germany and Sweden workers get their first inkling of planned dismissals from their representatives on a company's supervisory board of directors. After high-level agreement on the need to dismiss workers, the company must consult with, and in some instances, negotiate an agreement regulating dismissals with the plant-level workers' council. In some basic industries (the German steel industry), the government and the workers' council have developed a comprehensive social plan to expedite the reemployment of displaced workers. Britain's 1975 Employment Protection Act requires all public and private employers to give two to three months notice to affected unions and the Department of Employment. Advance notice normally allows a company to begin reducing its work force through attrition and subsidized early retirement, but rarely do the parties notified succeed in reversing a company's decision to close a plant partially or totally.

Advance notice is usually given to the workers' council, the union, the government employment service, and the individual worker. The minimum notice required depends on the number and characteristics of workers involved. In Sweden, at least two months notice is required to dismiss up to twenty-five workers, but six months for more than one hundred. If workers 45 and older or handicapped persons are to be dismissed, at least six months individual notice is required. French law requires at least one month's pay or one month's advance notice for all dismissed workers with at least six month's service and

Philip L. Martin, *Labor Displacement and Public Policy* (Lexington, Mass.: Lexington Books, 1983), pp. 83–90.

the approval of the Inspectorate of Labor. Since the Inspectorate can (and does) withold dismissal permission until workers find new jobs, French employers often try to find alternative jobs for redundant workers.

German law prohibits socially unjustified dismissal after a six months' probationary period of employment and requires companies planning mass dismissals (defined as more than 25 percent of the work force of small employers and 10 percent or more for large employers) to give advance notice to the regional office of the Labor Department. German companies are also obliged to notify their plant-level workers' council and to negotiate the number and timing of dismissals and the assistance to be offered to displaced workers. The workers' council opinion on the need for dismissals must accompany the employer's advanced notice petition to the regional labor office.

Advance notice is best suited to the production planning of large multi-plant firms. Smaller firms supplying large companies are often at their mercy. If a small auto supply firm is not informed of an impending auto assembly slow down, it can find its market gone with less than six months notice. The requirement to pay workers' wages at least six months sometimes bankrupts smaller firms. Although advance notice to supplier companies has been discussed, no country yet requires one company to inform another of its production plans. Companies continue to rely on word-of-mouth to stay informed.

Few countries require severance pay for all dismissed workers. Britain's Redundancy Payments Act (1965) requires employers to pay dismissed workers who have had at least two years' service a minimum one and one-half weeks of severance pay for each year of employment with the firm. This lump-sum severance pay does not reduce a worker's normal UI benefit. Employers who dismiss workers according to statutory procedures recover 41 percent of total severance pay from the national unemployment insurance fund.

British unions can and do negotiate additional severance pay. The British Steel Corporation made $17,000 to $35,000 lump-sum payments to workers dismissed when it closed its Corby (England) steel mill. In addition, the local housing authority reduced rents in public housing by 60 percent for unemployed workers. A joint European Economic Community (EEC)–British Government fund either brings a displaced worker's wage up to 90 percent of its former level for seventy-eight weeks if he takes a lower paying job or continues to pay a displaced worker his full wage for one year if he enrolls in an approved training program. In most other European countries, severance pay is fixed by negotiation and is subject to appeal to labor courts. Belgium is generally acknowledged to have the highest severance pay; it can cost $250,000 to dismiss a worker.

Advance notice is usually coupled with the duty to justify dismissals. One response to advance notice of mass dismissals is a strategy of delay. If workers' councils, unions, government, or individual workers disagree with the dismissal decision, they can appeal the employer's decision to labor department offices and labor courts empowered to delay actual dismissals. These courts can order that dismissals be delayed and that the firm retain its workers by paying short work-week benefits or getting government aid to subsidize inventory accumulation.

The need to justify dismissals usually means that the affected workers and unions at least grudgingly accept the final decision. The time that elapses until justification is secured permits a variety of targeted public and private assistance efforts to be offered to workers displaced and persuades employers to provide additional assistance in order to discourage time-consuming appeals.

The advance notice requirement and appeal options give workers facing dismissal a limited job property right that employers are willing to buy out with severance pay to induce quits. Employers sometimes induce voluntary quits by offering golden handshakes or severance pay. Volkswagen, partially state-owned, announced plans to dismiss 25,000 workers in one area in 1974. Opposition to the dismissals prompted a golden handshake program that offered up to $12,000 each to voluntary quitters, a scheme intended to appeal to women and foreign workers, who left the country after quitting. The Ford and Opel plans to trim their work forces 10 percent in 1979–1980 were estimated to cost at least $15,000 per voluntary quit. A 1981 agreement in the German cigarette industry encouraged early retirements by giving workers aged 63 or older the choice of 75 percent of their earnings for not working, or full wages for working twenty hours weekly (older workers could also work forty hours for full pay). Even with these ostensibly generous bribes to quit, some German students of displacement protections believe current laws and rules fail to protect workers.

Workers receiving severance pay also get regular UI benefits. These UI benefits usually replace a higher fraction of an unemployed person's previous take-home wages for a longer period than is the case in the United States. U.S. UI benefits replaced an average 50 percent of gross earnings in 1975; most European countries replace more than 60 percent of previous net earnings. Further, the real value of European UI benefits is increased by the continuation of health insurance and pension contributions. European UI benefits normally continue for at least one year, after which still-unemployed workers are eligible for means-tested unemployment assistance.

European countries operate extensive and expensive active manpower programs. Sweden spends almost 10 percent of its national budget—$2 billion annually—to train and retrain workers (a roughly equivalent U.S. expenditure would be $50 to $60 billion instead of the $8 billion spent through the Comprehensive Employment and Training Act (CETA) in 1981). Germany spends $1.5 to $2 billion annually for retraining and encourages participation by supplementing normal UI benefits to bring the earnings of training program participants up to 90 percent of their previous net earnings. Thus, displaced workers have superior protection on three fronts:

1.  Advance notice requirements and appeal options that encourage employers to bribe them to quit voluntarily.

2. Relatively generous regular UI benefits that also include maintenance of health insurance and pension coverage.

3. Eligibility for training and retraining in readily available government programs with monetary encouragement to participate.

Displaced workers may find jobs with new or expanding employers who have been induced to create jobs by industrial and regional policies. The tripartite Swedish Labor Market Board (AMS) coordinates labor and industrial policies, permitting a single agency to weigh trade-offs between dismissal and a variety of investment and employment subsidies, on-the-shelf needed public works projects, and training coordinated with the needs of employers. German industrial and regional policies do not permit Swedish fine-tuning but do subsidize investments in specific areas. Most European countries try to coordinate labor and regional policies.

European countries spend relatively more on industrial and labor policies than does the United States. Most European countries believe their extensive programs are successful, even though none have made any significant effort to apply detailed cost-benefit analysis to them. A general fear of unemployment and a tradition of cooperation between business, labor, and government portend continuation of the current mix of labor and industrial policies.

The fact that both workers and employers are relatively well-organized expedites industry-wide adjustments to change. If an industry faces structural adjustments (for example, autos or steel), the machinery to work out an adjustment program is already in place. Legislation that permits government to extend agreements to all firms in the industry encourages firms to compromise.

In sum, most European employers are encouraged to avoid dismissals. Before putting workers on temporary lay-off, employers may:

1. put workers on short work weeks, paying them for twenty to thirty hours of work and having government pay the balance of the weekly wage;

2. build up inventories that are carried with low-interest government loans and/or accelerate the delivery of government purchases;

3. receive grants to train and retrain workers in a plant during low production times.

If employers want to dismiss workers, they must first notify the affected union, the plant-level works council, the government, and the individual worker. Any of the four notified parties can find the justification inadequate and delay actual dismissals.

Most employers wanting to dismiss workers eventually succeed. Dismissed European workers:

1. often receive severance pay that varies with length of service but is often in the $10,000 to $20,000 range;

2. generally have a higher proportion of their after-tax earnings replaced for a longer period by the regular UI system than do U.S. workers;

3. usually retain health insurance coverage and often find their employer or the government continuing to make (high) pension contributions;

4. are eligible to participate in extensive retraining programs and/or obtain relocation assistance;

5. may find local work if other employers move to the area to take advantage of the investment, tax, and training subsidies offered through industrial and regional policies.

There are no European-wide rules that standardize job protections, but the European Economic Community (EEC) does maintain a special fund to compensate workers dismissed from coal and steel jobs. The EEC funds help displaced workers readapt by supplementing cash assistance payments and providing relocation and retraining aid. EEC coal and steel aid varies by country and age of the dismissed worker; for example, in 1979 the average EEC payment to a British worker was 930 pounds ($2,232). EEC payments may continue for up to two and one-half years.

Pending EEC legislation would standardize job protection legislation. An EEC redundancy directive requires employers to give advance notice of dismissal to individual workers, unions, and labor department offices. It also requires justification and negotiation over planned dismissals and sets minimum standards for severance pay. It should be noted that most northern European job protection programs already exceed these minimum EEC standards. Severance pay requirements are collected from the parent company if a country's branch plants are closed. The American Badger Company of Massachusetts closed its Badger Belgium subsidiary and left only the subsidiary's assets to cover severance payments. European courts ruled that the U.S. parent had to provide eight more months' severance pay for each worker.

Investment subsidies to create new jobs may be standardized also. Ford's 1979 auction of an auto plant and its 4,000 jobs to the countries offering the fattest subsidies (Britain and Spain) prompted the European Commission to request that all countries offer similar job-creation subsidies to avoid competitions that only increase company profits. The Organization for Economic and Cooperative Development (OECD) urged its members (including the United States) to adopt positive adjustment policies that phase out inefficient firms and aid individuals, not subsidize companies to preserve jobs.

Despite comprehensive assistance programs, the need to reduce employment in basic industries, especially steel, autos, textiles, and mining, has prompted special adjustment assistance programs. The French government took over two money-losing steel companies in 1979 and bought out 20,000 redundant workers with early retirements and severance pay. Displaced workers 55 or older get 80 to 90 percent of salary for one year and 70 percent per year thereafter; younger workers get a $12,000 severance payment. Workers are offered new local jobs and have the right to reject the first two offers without losing benefits. The fact that the state has an ownership stake in many troubled French industries promises new pressures for these special displacement programs.

Will European governments continue adding SPPs and create two-tiered UI systems like the United States? No general trend is yet discernible. Henk Vredeling, the EEC Commissioner for employment and social affairs in 1980, proposed an EEC-wide rule that would require companies with fifty or more employees to prepare an annual report for workers that informed them of any company plans that might affect their employment. Some observers fear that if the EEC does not establish European-wide standards, partial government ownership may force individual firms to offer special adjustment assistance, especially if large numbers of workers face dismissal and dim prospects for recall. If so, then even mandatory advance notice, negotiation, and generous UI cannot overcome entrenched job property rights.

Canadian job protections parallel those in Europe. The Canadian Manpower Consultative Service sends teams to communities experiencing or threatened with mass dismissals. Tripartite committees are created to help cope with dislocation. This low-cost joint consultation is supplemented with a $290 million, six-town pilot program that subsidizes job creation and encourages worker relocation. Among the subsidies available is a $1.68 hourly wage payment to employers hiring laid-off workers over forty-five. Laid-off workers fifty-four and older unable to find another job are eligible for full unemployment benefits until pensions are available at age sixty-five.

Consultation and subsidies are being reinforced with tougher labor laws. The Canada Labor Code, which covers the 10 percent of the work force employed in federally regulated industries (for example, banks and airlines) and sets the standard for most provincial labor codes, requires companies to offer at least five days severance pay after one year of employment and an additional two days pay for each additional year of service. Companies with more than fifty employees must provide at least four months advance notice of lay-offs.

In Japan, worker protection from lay-offs during business cycle downturns is a function of place of employment. About one-third of Japan's work force is employed by large firms that promise lifetime employment. Small firms, often subcontractors for the larger firms, serve as work force shock absorbers by making their hiring and lay-off decisions contingent on their workloads, derived from orders placed by larger firms. The lifetime system threatened to unravel after the 1974 recession, so the Japanese government decided to subsidize large firms by paying one-half of normal wages for workers who would have been laid off if employers agreed to retrain them instead. If a majority of firms in a structurally depressed industry appeals for assistance, firms become eligible for innovation loans, and workers receive extended UI benefits and additional job search assistance. In 1980, thirty-nine industries were designated structurally depressed.

## Assessment

European job protections are more comprehensive than those available to the typical U.S. worker. Tripartite economic planning, a commitment to full employment, active trade unions, plant-level workers' councils, and a history of exposure to the dislocations transmitted via international trade have made programs to assist displaced workers an integral part of European economic policies.

Most European economies had labor shortages between 1960 and 1973, expediting reemployment for anyone displaced. Today displacement is minimized because firms have to justify lay-offs and dismissals to their own board of directors (including worker representatives in Germany and Sweden), to plant-level workers' councils, and ultimately to labor courts. Instead of dismissal, many governments attempt to keep workers at work for twenty to thirty hours weekly and pay partial wages for the nonworked hours. Some governments avoid lay-offs by subsidizing inventory build-up, stepping up government purchases, or starting public works projects held in reserve.

European workers who are dismissed usually have at least two-thirds of their after-tax earnings replaced, have their health coverage extended and their pension contributions continued, and often receive severance pay. Most European countries operate extensive retraining programs and many subsidize worker relocation. Employment policies are linked to regional and industrial policies that encourage firms to create new jobs in depressed areas with investment and training subsidies.

The European job protection system worked remarkably well so long as unemployment was minimal and some fraction of lay-offs were absorbed by foreign workers returning home. Since 1974, unemployment is higher and the alien guestworkers are no longer willing to return and be denied reentry. Some employers want to emulate U.S. lay-off policies, arguing that dismissal must be an employer right, not a privilege granted by labor courts. Furthermore, they argue that subsidized housing in depressed areas and housing shortages in regions with jobs make redundant workers reluctant to move. Subsidizing the entry of firms into depressed areas is too expensive. One British critic argues that the delays and rigidities of European job security systems that promoted labor peace in the past now mean that "governments for the most part have engaged in adjustment-resistance policies . . . rather than . . . adjustment-assistance policies."

European economies now confront severe and persistent unemployment problems. This joblessness may persist because many of the experienced unemployed worked in the shrinking steel, chemicals, textiles, and shipbuilding sectors. Even after economic recovery, many experienced unemployed workers will not be recalled to their old jobs. However, European workers are reluctant to move to areas with expanding industries because of their sense of community, housing shortages, and generous unemployment benefits that do not require mobility. If job security and mobility systems are not revised, the duration of unemployment in Europe is likely to lengthen. For example, in France and the United Kingdom, about one-half the jobless were unemployed six months or more in October 1981.

The International Labor Organization (ILO) has joined the discussion on the side of those who argue that more flexibility and "anticipatory adjustment assistance" must replace "widespread adjustment assistance." The ILO finds that the constitutional right to work and the difficulty of securing dismissals encourages employers to retain redundant workers. However, the ILO is expected in June 1982 to endorse a proposal that companies closing plants consult with affected workers. Job security is even more important in Eastern Europe. The ILO estimates that only 10 to 20 percent of all Soviet workers displaced by technological change are actually separated from their jobs. To the extent that worker displacement results from Third World imports, some European countries propose a linking of free trade and fair employment standards.

Developing countries realize that positive adjustment policies in industrialized nations can promote the adjustment of dislocated workers and stave off pressures from protectionism. One former trade negotiator argues that continuous efforts to liberalize trade in the United States are necessary or else impacted firms and workers are likely to succeed in imposing further restraints on trade. On these grounds "TAA could be considered a foreign assistance program because, by promoting freer trade, it facilitates a transfer of resources to the developing countries."

High unemployment, on the other hand, demands attention before it wreaks irreparable damage on Europe's social and political fabric. The 10 million unemployed Europeans in a work force of 110 million mean Europe is experiencing its highest postwar unemployment rate (9 percent). The worrisome upward trend in youth unemployment is prompting more calls for reduced hours, work sharing, on-the-job training, and inventory accumulation to create jobs. However, employers complain that more labor market interventions will undermine their competitiveness in international markets.

# 19
# Comparison of the United States, Sweden, and France

*Gregory Hooks*

T he economic stagnation of the past decade led to a wave of factory closings throughout the United States, with thousands of workers permanently losing their jobs. These events have sparked a debate in this country over the responsibility of the government in two distinct policy areas: (1) promoting economic growth; and (2) protecting the economic security of workers, their families, and communities. Although the definitions of the problem and the proposed solutions to it are numerous, these can be condensed into three options:

1.  Free market. Conservative advocates of the free market argue that job security can be ensured, if at all, only by permitting the admittedly harsh discipline of the market to perform its function. The government's attempt to contravene the market through social programs only delays economic recovery, hence prolonging the suffering, whatever the intentions of policy-makers. For conservatives, including much of the Republican party and the present administration, the solution is to get the government out of the way to permit a purging of the economy and a restoration of the conditions for growth.

2.  Industrial policy. Industrial policy advocates believe that the government not only can play an active role in restoring economic well-being, but assert that its failure to intervene, especially in capital markets, is part of the problem. Because the U.S. government has failed to shape the flow of investments, capital has gone to inefficient and even socially harmful activities and has resulted in high unemployment. From this liberal point of view—one publicly identified with Felix Rohatyn and prominent members of the American Federation of Labor and Congress of Industrial Organizations (AFL-CIO) and the Democratic party—the solution requires the

Gregory Hooks, "The Policy Response to Factory Closings: A Comparison of the United States, Sweden, and France," *The Annals*, American Academy of Political and Social Science, 475 (Sept. 1984), pp. 110–124.

creation of an agency modeled after the Reconstruction Finance Corporation as a means of bringing a longer-term view to capital markets.

3.  Social welfare. This alternative, a self-consciously leftist one, agrees that the government should be doing more, not less, but argues that the interventions must concentrate on social welfare, not investment flows. That is, the proper role for government is not to provide still more subsidies to those already well-off; rather it is to buffer the workers and communities hardest hit by economic dislocation.

Although spirited and informative, these debates have not taken full advantage of solutions offered in other industrialized nations. Among industrialized nations, the United States has been the least reliant upon government interventions throughout the postwar period. Predictably, the second and third alternatives have been evaluated through lenses inherited from Adam Smith and judged to be major and untried changes in this nation's approach to economic policy. But, in point of fact, they are not merely the dreams of reformers; they have actually been implemented in other industrialized nations.

In France, economic planning based upon the state's influence over capital flows has been pursued, while in Sweden social welfare policies have served as the cornerstone of economic growth. Although it is clearly impossible to apply the lessons of France and Sweden directly to the U.S. case, valuable insights can be gained by considering and comparing the experiences of nations facing similar challenges. This article provides such a comparison. The bulk of the article explores the response to factory closings in the United States, France, and Sweden. In each case the discussion addresses institutional developments of the government, especially the ability of the government to implement industrial and welfare policies, and the activities of the two social groups most affected by these developments—business and labor. In turn, the conclusion identifies the lessons to be drawn from this comparison for current debates in the United States.

## The United States: Nonresponse of a Domestically Weak State

Listening to the debates over factory closings and economic policy more generally, one gets the impression that the government is able to intervene in any manner it chooses, as long as it makes a firm commitment to do so. But the reality is that the government's choices are limited by its own institutional capacity—or lack thereof—and the resistance or assistance of groups in society. And this point is particularly salient for the United States, where the government is exceptional among the Western industrialized nations for its weakness and the incoherence of its domestic economic and industrial policies. The federalist U.S. Constitution explicitly vests the 50 states with their own taxing capacity, the ability to maintain a staff independent of the central government, and gives one of the houses of Congress, the Senate, to answer to their interests. Not only does this contribute to confusion in the policymaking process, but implementation of federal programs becomes a bargaining process among numerous autonomous decision makers. The Constitution further divides power by mandating a system of checks and balances that pits one federal agency against another, thereby offering still other impediments to the emergence of far-reaching and coherent domestic economic policies.

It is not the case that economic planning has been unthinkable in the U.S. context. Throughout this century, reformers have engaged in state building in an effort to build a

> new democracy organized around executive leadership, . . . a new political economy based upon central planning in cooperation with industry, . . . and grounded in scientific principles of public administration.[1]

These principles have been shared to a greater or lesser extent by the reformers of the Progressive, New Deal, and post-World War II eras. But that state building has been only partial and contradictory in character. As a consequence, the attempts at planning, from Roosevelt to Carter, have been incoherent in conception and nearly always thwarted in execution.

In fact, even more modest forms of economic management in this country are hampered by the bluntness of the federal government's fiscal and monetary tools. Outside of the defense sector, the government usually makes fiscal interventions at the aggregate level—that is, interventions at the sectoral or individual firm level are precluded. Only since 1960 has the president gained any flexibility in making timely fiscal policies, and that remains sharply restricted to this day. At the same time, monetary policies are largely controlled by the quasi- (some say completely) autonomous Federal Reserve Board, and individual banks have the capacity to create money on their own. With blunted and only partial control over fiscal and monetary policies, the U.S. government is unable to shape capital markets and investment flows, nor has it ever gained significant control over labor markets; beyond jawboning, its major role is to ensure that business and labor bargain in good faith. Taken together, the lack of powerful policy tools has severely constrained the possible policy responses to factory closings.

While the bluntness of U.S. economic policies can in large part be traced to tendencies intrinsic to the federal system, the power of the business community has also been important. During the first decades of this century, U.S. business was cartelized into major corporations dominating multiple markets, employing thousands of people, and engaging in economic activities around the country and the world. While big business was emerging, however, the federal government remained small and its influence constrained. There were no federal laws regulating security exchange; income taxation was still unconstitutional; and the federal bureaucracy was small, transient, and weak. In other words, big government in the United States, unlike Western Europe, grew up in the shadow of big business.

The timing of the growth of the federal government, especially the resistance of an already powerful and entrenched business community, meant not only that the government's economic policy tools were blunt, but that the levers of control of economic activity would be vested in the institutions of the business community—corporations. The securities market, though heavily regulated, provides U.S. corporations the opportunity to acquire external financing in an impersonal finance market, thereby reducing the influence of a central bank as in France and Japan or private universal banks as in West Germany. In addition, negotiations with labor, whether organized or not, take place within individual firms or plants. No national tripartite corporatist institutions have been created, nor are any proposed. The subjects labor may address in these negotiations are specific to wages and fringe benefits, with working conditions carrying less importance. Only in the rarest of circumstances are broader questions of the organization of production or investment decisions addressed.

Relatedly, workers in the United States are in a relatively weak position, economically and politically, and that position is deteriorating. Less than 25 percent of the labor force is unionized, and the proportion falling. In part this decline is a direct result of the past decade's wave of factory closings. Further, the plant-by-plant bargaining procedures and the concentration of the organized work force in the closed-shop states of the North and Great Lakes region has reinforced both the unions' weakness and their concentration on bread-and-butter issues.

The difficulties confronted by the U.S. labor movement in the economic sphere are exacerbated by political constraints. Labor has been obliged to throw its support to the multiregional, multi-interest group Democratic party due to the lack of a viable labor or social democratic party in the United States. This has meant that the favored party is not directly responsive to the interests of labor in the policy formation process, while the ineffectiveness of the government's policy tools means that policy execution is hampered during implementation. As a result of these multiple economic and political limitations of organized labor in the United States, the social group most interested in controlling the tempo and severity of factory closings has had little success in doing so.

This, then, is the context in which U.S. policymakers developed a nonresponse to factory closings. As noted, our economy has never been planned during peacetime. The postwar consensus has been built around growth and productivity in private sector corporations, with a minimum of direct government intervention. When policymakers in the 1970s considered the development of a coherent industrial policy as a means of controlling factory closings, they confronted the prospect of a vast overhaul of the core institutions of the federal government that affect economic activity—namely, the Treasury, Federal Reserve Board, and Department of Commerce—as a first step. Further, the inevitable resistance of the powerful business community would have to be overcome if such policies were to work. On both counts, such a sudden shift in the direction of policy would represent a major rupture, a rupture that was never risked. Though often articulated and promoted by prominent union officials and Democratic politicians, this policy response has never emerged as a serious policy alternative, nor is it being considered by either the current administration or Congress.

If factory-closing legislation were to emerge in the United States, it would have to come in the form of social welfare provisions and regulations. However, the underdevelopment of the U.S. welfare state has worked against such an outcome. Social welfare reformers have lacked the support of either a powerful aristocracy or a socialist movement. As a consequence they have met with only partial successes and many failures. This leaves private sector institutions such as corporations with the central role to play in the allocation of labor and in providing the means of subsistence; the state intervenes only to ensure that individuals get a fair shake and to provide custodial care for the victims of the market. Though expensive, U.S. welfare policies do little or nothing to prevent victimization in the first place and have had little success in rehabilitating those in need. Further, these sizable expenditures are nonproductive in the sense that they do not systematically improve the quality and availability of the labor force and represent a net drain on the funds available for investment.

## Extant U.S. Responses

The incoherence of U.S. industrial and social welfare policies imposed powerful constraints upon this nation's response to the most recent wave of factory closings. Only after capital disinvestment was well under way in the northeastern and Great Lakes areas was such legislation even considered. By this time, the productive infrastructure was already outmoded and reindustrialization posed an expensive and difficult challenge. Further, the policy proposals never challenged private control of the investment process; their intent was to impose political constraints on these decisions for the purpose of protecting workers and communities.

The only serious attempt at national factory-closing legislation came in 1974 and was made by then-Senator Walter Mondale and Representative William Ford. These legislators were Democrats from Minnesota and Michigan, respectively, states in which unions are strong. The proposed legislation would have provided tax subsidies and other financial aid to factories at risk of closing. The bill proved unpopular to all concerned. The business community was hostile to any intervention affecting their rights to reinvest as they saw fit, while labor supporters and others on the left objected on the grounds that it represented a tax giveaway to any firm willing to declare they were thinking about closing a plant. The bill was never reported out of committee.

More recent national legislation has treated factory closings in an oblique manner. The Trade Adjustment Act (TAA) provides some severance and retraining benefits to workers losing their jobs due to international competition; and Title III of the Job Training Partnership Act (JTPA; 1982) is earmarked for the retraining of redundant workers. Not only do these more recent laws do nothing to prevent factory closings, but their ability to buffer the workers affected has been minimal. The TAA benefits were earmarked for a minority of the unemployed in the first place—that is, those losing their jobs as a direct consequence of international competition—and in implementation, a strict interpretation has been employed. As a result, very few workers have received these benefits. At the same time, Title III has been allocated only $0.1 billion of the JTPA's overall budget of $3.7 billion.

In a perverse sense, the policy response directed toward preventing factory closings has come at the state and local levels. Instead of buffering workers or attempting to redirect capital flows, the federal government leaves states and communities to compete against one another in attracting employers—smokestack chasing. This no-win competition results in a further drain on already strained public budgets and has little success in creating or retaining jobs. Several states, including Ohio and Michigan, have given serious attention to legislation to impede plant closings. Early warnings of anticipated closings and financial buffering of affected workers and communities were central concerns in these bills. Businessmen criticized the proposals because they would undermine the business climate; and these bills were ultimately defeated in the late 1970s.

With the defeat of these more extensive factory-closing laws, state-level legislation was directed toward minimal protection of workers and communities. Maine and Wisconsin require advance notification. California has created the California Economic Adjustment Team (CEAT), which works with management and labor on the one hand, and federal, state, and local agencies on the other to respond to individual closings. Though laudable first steps, these various state programs cannot prevent factory closings and have only scratched the surface of easing the transition problems faced by workers and communities.

And this is the situation at present; by default the United States has adopted a laissez-faire response. Though, as Gordus documents, a few communities have taken important steps to buffer the transition of laid-off workers, most workers are obliged to bargain with individual firms for job security, and the prospects for protecting a large proportion of the work force using this approach are dim. The likelier outcome is that more powerful monopoly-sector unions will succeed in gaining some job security from the larger corporations in exchange for major concessions, but the nonunionized and competitive-sector work force will be left without protection of any sort. The recent agreements between the United Automobile Workers and the big automakers—Ford and General Motors—represent an example of this trend. Factory closings remain a private decision despite their far-reaching and deleterious consequences for workers, their communities, and the larger economy.

## Sweden: The Social Democratic Response to Factory Closings

If the United States represents an example of a free-market approach, then Sweden provides an example of a social-welfare–based response to factory closings emerging from a long-standing commitment to those goals. The Swedish state has evolved into a highly centralized and coherent organization. After some striking defeats in the eighteenth and nineteenth centuries, the Swedish monarchy was obliged to forgo its penchant for military adventures and to focus primarily on domestic developments. In this vein, the state played a very active role in the transformation of Sweden from the "poorhouse of Europe" in 1880 into the ideal-typical social welfare state of the postwar era. From the turn of the century onward, the state financed and controlled key infrastructural projects while establishing and expanding tripartite corporatist institutions to coordinate labor markets.

During the same period that it intervened from above to promote industrialization and rationalization of the economy, the Swedish state proved permeable to movements from below. Through a series of reforms, universal suffrage and recognition or organized labor in the realm of collective bargaining and in national economic policy-making were secured. Also a number of social welfare programs were created before the Great Depression, and the central bureaucracy established a tradition of efficiency and centralization in administering these.

This tradition of buffering Swedish citizens from the anarchy of the market was inherited by the Swedish Social Democrats (SPD) when they became the ruling party during the depression. Also important in this discussion is the power of the Swedish labor movement and its close relations with the ruling Swedish Social Democratic party. Not only is over 90 percent of the labor force unionized, but the Landsorganisationen (LO) has managed to fuse much of organized labor, especially the bluecollar unions, into one organization. In turn, the labor movement has provided the SPD with a consistent plurality in the elections, hence a respected claim for influence over policies chosen. Consequently, the Swedish state is not only coherent and centralized enough to formulate and execute ambitiously reformist policies, but due to its accessibility to organized labor and the SPD, it has been used for precisely those ends.

The institutional developments outlined above contributed to an innovative policy response to the depression and to postwar economic development. The Swedish response to the depression became one of deliberate and far-reaching aggregate demand management through the construction of an extensive social welfare state. Sweden's welfare policies go beyond caring for the victims of the market; they are directed toward the prevention of victimization in the first place, sustained economic growth, and full employment. But the Swedes have faced difficult challenges in pursuing this approach to economic policy. In the 1950s an inflation rate that was higher than desired threatened these social welfare programs.

Instead of retreating, the Swedes adopted a proposal offered by LO economists Gösta Rehn and Rudolph Meidner. Beginning in the mid-1950s and continuing into the 1970s, they expanded social welfare expenditures and job security while making the economy more efficient and productive. Essentially the plan called for a deliberate squeeze on profits through taxation and uniformly high wages so that only the more efficient firms could survive. A sizable portion of the revenues generated were controlled by the Labor Market Board, where the LO and the SPD have a strong voice in making allocation decisions. Primarily, the funds have been used to provide (1) relocation benefits, extensive retraining, and severance benefits for workers affected by redundancy; and (2) investment capital for projects deemed worthy by the board. While forcing a rationalization of the economy, these active labor market policies also improved the quality and the mobility of the work force.

Sweden, like other industrialized nations, faced economic stagnation in the 1970s. However, the response to the situation reflects these institutional developments and an ongoing commitment to social welfare. In 1972 the employers' organization and organized labor negotiated a national agreement dealing with shop-floor rationalization. Both parties recognized a common interest in increasing productivity and remaining competitive, and the agreement reflects this consensus. In return for facilitating the rationalization of production, representatives of the LO are given prior knowledge of and input into changes that affect job security.

During the same period, legislation passed in 1974—updating an act passed in 1971—specifically dealt with factory closings. This legislation requires:

Full earnings paid to displaced workers for two years after layoffs, provided the Labor Market Board cannot find suitable alernative employment.

At least one month's notice of termination, six months' for establishments with 100 or more workers.

Leave with pay for workers searching for alternative employment during the notice period.

An enhanced role for the LO in selecting which and how many workers would be rendered redundant.

Additionally, the Codetermination Act of 1976 gives unions advance notification of plant closings and an opportunity to develop alternative plans to prevent job loss.

The Swedish response to factory closings outlined here exhibits the historical developments and policy streams cited earlier. These policies have successfully softened the blow of permanent layoffs, but in and of themselves they cannot prevent factory closings or ensure the creation of new employment opportunities. As a result, the Swedes are debating policies explicitly designed to foster economic growth, and these policies focus on the investment process. In part due to the high taxation and wage rates associated with the active labor market policies, and in part due to the limits inherent in private capital markets, needed investments, especially venture capital, have been in short supply. The Labor Market Board, the agency implementing these labor policies, began in the early 1970s to provide grants and loans to foster the creation of employment opportunities. This contributed to a significant expansion of the Labor Market Board's funding, and in the recession year of 1977 it amounted to 9 percent of the government's total budget.

Now the Swedes are exploring a gradual but vast expansion of the public sector's influence over capital flows, thereby ensuring adequate capitalization of desired projects. In other words, Sweden has recognized the limitations of social welfare policies as a means of fostering economic growth. But building upon their institutional abilities in the arena of labor and social welfare policies, the Swedish government and citizenry are considering expanding their capacity to influence investments, thereby complementing existing programs. By playing this dual role, the government hopes to ensure economic growth while maintaining the emphasis on equity.

## France: A Minimalist Response by an Interventionist Government

Though the French government is famous for intervening in the economy, unlike Sweden, those interventions have not concentrated on social welfare policies. Because laissez-faire policies have never characterized postrevolutionary France, Hayward finds it ironic that a French term is used to describe a doctrine that opposes governmental involvement in economic affairs. Not only is the political system highly centralized, being concentrated in or answering to Paris, but there are powerful economic policy tools at the disposal of the central government. The French banking system is directly controlled by the state, and using its control over capital flows, tariffs, and a variety of other policy tools, the state has consistently protected its client groups, the business community and peasantry, from the market. The uniqueness of the postwar era is that the French state used a system of indicative planning, not to simply buffer client groups, but deliberately to centralize and modernize key industries and to make the entire economy more vibrant. In doing so, the Commissariat du Plann relied on the coordination of the state-controlled banking system as well as sectoral and firm-level state interventions to encourage compliance with the plan. In what is in large part a reflection of a successful planning effort, the French maintained a high growth rate throughout the postwar period and doubled the United States's productivity increases.

While the French planners were insulated from society's pressure groups, the goals and outcomes of the planning effort reflected a bias in favor of the bureaucracy's traditional client group—the business community. The stress was on profits, investments, and productivity, not equity and democracy. A high rate of inflation was tolerated for most of the postwar period, while gains in the workers' wages were constrained. As a result, the rise in the real income of workers was quite modest despite impressive economic growth and productivity increases.

This bias is equally evident in the political sphere. While employers were regularly consulted in the planning process, organized labor was consistently ignored. In other words, no tripartite corporatist institutions to incorporate labor were created despite the interventionist nature of French economic policies. And the ability of French labor to negotiate with individual firms has been hampered by the fact that only 25 percent of the workers are unionized, and relations among the Catholic, Communist, and Socialist unions are often hostile. Further, the government permitted employers gradually to back out of collective bargaining beginning in the 1950s and to refuse to negotiate at all in the early 1960s. As will be described subsequently, only in response to the events of May 1968 did the government push employers back to the bargaining table.

In parallel fashion, the French welfare state has only recently been as extensive as it has been elsewhere in Western Europe. A whole range of statistical indices—for example, the Gini coefficient, net redistribution of wealth through taxation, and so forth—reflect this inegalitarian feature of French society. In the areas of health, welfare, and education the French state has only begun to take on an active role in buffering its citizenry from the market. With regard to the issue at hand, work-force training and compensation for workers displaced in the process of rationalizing the economy were given scant attention by French planners prior to 1968. Before the mid-1960s vocational training was archaic and covered by legislation passed in 1919. Despite rhetoric to the contrary and the rapid modernization of the economy throughout the postwar era, state planners and administrators were willing to tolerate an inefficient and privately controlled labor market and the virtual absence of vocational training.

The heightened pace of mergers, conversions, and redundancy in the 1960s, many of these occurrences being a direct result of the government's planning initiatives, brought on a period of labor tension, a tension compounded by the employers' refusal to negotiate. The state pushed for a resumption of collective bargaining in hopes of achieving the plan's goals and maintaining political stability. It was to this end that the state revamped vocational training in 1966. The Vocational Training Act established the National Employment Fund and made retraining available to displaced workers. Simultaneously, the state pressured the employers to begin bargaining with labor once again and to deal with the issue of job security. The Lorraine agreements were reached in 1967; they dealt with a few of the steel industries and provided for notification to workers facing impending job loss, established a set of priorities in choosing those to be laid off, and promoted intrafirm transfers. Retraining of affected workers was to be supplied by the state. This agreement was weak in any case, and few firms agreed to these provisions prior to May 1968.

The issue of job security, in addition to wage increases, was, however, central to the concessions given to labor in the return to normalcy after May 1968. The Vocational Training Act was updated to provide more extensive training and to make it more available. Further, the state mounted pressure on the employers to engage in national and multi-industry collective bargaining. In February 1969, the Grenelle Agreements not only provided the warning period established under the previous bargain but did so at the national level and for most industries. Further, workers notified of an impending layoff were awarded full pay while being retrained in the government's vocational training facilities.

Economic planning became less rigorous during the 1970s for several reasons. Internal to the state, the Treasury, the key agency in making the planning of the 1950s and 1960s possible, became both less willing and less able to coordinate the other agencies that participated in the implementation of the plan. Furthermore, the very success of previous planning efforts spawned major French corporations that were fully competitive in the international market. They were less dependent upon the state for their survival, hence less amenable to cooperating with government planners.

During this same period, the approach to job security was built upon the post-1968 Grenelle Agreements. These proved durable, surviving in roughly the same form to the present, but neither labor nor business people are satisfied with them. The policies provide only partial protection and retraining, so workers still face the threat of permanent layoffs and lack guarantees of alternative employment. With unemployment hovering near 10 percent over the past decade, that threat has proved all too real. On the other hand, the business community protests the impediments to the reorganization of production that the job security provisions have caused. Ultimately, employers retain the option to hire and fire without consulting workers, but the necessity of providing a warning period and a paid leave of absence for the workers' job search and retraining has meant that exercising this option is both more expensive and more circumscribed.

The strong institutional traditions of the French government have ensured a continuation of these basic tendencies even under the Socialist party (PSF) government elected in 1980. The PSF's growth policies marked a return to centralized planning—with nationalizations and strategic use of credit used to increase the government's influence over economic activities in the private sector. Though the PSF emphasizes the importance of full employment and job security, the policies to date have been aimed at overall employment rates and increasing the earning power of employed workers.

The government has been successful in these larger efforts, but the security of individual workers has not been protected. In fact, the Socialist government has evidently felt less constrained by the Grenelle Agreements than its Gaullist predecessor. The government has closed a number of its own plants in core industries and has pushed the private sector to do likewise, all in an effort to modernize. Though there are some severance benefits, unemployment insurance, and partial retraining of workers rendered redundant as mandated in the Grenelle Agreements, the brunt of the burden imposed by these government-backed layoffs is borne by individual workers. Further, newer immigrants to France, older workers, and those with handicaps face a particularly difficult time finding alternative employment.

Recent events have tested this basic response to permanent layoffs; in fact, it has led to a legitimacy crisis for the PSF government. The unemployment insurance fund became less solvent as unemployment mounted. By the end of 1983, the government was negotiating with the unions to lower the benefits during the same period that it was contributing to a growth in permanent layoffs. The problem exploded into crisis with the 1983–1984 strike and riot at Peugeot. At issue was the government-backed restructuring that resulted in a permanent job loss for hundreds of employees. In a manner characteristic of French politics, workers excluded from decision making adopted extraparliamentarian tactics to attack private and public sector administrators of unpopular policies.

The long and violent strike brought bitter infighting among the unions and to the PSF government. The government began negotiating a wide range of issues with the unions, not just the level of unemployment benefits. In February 1984 an agreement was reached. While continuing to promote modernization of core industries, even providing massive capital infusions to the firms involved, the workers' interests were addressed as well. For the first time, a systematic training and transfer of workers to new employment opportunites was incorporated into industrial restructuring. If the accompanying rhetoric is indicative, the agreement will be more than window dressing used for crisis management. Instead, it is offered as a first step toward the incorporation of human capital into the planning equation. Out of this crisis, then, is the French promise to complement industrial policies with those addressing social welfare. In doing so, not only are they incorporating equity into the planning equation, but they are ensuring a more efficient transfer and growth of human capital.

## Conclusion

In the introduction, the three basic policy options for dealing with factory closings were summarized: the conservative free-market approach, the liberal industrial policy, and the social welfare alternative. For all intents and purposes, the first of these has been adopted in the United States. By most accounts this policy is not working. Disinvestment continues, especially in the regions already hardest hit. The workers affected by these economic disruptions receive very little protection, and the human capital they embody is not efficiently transferred to new economic activities.

As explored in this article, viable policy options exist and have been tried in other advanced industrialized countries, but the proponents of social welfare and industrial policies seem not to have learned all the lessons these other nations provide. Simply put, neither approach, by itself, can solve the problem of factory closings. Sweden is attempting to complement social welfare measures by increasing the government's control over investments. Extensive retraining and severance benefits become much less valuable if there exists an absolute insufficiency of employment opportunities. Conversely, the French government is facing a crisis of legitimacy because it has failed to complement industrial policies with those providing for the economic security of workers. Even the French government's growth policies are threatened by the labor tensions resulting from unidimensional policies.

In the United States, reformers face even greater obstacles to achieving either social welfare or industrial policy reforms. Passing and implementing industrial policies requires more than the creation of a present-day Reconstruction Finance Corporation. To be successful, these industrial policies must be coordinated with and, therefore, lead to a significant change in the roles played by the Federal Reserve Board, the Treasury, and the Department of Commerce. The political obstacles to accomplishing this much are daunting, but the resistance of a powerful corporate sector that jealously guards its autonomy makes the task still more difficult. The liberals advocating industrial policies are not strong enough to bring about such an extensive transformation of governmental institutions by themselves, and they certainly cannot look for support to conservatives fearful of all government interventions. Instead, proponents of industrial policies must build a coalition with those concerned with social welfare, and these concerns must be reflected in the policies that emerge. As such, there would have to be less emphasis on the concessions exacted from labor and more on the options opened to workers and the job security such policies can provide.

Proponents of social welfare policies must also learn from other nations. Not only are social welfare policies inadequate in and of themselves to ensure full employment, but the social and political conditions for putting social welfare policies into effect are much less promising here than in Sweden. The passage and implementation of social welfare policies would likewise require a broad and powerful coalition. But building such a coalition is not facilitated by identifying industrial policies as antithetical to advances in social welfare. It is simply not true that industrial policies administered by a government bank are always nondemocratic; their democratic nature and relationship to social welfare policies are contingent upon the specifics of the enabling legislation and the manner of implementation.

The basic point is this: proponents of social welfare and industrial policies have more in common than recent debates suggest. Moreover, they need one another if either set of policies is to come into being. The most important obstacles to each are the inertia of existing governmental institutions, the conservatives fearful of all state interventions, and a powerful corporate sector. As such, there is much to be gained by building a broad coalition of reformers instead of making enemies out of potential allies.

# Note

1. Stephen Skowronek, *Building a New American State: The Expansion of Administrative Capacities, 1877–1920* (Cambridge, England: Cambridge University Press, 1982), p. 286.

# 20
# European and American Experience

*Bennett Harrison*

In all of the Western industrialized countries, laws have been passed or are being actively considered that call upon business managers who intend to close or relocate some facility within an otherwise ongoing company to provide their employees and the government with some degree of notification in advance of the actual shutdown.

Advocates of prenotification (and, more generally, of the periodic public disclosure by companies of information previously reserved to management) justify this demand by appeal to both efficiency and equity considerations. Advance notice of major investment, production, and employment decisions by firms is intended to facilitate capital, labor, and product market adjustments to structural change. Government agencies will be given valuable planning time with which to bring subsidies or services into play. Workers will be provided a greater opportunity to search for new jobs, to reorganize household arrangements, or to make connections with retraining programs. Especially in Europe, disclosure and prenotification are also justified by appeal to the right of workers and communities to expect "responsible behavior" from private corporations.

The demand for and the widespread interest in plant closing legislation in the U.S. are certainly understandable. During the period 1978–1982, the U.S. economy lost over 900,000 jobs a year to closures among those (generally corporate-owned) manufacturing firms having 100 or more employees. That turnover amounted to more than one-fourth of the stock of jobs in such companies at the beginning of the period. The mobility of especially middle-aged blue-collar workers out of industries and regions undergoing such deindustrialization is often hampered by mismatches in terms of skill, or by managerial discrimination on the basis of credentials, race, gender, or antiunion animus. As far as voluntary prenotification is concerned, fewer than one in five American workers are covered by collective bargaining agreements, and of those contracts,

Bennett Harrison, "Comparing European and American Experience with Plant Closing Laws," in *Proceedings of the Thirty-Sixth Annual Meeting of the Industrial Relations Research Association,* Barbara D. Dennis, ed. (Madison, Wisc., 1984), pp. 120–128.

only about 15 percent contain provisions for advance notification of shutdowns or of the introduction into the workplace of potentially labor-displacing new technology. Moreover, fewer than one-fifth of those contracts containing pre-notification language provide more than one week's advance notice of closure.

## European Experience

The European economies experienced even greater and more generalized employment loss during the 1970s than did the U.S. Among the largest member countries of the European Community (EC), negative rates of growth across the private sector prevailed everywhere. It was during these years that the Europeans began to introduce comprehensive legislation to regulate the plant closure process.

Local plant managers were required (or encouraged) to provide "timely" advance notification of plans currently under discussion concerning the implementation of new technology, possible work reorganizations, and changes in company structure such as proposed mergers and closures. With respect to closures per se, nationally negotiated or legislated prenotification periods vary from ten days (in the Italian Metalworkers' Federation national agreement with Intersind, the national employers' organization) to twelve months in the Federal Republic of Germany, "at most sixty days" in Belgium, three months in Britain and Holland, and six months in Sweden (for plants with 100 or more employees). In some cases, compliance is voluntary, although important public incentives to the firm may be triggered by that advance notification. There are no formal program evaluations of the extent of corporate compliance with these individual country-by-country laws in Europe. We do know that these laws (or negotiated contracts) are not easily enforceable when the local plant is a branch or subsidiary of a multinational corporation headquartered in another country.

The trade unions' response in Europe to what their members believe to be the unequal bargaining power of multinational corporations vis-à-vis a nation-based labor movement has been twofold. First, the unions engage increasingly in extensive international interunion consultation and coordination. Second, the unions have urged their political representatives at both the national and European Community levels to pursue legal remedies. Specifically, the European labor movement has for a dozen years sought to get the EC to issue regulations that would apply across national borders (and with greater force within each member state as well). Between 1975 and 1980, advocates managed to push through into European Community-wide law modest provisions with respect to advance notice of layoff, successorship, and workers' rights under bankruptcy.

In October 1980, the Commission of the EC took up the Vredeling Directive, named after the Commissioner of Employment, Social Affairs and Education, a

Dutch labor leader named Henk Vredeling. At the heart of the proposed measure was the requirement that the home office of any corporation doing business anywhere in the Community would provide periodic information to the managers of all of its EC subsidiaries concerning the parent corporation's Europe-wide and world-wide operations. The local managers of any plant or facility employing 100 or more persons would be required to forward this information to workers' representatives in the plant, region, or even at the national level, "without delay." If the local managers would not or could not forward the information, workers' representatives were authorized to appeal directly to the home office of the corporation. If the home office was located outside the EC (in the United States or in Switzerland, for example), then that parent firm was asked to designate an "agent" inside the EC. If it did not do so, responsibility would fall on the parent corporation's single largest plant or subsidiary presently doing business inside the EC.

Moreover, whenever home office management was seriously considering major decisions about some structural changes in the operations of any of its subsidiaries, especially a total plant shutdown, then the headquarters was required to notify the managers of all of its local subsidiaries in Europe within forty days. Local management would then be given ten days to inform the workers' representatives in every subsidiary. If the local managers would not or could not meet with the workers' representatives in the plant to be closed, then the latter were authorized to "directly approach the dominant undertaking"— in other words, the home office, provided that the office was located within the EC. If that were not the case, then, as before, the workers' representatives could approach either the non-EC multinational corporation's agent in Europe or its single largest European subsidiary. If that effort were rebuffed or if the nature of any response seemed to violate the intent of the law, then the workers' representatives could go to court following procedures which would have to be established by the member States. In any such action, it would be the parent organization's agent or its largest subsidiary that would be legally liable, not the local management in the plant or subsidiary from whence the local action would have been initiated.

The draft directive was debated throughout 1981 and 1982 in the various bodies of the European Community. Intensive lobbying by corporate groups, led by such American organizations as the U.S. Council for International Business and by a coalition of British companies, completely overwhelmed the advocates of the legislation. On December 14, 1982, the European Parliament voted to approve Vredeling, but only together with dozens of amendments that effectively gutted its most potent provisions. A substantially revised version was resubmitted to the 14 commissioners of the EC in the late spring of 1983. On June 15, they approved the revision and again sent it on to the Parliament for discussion. The new version contains major changes, most (although not all) of which substantially weaken the main intention of Henk Vredeling's original draft: namely, to regulate internationally the industrial relations practices of multinational corporations doing business in Europe.

What next for Vredeling? Certainly, business associations, especially in the U.S. and Britain, continue to strongly oppose it, even in its watered-down form, fearing it to be a step toward the eventual internationalization of collective bargaining. The Council of Ministers of the EC will probably vote on Vredeling II sometime in 1984. Even if it fails of passage this year, the European labor movement and its parliamentary representatives are determined not to let the question go away.

## Debate over Plant Closing Laws in the United States

Over the past dozen years, some American unions have succeeded in negotiating collective bargaining language concerning so-called "runaway shops." Some contracts call for employer neutrality in any union organizing drive launched in a new nonunion facility. So-called accretion agreements automatically extend union representation to new branch plants whose operations are logically an extension of those in the older (unionized) locations. Transfer rights and severance pay provisions have also been negotiated in connection with closures and relocations. But these examples remain few and far between.

Since 1979, in exchange for wage concessions, some unions have gained quid pro quos from management on a number of job security issues. However, as far as prenotification of closure is concerned, concession bargaining seems not to have yielded much new practice. At the level of the firm as a whole, even when central management was prepared to trade job security for wage concessions, it was typically a matter of the company agreeing not to go ahead with a previously planned closing or layoff. For the entire second half of 1982, Peter Cappelli could detect only three examples of concession bargaining yielding commitments by companies to any sort of procedure regarding advance notification of future shutdowns. MIT economist Harry Katz informally reports similar results from his own perusal of 1982–1983 contract renewals.

In recent years, the managements of some companies planning a shutdown have met with their unions to discuss outplacement or retraining and to plan for the redeployment of the displaced employees, although almost never to reconsider the decision to close. A small body of best-practice case material is gradually emerging on the ideas and experiences of these managers and their consultants.

Unhappily, there are far more stories of firms that display no consideration for their employees whatsoever. Thus, for example, a recent article in *Forbes* describes the 1983 Atari shutdown in California's Silicon Valley: "Consider the way Atari laid off 600 of its employees at one plant in California earlier this year. The company called workers off their jobs and told them they weren't needed anymore. Some were led off the premises through fire exits. [Additional] security guards were [placed] on duty."[1].

Advocates of national legislation point to these data as evidence that voluntary (including collectively bargained) arrangements have been insufficient in their coverage to adequately provide the protection they feel is needed. That vulnerability has been enhanced by recent U.S. Supreme Court and NLRB rulings. Thus, there is a growing interest in America in developing specific European-style legislation to deal with the problem of closures. As far back as 1974, then-Senator Walter Mondale from Minnesota and Congressman William Ford from Michigan introduced the National Employment Priorities Act into Congress. This was the first "plant closing bill," calling for mandatory prenotification of intended closures. When Ford-Mondale failed to gain support at the congressional level, the labor movement's effort to gain mandatory and universal prenotification and severance arrangements in the event of a plant shutdown or relocation shifted to the states. In July 1977, a new plant closing law was introduced into the Ohio legislature, which was to set the pattern for virtually all of the state and federal legislative attempts that would follow over the next seven years, by proposing specific legal language around three basic principles: advance notification, income maintenance, and job replacement. Together with a detailed set of proposals drafted by the United Auto Workers, following a three-union study tour of Western Europe in June 1978, the Ohio bill was taken up by Congressman Ford and Donald Riegle, U.S. senator from Michigan, and reworked into the second version of the National Employment Priorities Act. NEPA-II predictably ran afoul of the conservative political tide of 1980 and again failed to reach the floor of Congress for a vote.

In May 1983, 59 members of Congress co-introduced NEPA-III (H.R. 2847), which calls for (among other things) mandatory prenotification of from six to twelve months, depending on the size of the plant being shut down. The bill was drafted by the Industrial Union Department of the AFL-CIO, and this time was successfully reported out of the House Subcommittee on Labor Management Relations, on October 5, 1983. Even its strongest supporters are not optimistic about early passage on the floor of Congress, although it seems likely that plant closing regulations will be widely discussed over the next several years as part of the general debate on "industrial policy."

By the end of 1983, plant closing legislation has been passed or was being considered in 17 states and two cities. Maine has had a modest law on its books since 1971. Last year the Wisconsin legislature repealed its 1975 law and substituted another, replacing provisions for mandatory notification with voluntary guidelines combined with incentives in the form of so-called "positive adjustment" assistance. In 1983 Connecticut passed a law requiring a modest continuation of the health benefits of certain eligible workers displaced by plant closures. Philadelphia passed a law in 1982 mandating a sixty-day prenotification period; a recent evaluation of its immediate after-effects notes that a number of large service firms (in petroleum distribution, communications, and insurance) have moved into the city since the passage of the law. In July 1983 the city council of Pittsburgh passed a three to nine-month advance notice bill over the veto of the mayor, only to have it disallowed in August by two local judges. Fourteen other states and Connecticut are currently debating advance notification and positive adjustment legislation, including California, Illinois, Indiana, Iowa, Kentucky, Maryland, Massachusetts, Michigan, Minnesota, Missouri, New Jersey, New York, Pennsylvania, and Rhode Island. These bills call for an average prenotification period of from two to six months, depending on the size of the facility.

There is no question that employers' associations have been actively opposed to these government initiatives. Nevertheless, the business community is by no means monolithic on the question of prenotification of closures. For example, in a May 1980 survey of more than a hundred Forture 500 companies, the editors of *Forbes* magazine discovered that three out of five executives thought a prenotification period of at least three months was quite feasible, while "over a third of the respondents considered six months to a year to be the ideal period."[2]

Nevertheless, as public officials become increasingly fearful of doing anything that might pollute their "business climates," the original emphasis on mandatory regulations at the state and local level is giving way to a focus on experiments in the use of government resources and negotiation to facilitate so-called positive economic adjustment to the structural dislocation created by plant closures. Thus, California, Rhode Island, Massachusetts, Wisconsin, Ohio, and Michigan have all created or are actively considering the creation of "economic adjustment teams" or "industrial extension services." The policy instruments being discussed consist of subsidies, incentives, targeted government procurement, and moral suasion by governors, but minimal coercion on employers. The goals of these experimental programs are to use state power to help firms, employees, and local governments work out plans for restructuring businesses in trouble, for finding new buyers or assisting workers in considering whether to buy the plant themselves, or for effecting an orderly redeployment of the displaced labor if a shutdown cannot be avoided.

## Comparative Findings

In the United States, the public policy debate has become extremely confused, if not actually polarized, in terms of objectives. The unions have tended to be fairly oblivious to the efficiency questions, stressing instead the workers' right to know. Business and government officials talk almost exclusively about efficient adjustment—whether of labor, capital, or product markets. When unions raise the question of fairness, they are invariably criticized by the others for advocating protectionism which will impede efficient adjustment. In Europe, the language of the debate is much more likely to embody—indeed, to inextricably intertwine—both efficiency and equity objectives.

U.S. legislation has never called for European-style periodic disclosure of company operating information, although a bit of this has begun to emerge through the concession bargaining process, for example, in the 1983 agreement between Eastern Airlines and its three unions. Instead, in this country, plant closing laws are invariably seen as a mechanism for dealing with emergencies. Their provisions kick in only when management has already made a decision to shut down.

In the United States, prenotification requirements and positive adjustment assistance are coming to be treated by many as though they were very nearly mutually exclusive. In Europe, by contrast, subsidies and other aids to business are often tied to or triggered by prenotification.

A number of European countries (notably Sweden) complement public policies concerning closures with other policies geared to geographically targeted job creation—in other words, economic development. In this country, most existing economic adjustment programs consist of fairly traditional human resource approaches, emphasizing the diffusion of job vacancy information and the publicly subsidized provision of individual relocation and retraining assistance.

The very demand for notification and consultation puts an enormous responsibility on the unions themselves to learn how to combine advocacy and direct pressure with the acquisition of technical expertise in the evaluation and use of the information that they seek. This, in my judgment, is the issue that is most common to the situations in Europe and the United States. Whether for the purpose of facilitating knowledgeable participation in formal, high-level, neocorporatist codetermination policies, to enable a union local or works council to assess the chances that a firm is bluffing when it threatens a closure if wage concessions are not granted, or to plan a strategy of positive resistance to closure, workers and their unions are going to have to learn a lot more about the financial, legal, and even engineering aspects of investment and production decision making than they know at the present moment. If progress is to be made in the extension of workers' rights to protection from the consequences of unannounced corporate investment decisions, then the labor movement is going to have to learn how to combine direct political action, increasing sophistication in legal bargaining, and greater technical competence in making productive use of the information whose disclosure they are seeking.

## Notes

1. John A. Byrne, "Turning Off the Lights" *Forbes* (Nov. 7, 1983), p. 245.

2. Morris L. Sweet, *Industrial Location Policy for Economic Revitalization: National and International Perspectives* (New York: Praeger, 1981), p. 148.

# 21
# Policies in Japan

*Tadashi A. Hanami*

T here is no rule concerning the amount of advance notice of workforce reductions that management has to give to the unions. In practice, however, the policy of management is to notify them early enough to give itself adequate time to persuade the union, as well as the workers to be dismissed, of the necessity of the measures envisaged. (Table 21-1 provides data on the extent of union participation.) Before attempting to dismiss workers, management often very patiently tries to persuade a number of workers to retire voluntarily.

Management usually gives rather detailed information on the economic, technological, or financial reasons for reduction; the number of persons to be affected; and the occupational categories concerned. In the process of consultation or negotiation, measures are taken to limit the extent of the workforce reduction. The criteria for the choice of persons to be dismissed are negotiated. In the early stage of proposals for voluntary retirement the management usually offers favorable conditions, such as increased severance pay, arrangements to find jobs in the relevant other enterprises, and so on.

Consultation or negotiation concerning workforce reductions are mostly conducted at the enterprise level rather than at the plant level. Industrial federations or national organizations of employers and workers often assist and may even be involved in the negotiations, especially in times of deep recession, when the effects of large-scale workforce reductions on the workers who are laid off can be particularly severe.

In practice, the typical process of negotiation on workforce reductions includes the following steps: (1) negotiation takes place regarding the necessity of a workforce reduction; (2) if the necessity for such a reduction is agreed upon, management and the union negotiate the extent of the workforce reduction; (3) management often offers special retirement allowances for workers

Tadashi A. Hanami, "Japan," in *Workforce Reductions in Undertakings: Policies and Measures for the Protection of Redundant Workers in Seven Industrialized Market Economy Countries*, Edward Yemin, ed. (Geneva: International Labour Office, 1982), pp. 174–176, 180–185.

**Table 21-1**
**Japanese Union Participation in Employment Adjustment Measures,**
**by Numbers of Enterprises and by Percentages**

| Union Participation | Transfer | Detachment | Rest Days | Reduction of Workforce |
|---|---|---|---|---|
| Union acceptance | 117 | 110 | 120 | 128 |
| | 19% | 20% | 26% | 25% |
| Consultation with union | 306 | 257 | 257 | 281 |
| | 49% | 50% | 55% | 55% |
| Union opinion sought | 48 | 27 | 17 | 20 |
| | 8% | 5% | 4% | 4% |
| Advance notice to union | 65 | 46 | 22 | 28 |
| | 10% | 9% | 5% | 5% |
| Notice to union afterwards | 25 | 20 | 7 | 11 |
| | 4% | 4% | 2% | 2% |
| No union participation | 67 | 57 | 45 | 47 |
| | 11% | 11% | 10% | 9% |
| Total | 628 | 517 | 468 | 515 |
| | 100% | 100% | 100% | 100% |

*Source:* Ministry of Labour: *Summary of a Report on a Survey of the Present Situation of Collective Agreements and Other Practices: A Tentative Report* (1976).

who voluntarily resign; and (4) if dismissal is still necessary after voluntary resignations, there are negotiations concerning the conditions of dismissal.

Often special allowances for dismissal are agreed upon and the agreement may include provisions to help those dismissed to find new jobs.

Sometimes a dispute arises during the first or second stage of negotiations: unions may refuse to accept workforce reductions. This refusal by unions was one of the most important causes of serious labor disputes after the Second World War; disputes arising from workforce reduction negotiations continued for a long time and often led to violence. The reason for the rather stubborn attitude of the Japanese unions in their opposition to any workforce reduction may be explained partly by the form of organization of Japanese unions. The central core of union organization is the enterprise or plant union, which organizes the employees of a particular enterprise or plant. For a union organized in that way, any significant workforce reduction is a serious threat to its very existence. Any diminution in the number of employees means a direct fall in the number of union members. Sometimes unions even oppose appeals by management for voluntary retirement because, apart from a loss of members, such appeals will weaken the will of union members to fight against the workforce reduction. The negotiations can thus easily reach a deadlock. After a while the management will give up trying to get the union's agreement or it may be that the union itself will make some concessions and accept a compromise limiting the extent of the reduction. In either case the criteria for selecting the workers to be dismissed will be established either unilaterally by the management or by agreement with the union.

## Choice of Persons to be Affected by a Workforce Reduction

*Criteria*

There are no legislative provisions stipulating the criteria to be used for the selection of persons to be affected by a workforce reduction. In the event of an appeal to the courts, the decisions of the courts are often based on criteria such as the ability and efficiency of the workers and the economic and other effects of dismissals, detachment. or transfers on the lives of the persons concerned. Only a few collective agreements lay down the criteria to be used. In practice, management and unions usually negotiate on the criteria to be used, in each case of a workforce reduction, in choosing the workers to be dismissed or detached or transferred.

The criteria applied usually relate to both the personal situation of the affected workers and the interests of the enterprise. The personal factors relate to the degree of economic difficulty in which the worker will be placed as a result of the workforce reduction measures, including the worker's family situation. The interests of the enterprise relate to the worker's efficiency, ability, and adaptability to transfer and to other means of avoiding dismissal. The personal factors and the interests of the enterprise are mutually incompatible in that the former tend to work in favor of older workers while the latter act in favor of younger and more flexible workers.

The proper application of the criteria is subject to subsequent control by the courts. Control by the union prior to implementation of the workforce reduction is not very efficient since management is usually reluctant to negotiate the grievances of the workers affected and does so only after the decision on selection has been already put into effect.

Grievances about selections may be submitted to a grievance committee within the enterprise as well as to the courts. The former is not very effective because, as indicated above, once the decision on selection has been made, the interests of workers affected and of those not affected diverge.

*Special Situations*

If the management intentionally applies workforce reduction measures to union leaders or to workers involved in other union activities, the workers affected may appeal either to the courts or to the Labor Relations Commission for protection against infringement of their right to organize. Dismissals of pregnant women and of women on maternity leave are prohibited under article 19 of the Labor Standards Law. Dismissals of workers on sick leave caused by occupational illness or injury are also prohibited.

Article 3 of the Labor Standards Law guarantees equal treatment of workers regardless of their nationality. Dismissals in violation of that provision are declared null and void.

## Rights of Dismissed Workers

*Advance Notice*

Article 20 of the Labor Standards Law requires the employer to give thirty days' advance notice to a worker to be dismissed except where continuance of the enterprise is precluded by a natural calamity or other inevitable cause or where the employer has dismissed a worker for reasons for which the worker is responsible. In such cases, the employer must obtain the official approval of the reason for not giving notice. There is no special notice period for workforce reductions. The law does not provide for time off to seek other employment during the notice period.

The employer may pay average wages during the notice period in lieu of giving notice or may reduce the period of notice by paying the average wages for the days reduced. In such cases, of course, the workers can seek new employment without any obligation to their present employer.

In practice, if the worker finds a new job before the period of notice has expired, he may leave his job without losing entitlement to, for example, severance allowances.

*Compensation for Loss of Employment*

**Severance Allowance.** In general, it is usual practice for the employer to pay a lump-sum retirement allowance to workers who have been dismissed or who have resigned or retired after certain periods of service in the enterprise. Provision for such allowances is made in collective agreements or work rules. The amount of the allowance is usually calculated by mutiplying the last basic monthly wage by a specific factor related to the length of service. According to a survey made by the Central Labor Relations Commission in June 1977, the factor varies, in the case of retirement at retirement age, between 0.8 for one year's service and over 41 for 35 years' service, and, in the case of resignation, between 0.1 for one year's service and over 35 for 35 years' service. Those who accept voluntary termination in connection with a workforce reduction often receive a higher allowance.

The law on retirement allowances for small and medium-sized enterprises provides for a mutual benefits system to guarantee the payment of retirement allowances to workers of such enterprises. The enterprises contribute fees to associations which pay the allowances to the workers. The government provides those associations with financial assistance and is responsible for their administration.

**Unemployment Benefits.** There is no special unemployment benefit for workers dismissed as a result of a workforce reduction; they are entitled to the general unemployment benefits. The amount of such benefits is 60 percent at minimum and 80 percent at maximum of the previous wages, favoring low wage earners. The duration of the benefits is set according to age and reemployment difficulties.

## Assistance in Obtaining Alternative Employment

### *Promotion of Occupational Mobility*

The Employment Insurance Law provides for payment to workers of a skill acquisition allowance, as well as a lodging allowance where the worker has to reside away from home in order to take a course of public vocational training. Also under that law, the government may provide assistance to an employer compelled to reduce business activities and who requires members of the workforce to undergo vocational education or training.

The law relating to displaced workers in specified depressed industries provides for various forms of financial assistance to workers or enterprises with a view to promoting the training of workers covered by the law.

### *Promotion of Geographical Mobility*

The Employment Insurance Law provides for the payment of expenses to workers referred by the Employment Security Office to employment opportunities outside their own area.

The law relating to displaced workers in specified depressed industries provides for the payment of transportation and lodging expenses to workers who seek new jobs in a wider work area on referral by the Employment Security Office, as well as for the payment of a removal allowance to workers who have to change their place of residence in order to take up new employment or to obtain vocational education or training.

## Conclusions

The existing system governing workforce reductions is characterized by its passive nature. The main efforts to cope with the unemployment problem are focused on protection of the workers about to be dismissed and on the various kinds of assistance provided to enterprises as well as workers, especially in the industries and trades which are and will be suffering from the present and anticipated economic recessions. For these purposes, the system is carefully designed and will be efficient enough if the economic situation does not change drastically for the worse.

Nevertheless, by comparison with the recent trend in Europe in policies or measures to cope with workforce reductions, the Japanese system is characterized by a lack of positive restrictions on dismissal and other measures used by management for the purpose of workforce reduction. It aims to alleviate the impact of the dismissals on the workers but does not interfere with the decision-making of management on workforce reductions. As shown in this report, the unions are not very effective in protecting the workers' interests in this particular regard.

It has also been indicated that the courts have been trying to protect the workers' interests by examining whether the dismissals are justifiable or whether there is any abuse of the right to dismiss. However, the remedy given by the courts is always *ex post facto* and takes a long time. There is thus an urgent need to introduce legislation such as exists, for example, in Great Britain, France, and the Federal Republic of Germany, where employers are obliged by law to consult unions or workers' representatives, to disclose information and to follow some other procedures prior to carrying out a workforce reduction. There would be a need also for government intervention requiring the submission of plans for workforce reductions to, and their authorization by, public authorities such as the public Employment Security Offices. Certain exisiting laws, such as the law relating to displaced persons in specified depressed industries, already do require consultation with the union and the submission of plans for workforce reduction to the authorization of the public Employment Security Offices. The application of that particular law is, however, limited at present to a specified field and the requirements in question are only for the purpose of receiving allowances under that law.

Unions have been keenly aware of the gravity of employment problems in recent years, especially because of the rather negative prospect of wage negotiation. They are now emphasizing the importance of union policy concerning employment problems. They are demanding shorter working hours, regulations on overtime work, extension of the retirement age, regulation of dismissals, and the introduction of a consultation system by law.

However, the shortcomings of the Japanese system have a foundation in present Japanese industrial relations. First of all, it should be emphasized that

Japanese enterprises have been voluntarily taking various internal steps within the enterprise and cooperative groups to avoid redundancy, such as transfer, detachment, and the creation of new jobs. The lifetime employment system not only makes it possible to transfer workers within a larger number of organizational units and job classifications but also contributes in some measure to protecting regular workers' interests though at the same time sacrificing the interests of temporary workers. However, the lifetime employment system does not work to protect older workers and is thus at odds with its original purpose. As we have seen, when an enterprise needs to dismiss a significant number of workers, including regular workers, the criteria used for selection for dismissal often work against the older workers. It is those who are economically less affected than older workers who should be chosen for redundancy in cases of workforce reductions. Although liable to be dismissed, younger workers more readily accept voluntary resignation and are better able to seek an alternative job. The government is well aware of this point and has already taken several measures to protect the older workers, as already indicated. The extension of the retirement age might be one of the steps to be taken in this connection, but for that purpose a drastic change in the wage system, in which the amount of wages is related to length of service, is needed.

It is true that there exists a trend in the recruitment policy of enterprises to resort to casual employment or fixed-term contracts under tight economic conditions. Recruitment is still an area in which the employer and the worker have more freedom of contract than in the case of dismissals. Even so, there are certain limitations to this trend resulting from the lifetime employment system and legislation. Japanese management prefers contracts of indefinite duration, at any rate for the central core of the labor force, employing them as regular workers and thereby ensuring that a certain number of workers will be loyal and devoted to the cause of the company. Article 21 of the Labor Standards Law assimilates nonrenewal of fixed-term contracts to dismissal in a number of cases, thereby limiting employers' use of fixed-term contracts as a means of avoiding the protection resulting from contracts of indefinite duration.

The present system governing workforce reductions works more or less effectively in protecting workers' interests, at any rate in the present economic situation. It has, however, certain defects and might prove to be insufficiently protective, especially if the economic crisis were to become more serious in the future.

# Part V
# State Laws and Proposed Federal Legislation

Perhaps the most controversial issue concerning plant closure is whether to pass state and/or federal laws to protect affected workers and communities. Numerous arguments have been cited on both sides of the protective legislation question. The main arguments in favor of legislation are the following:

1. Advance notice and income guarantees are necessary to mitigate the great economic and psychological burden of closure on employees.
2. Businesses get substantial economic incentives and rewards from government; workers deserve the same protection.
3. Communities also deserve advance notice, because they lose a major source of tax revenue when plants shut down and face the added burden of social welfare payments to displaced workers.
4. Whether they realize it or not, businesses have a social obligation to their employees and the community.
5. Plant closure has a multiplier effect within a community, causing a greater loss of jobs than those in the plant itself.
6. Because businesses usually plan shutdowns far in advance, prior notification requirements pose no great burden.

The main arguments against legislation are the following:

1. State plant closure laws would place unconstitutional restraints on interstate commerce.
2. State plant closure laws would create more, not less, unemployment, because large firms would establish or increase operations in other states to avoid penalties.
3. Workers receive adequate economic protection through state unemployment benefits and job search services.

4.  Business is already overregulated; it needs incentives, not restrictions and penalties, to promote a healthy economy.

5.  Businesses need to be free to close inefficient plants with obsolete equipment and replace them with new facilities.

6.  Based on their educational qualifications and skills, manufacturing workers receive generous wages; as a quid pro quo they should assume the risk of closure.

In the initial reading, Antone Aboud and Sanford F. Schram provide a good overview of features contained in proposed plant closure legislation. They also compare existing laws in Maine, Wisconsin, and the city of Philadelphia. Three major questions policymakers have addressed are: (1) to whom would the legislation apply, (2) what are the specific obligations of employers, and (3) how will the law be enforced? As to coverage, existing and proposed laws typically limit coverage to larger employers, for instance, those with more than 100 employees. Employer obligations vary, with some proposed laws requiring only advance notice (from sixty days to two years), and others placing greater burdens on employers, such as severance benefits, continuation of health insurance, job retraining, transfer privileges, and financial contributions to communities affected. Enforcement provisions vary as well, from allowing voluntary compliance to mandatory compliance with imposition of criminal penalties. Aboud and Schram note, importantly, that the highly controversial nature of any such legislation is a major impediment to passage. Antone Aboud is an associate professor and director of the Graduate School of Industrial Relations at St. Francis College in Loretto, Pennsylvania. Sanford F. Schram is an associate professor of political science at the State University of Arts and Sciences at Potsdam, New York.

Next, Nancy R. Folbre, Julia L. Leighton, and Melissa R. Roderick analyze the Maine plant closure law. This selection is important because Maine's 1971 law was the first in the nation, and the authors were able to examine a decade of experience under the law. They conclude that the advance notification requirement substantially reduced local unemployment rates by allowing workers time to adjust to the labor market. They note that enforcement has been lax, and many employers have not complied with the law. Nancy R. Folbre is an assistant professor of economics at the New School for Social Research. Julia L. Leighton is a Research Associate with the Public Interest Economics Foundation in Washington, D.C. Melissa R. Roderick is a recent graduate of Bowdoin College.

In the following excerpt, Congressman William D. Ford of Michigan, a leading proponent of federal plant closure legislation, discusses ill-fated efforts to get a law passed in Congress. An early proposal, the National Employment Priorities Act of 1974, would have given federal financial assistance to individuals and communities. Later versions of the bill included severance payments and transfer rights for displaced workers. Ford's 1985 bill, known as The Labor-Management Notification and Consultation Act, not only would have required advance notification of major layoffs and closures, but would have included workers and unions in the decision making. Although Ford presented a strong argument in support of this legislation and it seemed to have a good chance of passage, the House defeated it in late 1985 by a vote of 208 to 203.

In the next essay, Bennett Harrison discusses the legislative history of proposed federal laws and also reviews the dynamics of proposed state and local legislation, comparing the United States policy debate with European experience. Harrison believes the problems of plant closure will intensify in years to come, leading to further initiatives for government regulation.

The concluding essay, by James J. Chrisman, Archie B. Carroll, and Elizabeth J. Gatewood, presents a strong case against plant closure legislation, reflecting the prevailing political consensus that the free market rather than government regulation would influence decisions on plant closure. Responses to their survey of managers' perceptions show managers' concern about the problems of plant closure, but express their misgivings over laws that would retrict their decision-making authority. In advocating a free-market approach, however, these writers do not preclude a government role. They see its role as stimulating an environment that would prevent plant closure or soften its impact, instead of imposing conditions that would limit its occurrence. James J. Chrisman and Elizabeth J. Gatewood are Ph.D. candidates and Archie B. Carroll is a professor of management at the University of Georgia.

# 22
# Overview of Legislation

*Antone Aboud*
*Sanford F. Schram*

**W**hen William G. Sumner testified before a Select Committee of the United States House of Representatives in 1878 concerning the depression of the 1870s, he was asked about the possibility that government might be able to assist individuals in combatting the debilitating effects of unemployment. In his response he stated:

> The only answer that I can give to a question like that would be the application of a simple sound doctrine and sound principles to the case in point. I do not know of anything that the government can do that is at all specific to assist labor—to assist noncapitalists. The only things that government can do are general things, such as are in the province of a government. The general things that a government can do to assist the noncapitalist in the accumulation of capital (for that is what he wants) are two things. The first thing is to give him the greatest possible measure of liberty in directing his own energies for his own development, and the second is to give him the greatest possible security in the possession and use of the products of his own industry. I do not see anything more than that that a government can do . . .[1]

In the 1980s many would agree with Sumner; however, as the recent recession intensified and plant closings increased, the adverse effects on workers and communities moved Congress and state legislatures, often under prodding from unions, to study more seriously the prospects of government regulation of the plant closing process. The rationale for such legislation was straightforward. William Schweke has summarized it succinctly:

> The problems of capital mobility and major job losses are real and growing. The major victims are the laid-off workers and their families. The massive job cuts often flood the labor market, overwhelming local employment opportunities. States and municipalities also face several fiscal difficulties, as their tax base erodes and public spending rises to pay for the social costs of economic dislocation, which include rapid increases in juvenile delinquency, crime, divorce, mental illness and despair.[2]

---

Antone Aboud and Sanford F. Schram, "An Overview of Plant Closing Legislation and Issues," in *Plant Closing Legislation*, Antone Aboud, ed. (Ithaca, N.Y.: ILR Press, Cornell University, 1984), pp. 33–43.

One form that regulation could take is what is commonly referred to as plant closing legislation, a combination of requirements involving notice and various types of compensation for employees and communities affected by the changes.

The U.S. Congress has considered plant closing legislation for a decade. As early as 1974, Congressman William Ford of Michigan introduced the National Employment Priorities Act, a bill that would have provided federal monetary assistance for communities and individuals adversely affected by plant closings and would have suspended certain federal tax advantages to firms violating the law. Later versions of the same legislation would have required firms to make severance payments directly to individual employees affected by the closing. Other attempts at federal legislation have been made with similar purposes, although details of each piece of legislation vary. The Employee Protection and Community Stabilization Act (1979), the Voluntary Job Preservation and Community Stabilization Act (1978 and 1979), and the Employment Maintenance Act (1980) are all examples of attempted federal involvement in regulating plant closings.

Later sessions of Congress provided a number of legislative approaches that would speak to many of the issues raised previously. For example, H 2847, introduced by Congressman Ford and thirty-one additional cosponsors, on May 2, 1983, included notice provisions for closings and certain layoffs, as well as provisions for assistance to affected workers. The bill was referred to the Education and Labor Committee and the Banking Committee. On May 18, the Labor-Management Subcommittee of the Education and Labor Committee began hearings. A similar bill, H 807, had been introduced on January 25, 1983, and was referred to the same committees as H 2847. A number of other bills dealt with the problems of continuing health insurance coverage of unemployed workers. For example, H 3021 was introduced on May 16, 1983, and referred to the Ways and Means and the Energy and Commerce Committees. It was subsequently referred to subcommittee and approved with amendments on May 18. Although broader in scope than legislation designed simply for the plant closing situation, the bill would cushion the economic impact of closing a plant, at least for the unemployed worker.

So far, few U.S. jurisdictions have enacted such legislation. Public policy in Maine requires certain employers to notify employees of planned closing sixty days in advance and make severance payments to individuals who have been employed for three or more years. Violators may be fined up to $500. Legislation in Wisconsin also requires a sixty-day notice in such circumstances and provides for a $50 fine for each worker affected by the change. The city of Philadelphia also requires employers to give sixty days notice of closings and relocations; the law also requires that extensive information be included in the notice.

There have been other events at the local level. The city of Pittsburgh passed an ordinance similar to Philadelphia's. A group of employers challenged the statute on three grounds: (1) the statute violated the home rule provisions of Pennsylvania's state law; (2) the ordinance violated the Fourteenth Amendment of the United States Constitution; and, (3) it regulated matters otherwise subject to the National Labor Relations Act and therefore was preempted from state or local legislative control. The state trial court ruled the ordinance invalid with respect to the home rule and preemption issues. The city of Vacaville, California, negotiated a plant closing agreement with labor and other community groups that would apply to employers who receive financial incentives from local sources to relocate to Vacaville.

## Legislative Proposals

Existing legislation hardly represents the widely varied proposals that many jurisdictions have considered. The purpose of this section is to describe both the common and unique characteristics of the provisions in these bills. The bills from which we drew this description are those considered in 1981 by state legislatures. The U.S. Department of Labor made these proposals available to the authors; they are the ones on which a survey of state legislative activity in the January 1982 issue of *Monthly Labor Review* was based. Although other proposals have been introduced since 1981, this group provides a sufficiently large sample to illustrate the variety of ways in which the problem is being addressed.

### To Whom Would the Legislation Apply?

Virtually all of the proposed bills contain extensive definitions that, taken together, identify those employers who would be required to provide notice and payments when making the types of changes regulated by the proposals. The definitions usually involve the interplay of "employer," "establishment" or "affected establishment," and other qualifying clauses contained in the definitions of "closing," "relocation," "reduction in operations," or similar clauses. For example, Indiana's HB 1274 defines an employer as

> any person who employs at least one hundred (100) individuals and who operates or owns more than fifty percent (50%) of a commercial, industrial, or agricultural enterprise, but does not include the state or its political subdivisions or any organization or group of organizations, described in Section 501(c)(3) of the Internal Revenue Code (26 U.S.C. 1), as in effect on September 1, 1981, and exempt from income tax under Section 501(a) of the Internal Revenue Code . . . as in effect on September 1, 1981.

Clearly, the definition goes beyond an employer operating simply a "plant." Virtually any business enterprise that pays taxes and employs more than 100 individuals would be covered. The exclusion of state and local governments and tax exempt organizations is a common feature of most of the proposals.

The definition of establishment further narrows the number of business decisions that might trigger responsibilities under the proposal. Establishment is defined as

> any factory, plant, office or other facility that an employer has operated in the state for five (5) years or more, but does not include a construction site or other work place that was never intended as other than a temporary work place. . . . If the current employer has operated an establishment less than five . . . years but acquired it from a previous employer, the current employer shall be considered to have operated the establishment during the period in which the previous employer operated the establishment.

Although not all definitions require that a facility has been in existence for five years to qualify as an establishment, most create a grace period. Of greater importance than the length of that period is the part of the definition that credits a successor employer with the time during which the facility was operated by the previous owner. Again, this provision is a common feature of many proposals, one which would prevent the creation of "alter ego" employers for the sake of escaping responsibilities that would have vested but for the change in ownership. Also, the explicit exception in the case of a construction site is almost universal in the bills studied, although not always contained in the definition of establishment.

Finally, the definitions of closing, reduction in operations, relocations, or similar terms often contain exceptions to their application. Indiana's S 541 definition of closing, for example, contains an exception where an employer has filed for bankruptcy, perhaps the most common of all exceptions. Oregon's SB 830 contains a list of such exceptions, although in this case the exceptions do not include bankruptcy:

> Section 3. This Act does not apply to reduction in operations that:

1. Result solely from labor disputes;
2. Occur at construction sites or other temporary workplaces;
3. Result from seasonal factors that are determined by the Bureau of Labor and Industries to be customary in the industry of which the business operation is a part; or
4. Result from fire, flood or other acts of God.

The magnitude of the business change is another factor considered by many bills. For example, Maryland's H 1411 defines reduction in operations as "permanent shutting down of a portion of the operations at any establishment so as to reduce the number of employees employed at that establishment by at least 50 percent over any two-year period." Although the percentage of employees affected and the time period may vary in other proposals, such clauses would exempt a significant number of business decisions from regulation. Also, definitions of "relocation" in many cases contain qualifiers that limit the application of the proposal. For example, Minnesota's S 294 contains the following definition of relocation:

> transfer or series of transfers of part of an employer's operation from an affected establishment to an *existing establishment* located at an *unreasonable distance*, as provided by rules promulgated by the commissioner, from the affected establishment, and which results in at least *a ten percent reduction* over a *two year time period* in the number of employees at the affected establishment. Relocation does not include a transfer from one establishment to another establishment *in the same county* (emphasis added).

Each of the phrases emphasized creates a separate criteria for determining whether the bill's restrictions would apply to an employer at an establishment otherwise covered.

## What Obligations Would a Covered Employer Acquire?

At a minimum, plant closing proposals require that covered employers provide notice to employees and communities and sometimes to unions. The period of notice may vary from six months to two years. Many of the proposals list in varying detail the groups to whom the notice must be sent. Oregon's S 830 contains one of the longer lists. Notice must be sent to:

1. All affected employees;
2. Affected labor organizations;
3. Elected officials in communities in which affected workplaces are located;
4. Clerks of all taxing districts affected by the changes;
5. Bureau of Labor and Industries; and
6. Economic Development Department.

Most bills also require that some additional information be included in the notice. Common features of these provisions include locations and nature of the various facilities owned by the company and the reasons for the changes that caused the closing or relocation decision. In some cases companies must give financial data related to the size of the workforce and wages of the employees. To measure the impact of the change on communities, certain bills require a list of taxes paid to affected communities during the previous year. Minnesota's H 294 goes so far as to require the employer to estimate the financial impact of the change on the communities and businesses that the closing or relocation affects.

In addition to the notice and information requirements, the bills require payment of severance benefits to either the employees or the affected communities or both. Where benefits are owed employees in the form of severance payments, the calculation is almost always the average weekly wage of the employee during the preceding year times the number of years the employee was employed (for example, as in Massachusetts H 1541, California SB 114, and Maryland H 1411). Oregon's S 830 requires that employers pay income maintenance amounts equal to 85 percent of the employee's average pay for the one year prior to the change. An employee who enters a job training program would be entitled to a payment equal to 100 percent of the average wage. In both cases the amounts are reduced by any unemployment insurance received and cannot exceed $15,000 per employee. In essence, then, the options operate as supplemental unemployment benefits with built-in incentives for affected employees to be given additional job training.

Michigan's H 4330 is perhaps unique among proposals requiring severance or income payments to affected employees. At the time of announcing a closing, relocation, or reduction in operations, the employer must create an escrow account with sufficient monies in it to pay health insurance benefits to the employees for a period of twelve months after they lose their jobs. In addition, the employer is required to give extra paid leave of either one or two weeks (based on length of service) for the employee to interview for other jobs. The employee is entitled only to an actual severance benefit if the employer has violated either Sections 8 or 14 of the bill; Section 8 specifies notice requirements and Section 14 spells out the paid leave provisions.

Obviously, there are some collective bargaining agreements that provide for severance payments to employees when a firm closes or relocates. Virtually all of the bills studied indicate that if a collective bargaining agreement calls for benefits greater than those provided by law, employees would receive the negotiated benefits. Thus, the bills provide a benefit floor above which the parties are free to negotiate.

Perhaps the most complex of all provisions in these proposals are those related to the assistance of communities. The vast majority of proposals are not as simply drawn as California's SB 114, which merely provides that a payment equal to 10 percent of the total annual wages of employees who lose their jobs be made by the employer to the affected community. The more common arrangement is the set of policies and procedures outlined in Massachusetts H 1541. This bill provides for the establishment of the Community Jobs Assistance Fund from payments of employers covered by the law. Such an employer would pay a total of "15 percent of the affected annual payroll" into the fund administered by the state. Section 179(f)4 gives the purposes for which the fund may be used:

> The Commissioner shall have authority to make grants or loans from the Fund to affected municipalities and to non-profit corporations for the purposes of providing employment opportunities for employees directly affected by a mass separation and maintaining and restoring levels of employment in affected municipalities. Such grants and loans may be made to provide technical and planning assistance, matching funds to secure federal or other assistance in creation of new employment opportunities in affected municipalities, provisions of temporary financing or capitalization of new employment opportunities in affected municipalities.

There are a number of variations of this model, including the amount of money an employer would have to pay into the fund. In some cases the proposal makes clear that payments may be made to communities "to provide emergency tax relief where a political subdivision faces substantial loss of tax receipts from a business closing, relocation, or reduction in operations" (Maryland H 1411). In addition, many proposals create local bodies to administer the funds and may assign them duties other than merely dispensing the money. The Michigan bill (H 4330) would give local communities the right to establish Community Service Councils. In responding to pending changes, councils are empowered to do the following:

1.  Attempt to persuade or induce the employer to reduce the operations of the affected establishment rather than close or relocate.

2.  Offer assistance to the employer to promote operations.

3.  Evaluate the feasibility of a proposed employee-owned corporation.

4.  Compile and distribute a list of private, local, state, and federal agencies or programs that may be of assistance to affected employees and other persons similarly situated in the community with an explanation of their services and an accompanying address and telephone number.

5.  Encourage the development of in-plant and community counseling programs for affected employees and other persons in the community who are or who soon may be unemployed. The counselors may help these persons to anticipate, define, and identify their problems, and inform them about possible sources of assistance and information.

6.  Establish a job training program or cooperate with others in providing job training.

7.  Attempt to persuade or induce state and local officials to grant emergency tax relief to an affected municipality that faces substantial loss of tax revenue due to a closing, relocation, or reduction in operations.

8. Provide a grant or loan to an affected municipality.

9. Apply for state and federal grants and matching funds.

10. Provide for other programs as the council considers appropriate.

### How Would Plant Closing Laws Be Enforced?

With respect to the enforcement of various plant closing regulations, the proposals ranged from the imposition of criminal penalties to the virtual absence of any enforcement provisions. For example, California's SB 114 contains no provision that would penalize employers who failed to live up to its requirements, except with respect to giving notice. In that case, the state is authorized to seek an injunction where it determines that an employer's failure to give notice was "willful." On the other hand, Massachusetts's HB 1411 contains the following provision:

> Any employer who violates any provision of this subtitle is guilty of a misdemeanor and on conviction is subject to a fine not exceeding $500 or imprisonment not exceeding 3 months or both.

Between these extremes are a variety of enforcement devices. We have already noted that Michigan's HB 4330 would require payment of severance benefits only if an employer violates obligations under the act. A complaint alleging violation of that statute could be lodged by an employee or a union with the state department of labor within "1 year after termination of employment or 1 year after the affected employee or affected employee organization is aware of the alleged violation, whichever is later." The bill creates an elaborate set of administrative investigatory steps and appeals to the judiciary. In addition to the severance payments, employers would be liable for attorneys' fees and court costs should they lose in litigation; the state attorney general, upon request of the director of the labor department, is also free to seek civil enforcement of a final order of the department.

Many proposals contain a general clause prohibiting discrimination against any employee who seeks to enforce his or her rights under the act. Pennsylvania's SB 727 contains common language of this sort:

> No employer shall discriminate against any employee in the terms and conditions of employment on account of that employee's reporting of information concerning the possible or actual violation of this act or because that employee has filed any administrative or judicial proceedings to enforce the provisions of this act.

Finally, Minnesota's SB 294 contains a unique enforcement provision. According to Subdivision 1 of Section 21, "an employer operating an affected establishment who violates this act shall forfeit any tax benefits that the employer has received within the past five years due to the employer's status as an employer." Given the present state and local mood to extend such benefits in order to attract employers, the total of such benefits for any particular establishment may constitute a much greater penalty than statutory or court imposed fines.

*Conclusion*

That so few states and municipalities have passed plant closing legislation is attributable, at least in part, to the controversial nature of this type of regulation. The subject is controversial enough that some states have established study committees, by statute, to evaluate the desirability of such legislation. In one case the resulting recommendations of the committee reflected the balance that appears to exist in the current debate. The Connecticut Committee to Study the Issue of Plant Relocations and Layoffs split six to five in its deliberations. The majority of six issued a report generally critical of passage of a bill requiring notice and severance payments. The minority of five issued a report titled *The Positive Approach Report*, in which they recommended one year's notice for employees and the establishment of severance payments for employees and affected communities.

## The Emerging Debate

In addition to the debate at the legislative level, discussions of plant closing restrictions have flourished among academics and various interest groups. Some economists, such as Richard McKenzie of Clemson University, place the debate in a free market context, criticizing plant closing legislation as an unwise and, perhaps, dangerous infringement upon capital mobility. According to this point of view, not only will such infringement hamper the efficiency and efficacy of the market, but states and communities choosing to enact such legislation will hurt themselves in the long run because companies will choose to settle in locations that provide the greatest freedom of choice. Other economists, among them Barry Bluestone of Boston College and Bennett Harrison of Massachusetts Institute of Technology, examine the issue from a perspective of the systematic "deindustrialization" of the U.S. economy by corporate America, particularly by multinational conglomerates with loyalty to no community, state, or country. They argue that plant closing legislation must be instituted to ameliorate the social costs—loss of income to a community and workforce dislocation—of unrestricted capital mobility. The balance of this section will outline these positions in greater detail as they have been set forth by business and labor interests.

*Right to Mobile Capital*

While not denying that workers and communities are hurt by plant closings, business resists the temptation to treat the problem by doing something that, from the business perspective, amounts to no more than simply treating symptoms. For business, plant closing legislation addresses a real problem in a simplistic, superficial way that overlooks the fundamental questions about the market and the conditions necessary for ensuring economic growth. For business, plant closing legislation, while trying to band-aid over the problems of disinvestment, creates more serious obstacles to economic revitalization.

The typical business viewpoint suggests that the most fundamental question advocates of plant closing legislation overlook is, how will plant closing restrictions affect capital's right to mobility and the ability of the market to ensure the most profitable, productive, and efficient investments? For business, such restrictions can only introduce additional inefficiencies into the market and, in the long run, further hamper the economy's ability to grow, prosper, and offer increased economic opportunities for all. In the view of most business spokespersons, the right to capital mobility shall not be infringed. If it is, we all suffer the consequences of a less efficient economy. This cardinal principle for advocates of a free market leads them to oppose even plant closing legislation that does nothing more than require advance notice to affected workers and communities. Free marketeers argue that such notice reduces the ability of a firm to find new owners for their plant or other possible alternatives to a closing. Notice spreads the word unnecessarily; it creates a bad name for the plant and disminishes the propensity of others to become involved in keeping the plant open through purchase, transfer of title, buy-out, loans, and so on. Therefore, when it comes to plant closings, even the most rudimentary form of government regulation requiring adequate advance notice, according to business, constitutes a dangerous infringement on the right to capital mobility.

A second important question business says is overlooked by proponents of plant closing legislation is, do the short-run benefits to workers and communities really outweigh the long-run costs of plant closing legislation? Keeping inefficient, uncompetitive plants open will only intensify and delay the need for restructuring the economy. Penalizing firms for reallocating investment in the most efficient ways only reduces the ability of the market to create economic opportunities for workers in the future. And, most important, restricting the ability of business to disinvest may do very little to keep business where it is, while doing much to discourage new business from coming into the affected community. Irrespective of plant closing legislation, business will continue to seek the most profitable, productive, and efficient investments that simultaneously have the greatest liquidity and flexibility. Therefore, states or even countries with plant closing legislation will have one strike against them and business will grow reluctant to initiate new investments in these places, knowing it may have to meet the additional costs of plant closing laws some time in the future. McKenzie has written:

> States and communities that are mulling over business mobility restrictions may believe they would be protecting their economies by protecting their industrial bases, but in fact they would be hurting them—and themselves. What company would want to move into an area that had substantial economic penalties for moving out? What entrepreneur would want to start a business in a community or state that had penalties for changing locations? Companies interested in profits will always try to settle in those areas that leave them free to make the basic decisions of when to shift among products, when to close, and when to move. States or communities that do not impose restrictions will obviously have a competitive advantage over those that do.[3]

Finally, the business perspective suggests that proponents of plant closing legislation incorrectly answer the question: What is the social responsibility of a firm to its community? Business spokespersons doubt that firms owe more to their communtities than workers and like workers, in market economy, they should be able to move elsewhere without incurring a social debt to the community they leave behind. To restrict either business or workers mobility would be to diminish the meaning of living in a "free" society and reduce the efficiency of the market. Such restrictions would also reduce the ability of depressed communities that have already lost plants, such as Youngstown and Dayton, Ohio, and Mahwah, New Jersey, to attract companies. In other words, does corporate social responsibility lie with the immediate community or the entire nation? In the end, according to the business perspective, business best fulfills its social responsibility by being good business and allocating capital in the most profitable, productive, and efficient way.

### The Social Obligations of Capital

From organized labor's perspective, the right to capital mobility and the efficiency of an unregulated market can exist as priorities only to the point that they do not wreak havoc on the lives of workers and the stability of their communities. The massive pullouts of firms from communities like Youngstown and Dayton, Ohio, in the late 1970s must not be repeated elsewhere. The social costs of runaway shops increase unemployment and social pathologies that in the end generate instabilities in the market system itself.

Labor strongly supports plant closing legislation and prefers that such legislation be passed at the national level so that all states will be required to play at the same rules; states without restrictions should not be able to out-bid states with restrictions in the competition for new plants. Yet, labor is also willing to support state legislation as a first step and has done so in at least twenty states.

Labor's most fundamental argument for plant closing legislation is that government has an obligation to ensure that the market works for people rather than the other way around. The right of capital mobility and the efficiency of the market are not ends in themselves; they exist to create a quality of life for the people who live and work in a market economy. When the right of capital mobility leads to disrupting that quality of life, it begins to abuse its privilege. At that point, labor feels justified in characterizing it by epithets such as "disinvestment," "capital flight," "runaway shop," and "corporate irresponsiblity." For labor, the limits of mobile capital are the market's obligations to society; in the face of massive plant closings, government must regulate the mobility of capital to ensure a socially responsible market.

According to labor, the market itself benefits from plant closing legislation. Reducing the ability of capital to go wherever it is most profitable enhances economic stability. Knowing that business will probably not be leaving encourages additional investment in an area. Workers will be more motivated and productive once they gain a heightened sense of economic security.

From labor's perspective, the increase in plant closings is a symptom of a serious disruption in the economy, signaling a need for workers to gain new skills to reenter the labor market. Plant closing legislation would treat the problem at its source by offering displaced workers the time, resources, and training to acquire new skills and make the transition to a new economy. Without such assistance, the economy will falter as long as there is an inadequate supply of workers with the skills required for a restructured economy.

Labor also supports plant closing legislation because it believes disinvestment is often unrelated to the needs of restructuring the economy and represents nothing more than the flight of capital from unionized, high-wage labor to nonunionized, low-wage labor in the South and abroad. For labor, such closings represent an extreme abuse of the right of capital mobility for several reasons. First, such plants close not because they are unprofitable but because they are not as profitable as they would be if the workers where not unionized and were accepting lower wages. Second, such closings represent instances where firms draw their profits and reputation from the community and its workforce and then invest them elsewhere, deserting their benefactors in a quest for even greater profits. Third, such closings do nothing to improve the economy as capital flows out of the region or, more importantly, the country.

Finally, labor believes it possible to impose plant closing restrictions without duly infringing on the profitability and efficiency of the market. Labor draws its confidence on this point from the experience of Western European countries. Countries like West Germany and Sweden have been able to buffer the effects of the international economic down-turn through government regulation of plant location. Such regulation has been guided by a commitment to balance the priority of making productive and efficient investment with the need to ensure the work force economic security and shelter from the social consequences of mobile capital.

From labor's perspective, government can regulate capital mobility so as to ensure both efficient and productive investment and economic security for workers. Plant closing legislation would tame capital mobility and make certain it worked for the people without destroying the profitability of investment.

## Conclusion

The parties in the debate have set out their position. Business wants capital mobility to continue unfettered, labor wants the plant closing process regulated, and whole communities seek effective alternatives to protect them from what they see as economic disaster. Plant closing legislation of the type discussed in this paper is not the only way in which communities and governments have considered dealing with the effects of the disintegrating economies of the older industrial sectors. Concession bargaining, urban enterprise zones, and employee ownership are all phenomena that in different ways attempt to protect communities from the dislocation and hardship of unemployment due to failing industries. When one considers that there is no clearly correct alternative in dealing with economic change, one appreciates the complexities that face policymakers. What emerges, perhaps, is the recognition that plant closing legislation is not a panacea to current economic ills. It may, however, be a lightning rod for the debate about potential new directions for our economy.

## Notes

1. William G. Sumner, Testimony before the U.S. Congress House Select Committee Investigation. *The Causes of the General Depression in Labor and Business, Etc., 1878* [Hewitt hearings], reprinted in G. Gerd Korman, comp., *Labor History Documents*, vol. 1 (Ithaca: New York State School of Industrial and Labor Relations, 1974), p. I:002;19.

2. William Schweke, ed., *Plant Closings: Issues, Politics and Legislation* (Washington, D.C.: Conference on Alternative State and Local Policies, 1980), as quoted in Richard B. McKenzie, ed., *Plant Closings: Public or Private Choices* (Washington, D.C.: Cato Institute, 1982), preface.

3. Richard B. McKenzie, ed. *Plant Closings: Public or Private Choices*, p. 18.

# 23
# Legislation in Maine

*Nancy R. Folbre*
*Julia L. Leighton*
*Melissa R. Roderick*

T he dramatic decline in industrial employment in many regions of the
United States has contributed to a widespread awareness of the economic consequences of capital mobility. Plant closings, by no means a
new phenomenon in this country, have assumed a new significance in regions
where the growth of new employment is not sufficient to compensate for the
jobs lost. In New England, for instance, a 14 percent decline in manufacturing
jobs between 1967 and 1976 caused serious economic dislocation and aroused
substantial public concern.

A growing number of labor contracts and legislative initiatives have been
designed to regulate or restrict plant closings. In 1980–81, almost 10 percent of
1,550 collective bargaining contracts examined by the U.S. Bureau of Labor
Statistics required advance notice of plant shutdown or relocation. Legislation
requiring advance notice of plant closings is currently in effect in Wisconsin
and Maine, and Maine requires further that firms closing a plant with over 100
employees must provide severance pay to workers with three or more years of
seniority. Early notification was a provision of 11 bills regarding plant closings
that were introduced into state legislatures in 1981. In 1982, Connecticut and
California passed legislation designed to lessen the impact of plant closings on
workers by means of retraining or assistance programs, and in 1983, Connecticut passed legislation requiring employers of 100 or more laid-off workers to
continue paying health insurance premiums for those workers after their
termination.

These political initiatives highlight the importance of a clear understanding
of the economic impact of plant closings. Regulation of plant closings can be
justified as economically efficient if and only if its social benefits clearly exceed
its social costs. The existing economic literature fails to explore this issue
systematically. Opponents of regulation simply assert that restrictions on plant

Nancy R. Folbre, Julia L. Leighton, and Melissa R. Roderick, "Plant Closings and Their Regulation in Maine, 1971–1982," *Industrial and Labor Relations Review*, 37, no. 2 (Jan. 1984), pp.
185–196.

closings will impair the efficient allocation of resources and deter the growth of new sources of employment, while proponents of regulation often describe the social costs of unemployment without directly relating these costs to plant closings or explaining how particular policies might mitigate social costs. Thus, the debate has largely revolved around the issue of whether or not plants should be closed rather than the issue of how they should be closed. The failure to formulate the regulatory issue in rigorous theoretical terms has been compounded by the paucity of empirical research on the local economic impact of plant closings.

This paper represents an attempt to remedy, at least in part, that failure in the literature. It presents a detailed case study and analysis of plant closings and plant closing regulation in Maine during the period 1971–82.

## Social Costs and Benefits

Much of the controversy over plant closing legislation stems from fundamental differences in political philosophy. Those who believe that workers have, or should have, a right to substantial job security will obviously place a high value on policy measures that mitigate the effects of job loss. Those who stress employers' right to operate their businesses without any interference from the state will obviously value the absence of regulation. Partisans of both views must recognize, however, that their final assessment of the legislation rests, at least in part, upon an understanding of its objective effects. Regulation that delivers important gains to workers without imposing significant costs on employers is likely, of course, to be viewed quite differently from regulation that imposes significant costs without providing commensurate benefits.

Although social costs and benefits can never be fully quantified, a number of means can be used to explore their quantitative dimensions. The most immediate result of a plant closing is involuntary unemployment, and there is considerable evidence that this kind of unemployment has particularly serious consequences for workers. Bluestone and Harrison, for example, review a number of studies describing the loss of pension rights and medical insurance coverage and the deterioration of physical and psychological health that result from closings. Lipsky and Stern, among others, show that women and older workers experience particularly long periods of unemployment and particularly significant declines in earnings trajectories after a closing.

Neither the duration nor the consequences of unemployment following a plant closing can be completely attributed, however, to the closing itself. Labor market adjustment to involuntary job loss is largely a function of macroeconomic factors that determine the aggregate unemployment rate. Nonetheless, unemployment in any local labor market that has experienced a plant closing has a firm-specific component that can be conceptually and empirically distinguished from the underlying aggregate unemployment rate. This distinction between plant-specific unemployment and aggregate unemployment, seldom observed in the plant closing literature, is crucial to the issue of regulation. Plant closing laws can aim at little more than reducing the social costs of plant-specific unemployment.

Those costs are not captured by market transactions. Although individual employees may place a certain "price" on job security in agreeing to a specific wage that could include a compensating differential for the risk of job loss, individual workers have no accurate information on the probability of job loss. Furthermore, workers are normally subsidized to some extent by unemployment insurance and welfare programs, both of which shift part of the economic burden of unemployment to society as a whole. Shutdowns have a particularly disruptive effect on local communities, a party that has not participated in the wage bargain. Small businesses typically suffer from the sudden loss of buying power in their community, and local governments often experience tight budgets as the result of a sudden loss of tax revenue.

The imposition of these social costs raises equity and efficiency issues that could be addressed by transfer programs rather than by regulation. Several social programs in effect before the 1982 federal budget cuts clearly had this intent. The Trade Readjustment and Recovery Act, for example, provided economic assistance to workers losing their jobs because of plant closings caused by foreign competition. Similarly, the Comprehensive Employment and Training Act amendments of 1978 provided funds for special training programs for individuals unemployed as a result of plant closings or mass layoffs.

The case for regulation rests on the claim that it represents a more efficient way to lower the social costs of plant closings. By internalizing some of these social costs, regulation can diminish the discrepancy between the cost to firms and the cost to society as a whole. A number of specific policies regulating shutdowns have been effected in Europe and the United States, none of which completely prohibits closings. Regulation has instead been directed at the process of closing plants, with particular emphasis on the timing of management's decisions.

British law, for instance, requires that employees be given sixty to ninety days' notice in the event of a plant closing or mass redundancy. In West Germany, announcements and explanations of mass dismissals must be submitted to a regional labor department that may delay dismissals for up to two months. And here in this country, a Wisconsin law passed in 1975 requires firms with 100 or more employees to provide the state's labor department sixty days' notice of a shutdown. Maine law currently requires severance payments to workers with more than three years' seniority who are affected by a major plant closing, as well as advance notification to all affected workers if the plant is to be relocated out of state. Under certain conditions, such as relocation of the plant within 100 miles, firms are exempt from one or both requirements.

The imposition of such requirements upon individual firms may represent far more than a simple internalization of social costs. Opponents of regulation argue that plant closing regulation may harm the very group it is intended to help. To the extent that severance and benefit-extension payments are analogous to a tax on labor, they may be shifted to employees in the form of lower wages, while the prolonged operation of an economically inefficient plant may represent efficiency losses for society as a whole. Advance notification may hamper a firm's ability to make quick decisions and may lead to high labor turnover or poor productivity in the period immediately preceding the shutdown. Regulations of any sort, the argument continues, may damage the region's reputation as having a desirable business climate, thus discouraging firms contemplating new plant locations in the region.

There has been little empirical research on the magnitude of the costs that plant closing regulation imposes on business. McKenzie asserts that such costs significantly affect plant closing decisions. On the other hand, Bluestone and Harrison argue that the elasticity of response to regulation is quite low and that the states themselves effectively lower that response by competing with one another to offer exemptions and subsidies to firms choosing to locate within their borders. Similarly, Schmenner and others conclude that there is little empirical support for the view that a favorable business climate can attract new plants.

Since the effects of large tax or other subsidies on plant location decisions are difficult to discern, it seems unlikely that relatively low-cost plant closing regulations affect those decisions. There is some evidence, in fact, that the costs of compliance with short-term notification regulations are negligible. A study of thirty-two plant closings in several states in the early 1960s concluded that the median time then required to plan a plant closing was between seven and twelve months; that advance notice seldom led to higher quit rates or lower productivity; and that eliminating job uncertainty by providing an early warning actually increased worker productivity in some cases.

More empirical research on the costs of regulation is clearly needed, but the benefits of regulation—the extent to which it reduces the social costs of a plant closing—also need to be explored. In light of this need, the importance of the Maine experiment looms large. Despite the fact that Maine's plant closing regulations have been in effect for thirteen years, there has been no previous systematic analysis of plant closings or the relevant regulations in that state.

The policy implications of this lack of analysis became particularly apparent in the spring of 1981, when several Maine legislators, strongly backed by organized labor, introduced amendments designed to strengthen the state's existing law by increasing severance payments, lengthening the period required for notification, and requiring that notice be provided directly to workers as well as to the Director of the State Bureau of Labor. Representatives of Maine's business community argued that such legislation not only would deter the growth of new industry, but would also fail to deliver any significant benefits to workers. Virtually all those opposed to the new legislation tended to assume that relatively few major plant closings had occurred and that most closings had been associated with bankruptcies.

The results of this study show that several of those assumptions were incorrect. They also show that the potential benefits of regulation were much greater than the actual benefits. In practice, that is, Maine's plant closing law fell far short of its intent.

## Plant Closings in Maine

Analysis of plant closings has long been hampered by the lack of any comprehensive source of data. No government agency officially collects records of business shutdowns. In constructing a list of Maine firms that had closed in recent years, we combined two partial lists of plant closings compiled by the Central Maine Power Company and the *Maine Marketing Directory*. These data reveal that 107 Maine plants with more than 75 employees terminated their operations between January 1971 and June 1981, resulting in a direct loss of 21,215 jobs. Virtually all the workers laid off were in manufacturing industries, and the U.S. Department of Labor estimates that when one manufacturing job is lost in Maine, an additional one and one-third jobs are lost in the state's economy as a whole. If we use this multiplier, the estimated total direct and indirect job loss in the state resulting from the 107 closings was 49,219 over the ten-year period.

A breakdown of these closings by industry reveals a distinctive pattern. The leather industry was by far the hardest hit, accounting for 59 percent of the shutdowns. Almost two-thirds of the leather shutdowns occurred in the first five years, 1971-76; and although these closings began to abate in the later period, there was a sharp increase in the number of food processing plants closing down between 1977 and 1981. Of these latter closings, accounting for 11 percent of the total over the period, 83 percent occurred after 1976. Difficulties in the poultry industry during those years probably account for this sudden rise. The textile and apparel and wood products industries accounted for 8 percent and 6 percent, respectively, of the total closings, but these seem to have been fairly evenly distributed over the period. The remaining 14 percent of the total closings occurred in several miscellaneous industries—petrochemicals, paper products, instruments, and electrical machinery.

Only 13 percent of the closings over the 1971-81 period were caused by bankruptcy. Although 33 percent of Maine's manufacturing labor force were unionized in 1978, only 16 percent of the work forces of the shutdown plants were unionized. In general, closings were concentrated in low-wage industries with a relatively high percentage of female workers: women comprised about

34.1 percent of the total manufacturing labor force in Maine between 1979 and 1980, but they accounted for almost 59 percent of the work force in the hard-hit leather and leather products industry. In fact, women's work-force participation was higher than the average in all five of the major industries most affected by closings.

Poor economic conditions in a state as a whole mean that workers are particularly vulnerable to the ill effects of sudden job loss. Among the 50 states, Maine currently ranks forty-sixth in per capita income, and in 1979, its average manufacturing wages were 81 percent of the national average. Moreover, Maine's unemployment rates in the years 1970–79 averaged 15 percent higher than the national average.

State policymakers have long been aware of these facts and of the economic burdens caused by plant closings. Maine's first plant closing law was enacted in September 1971. The law covered all businesses with more than 100 employees and required that one month's notice of an intended closing be given to all employees. The penalty for failure to give notice was one week's pay for every year worked, up to a maximum of one month's pay, for every employee who had been employed at least one year. This legislation was amended in 1973; the worker-notification requirement was dropped, but severance pay to employees who had worked a minimum of five years became mandatory.

The 1971 law was further amended in 1975 to require businesses to give notice to the Director of the State Bureau of Labor sixty days before a closing. The intent of this amendment was, and is, unclear, but it may have represented a compromise between legislators who favored notification of workers and those who did not, since the Director could, at his discretion, treat the notification as confidential.

The law directed firms to provide severance pay amounting to one week's current wages for every year worked to any employee who had worked at the plant for more than three years. Adjudicated bankruptcies, relocation within 100 miles, and closings caused by physical calamities (such as fire or flood) were exempt from the severance-pay requirement. Further, under the law, no company was liable for severance pay to any employee working under a union contract already providing severance pay, nor was any company required to pay severance to any worker reemployed elsewhere by that same company. With the exception of the notification penalty, the law stipulated no penalties for failure to comply with these requirements.

In 1981, organized labor in Maine, disturbed by the law's loophole allowing the Director of the State Bureau of Labor to regard the notification of a closing as confidential, lobbied for several amendments to the legislation, including direct notification to workers, lengthening the period of advance notice, and an increase in severance pay. The legislature passed an amendment requiring that sixty days' notice be given directly to employees of a plant closed as a result of relocation out of state. Thus, in late 1983, the law's provisions were the same as those adopted in 1975, except for the 1981 change in the notification provision.

Maine's plant closing legislation, in effect in one form or another for over ten years, has had some influence on the manner in which firms have shut down plants, if only because it has heightened public awareness of the problem. But the efficacy of the law has been severely limited by the confidentiality of the notice requirement, by the lack of explicit penalties for failure to provide notice, and by the consequently low level of compliance with the regulations.

Between October 1, 1975 and June 17, 1981, thirty plants employing more than 100 workers closed in Maine, but only seven notified the State Director of Labor. Twenty-three companies apparently failed to comply with the advance notification requirement of the law. Of the thirty plants that closed, fourteen were clearly exempt from the severance pay requirement because of bankruptcies, relocations, or union coverage. Of the remaining sixteen companies, nine paid severance and seven did not. In sum, only 23 percent of the companies that closed during this period complied with the notification requirement, and only 56 percent of those apparently covered by the severance-pay requirement did in fact provide such benefits.

This pattern of noncompliance invites comparison between the effects of plant closings that conformed to the law and the effects of those that did not. Unfortunately, confidential notification of a closing to the Director of the State Bureau of Labor cannot have had any significant impact, and an analysis of the effects of the severance pay requirement is beyond the scope of this study. Nonetheless, it is possible to estimate the potential impact of the requirement to notify workers directly and in advance by comparing the impact of publicly announced plant closings with the impact of unannounced closings.

Of thirty-seven major plants that closed between August 1974 and January 1982, fourteen voluntarily gave at least one month's notice directly to their employees before the 1981 direct-notification requirement. Notice was formally given to workers, union representatives, or town officials and often appeared in a local newspaper. (These notifications were verified by telephone interviews with employees.) We used this information, together with data from the Maine Department of Manpower Affairs on monthly unemployment rates by labor market areas, to estimate the effects of notified and unnotified closings on local unemployment, controlling for macroeconomic trends.

## The Effect of Shutdowns on Local Employment

The closing of a plant with 100 or more employees is likely to increase local unemployment rates significantly for the obvious reason that laid-off workers normally become unemployed for some period of time. The extent and duration of unemployment associated with a plant closing are of particular interest.

It should be noted, however, that changes in local unemployment rates do not capture the total impact of plant closings on employment, part of which is reflected in out-migration, early retirement, or discouraged withdrawal from the labor force because of poor job prospects. We will discuss separately the effects of plant closings on local unemployment rates and their effects on the size of the local labor force.

Advance notice lengthens the period of job search and can be expected to lessen the unemployment that would otherwise result from a major plant closing. The data set constructed to test this prediction consists of observations on the fifteen labor market areas (of the thirty in the state) in which at least one major plant closing occurred between August 1, 1974, and March 31, 1982, constituting a pooled set of cross-section and time-series data. Data for 1971 and the first seven months of 1972 were not included because it was difficult to ascertain the months in which closings took place during that period. Major closings were defined as closings of plants with more than 100 employees, a number consistent with Maine's plant closing legislation.

Our model treated any given local monthly unemployment rate (whether or not a plant closing had occurred in that month) as a function of the percentage of workers laid off by a major plant closing, if any, in that month and in each of the previous seven months; the notification or nonnotification of these workers; the monthly unemployment rate in the state; and local labor market characteristics. The estimating equation was specified as follows:

$$LUNEMP_t = \beta_1 + \beta_2\,SUNEMP_t + \beta_3\%LO_t + \ldots + \beta_{10}\%LO_{t-7} + \beta_{11}\%LO_t \cdot$$

$$NOTICE_t + \ldots + \beta_{18}\%LO_{t-7} \cdot NOTICE_t + \sum_{i=19}^{32} \beta_i\,LMAREA$$

where $LUNEMP_t$ = the local unemployment rate in month $t$, seasonally adjusted from August 1974 to March 1982;
$SUNEMP_t$ = the state unemployment rate in month $t$, seasonally adjusted from August 1974 to March 1982;
$\%LO_t$ = the number of workers laid off in a major plant closing in month $t$, as a percentage of the local labor force in month $t$;
$NOTICE_t$ = a dummy variable equal to one if advance notice was given at least one month before the closing that occurred in month $t$, and equal to zero if otherwise; and
$LMAREA$ = dummy variables for fourteen labor market areas.

The dependent variable was the unemployment rate in these labor market areas for every month, 1974–82, adjusted for seasonal variation. The independent variable used to control for aggregate macroeconomic trends was the unemployment rate for the state as a whole, adjusted for seasonal variation. Dummy variables were included for the fifteen labor market areas in which a plant

closing had occurred, to control for area-specific factors other than plant closings that might lead to differences in the level of unemployment. The effect of
plant closings on local unemployment rates was estimated by the coefficient of
the number of workers laid off as a percentage of the local labor force, and a
dummy variable was set equal to one in the event that advance notice was given.

A lag of at least one month in the percentage of the local labor force laid off
was necessary because the date of closing was known only to the month. Lagged
values for at least six months were desirable because the timing and duration of
unemployment were of particular interest in the analysis. Seven lagged values
were therefore included in the estimating equation. There were no instances of
multiple plant closings in the same month in which notification was provided
for one and not for the other plant's employees.

Normally, a simple lag structure of the sort specified above is inappropriate, since there is a high degree of multicollinearity among the lagged values.
This concern is not important here because of the somewhat unusual structure
of this data set. Since there were very few observations (6 among 1,160) in
which more than one plant closing had occurred in a given month or in the
preceding seven months, multicollinearity was not a problem. A geometric or
polynomial lag structure could have been specified, but this would have greatly
complicated interpretation of the coefficient for advance notice.

Because the dependent variable chosen was monthly local unemployment
rates, many of the observations had zero values for the percentage of workers
laid off. In other words, in any given month, there were some labor market
areas in which no major plant closing had occurred during the month or during
the previous seven months. A second equation was therefore estimated not only
for the entire period, but also for a subset of the observations in which a closing
occurred during the period $MONTH_t—MONTH_{t-7}$, as well as for the whole
sample.

The empirical results are displayed in columns 1 and 2 of table 23–1. The
estimated coefficients of $\%LO_{MONTH_t}—\%LO_{MONTH_{t-7}}$ indicate, as one might
expect, that the closing of a plant with more than 100 employees had a positive
and significant effect on local unemployment rates for seven months, peaking in
the month after the closing occurred. In the month of a closing, for every
worker laid off, 0.814 workers who would otherwise have been employed in the
local area were unemployed. The coefficient of $\%LO_{MONTH_{t-1}}$, which is slightly
greater than one, reflects significant indirect job loss in the local area. The sum
of the coefficients of $\%LO_{MONTH_t}—\%LO_{MONTH_{t-6}}$ in column 1, approximately
equal to 5, represents the estimated total number of worker-months of unemployment experienced in the local labor market area per worker laid off. These
are average aggregate worker-months of unemployment resulting from a closing, not the average months of unemployment experienced by the laid-off
worker, and therefore include local indirect as well as direct employment effects
of major plant closings.

**Table 23–1**

**The Effect of Closings of Plants in Maine with More Than 100 Employees on Local Unemployment Rates, 1973–82[a]**

*(t-statistics in parentheses)*

| Variables | Model 1: Uncorrected for Serial Correlation | | Model 2: Corrected with Estimates of Rho | |
|---|---|---|---|---|
| | Months in Which a Closing Occurred During the Period | | Months in Which a Closing Occurred During the Period | |
| | All Months | $MONTH_t\text{-}MONTH_{t-7}$ | All Months | $MONTH_t\text{-}MONT$ |
| | (1) | (2) | (3) | (4) |
| $SUNEMP_t$ | .864[b] | 1.178[b] | .790[b] | .628[b] |
| | (32.053) | (13.949) | (29.259) | (7.179) |
| $\%LO_t$ | .814[b] | .561[b] | .536[b] | .583[b] |
| | (3.748) | (2.177) | (3.175) | (3.034) |
| $\%LO_{t-1}$ | 1.032[b] | .731 | .888[b] | .930[b] |
| | (4.746) | (2.819) | (4.642) | (4.018) |
| $\%LO_{t-2}$ | .827[b] | .529[b] | .688[b] | .711[b] |
| | (3.803) | (2.042) | (3.479) | (2.883) |
| $\%LO_{t-3}$ | .827[b] | .543[b] | .672[b] | .686[b] |
| | (3.796) | (2.100) | (3.366) | (2.720) |
| $\%LO_{t-4}$ | .567[b] | .262[b] | .464[b] | .484[b] |
| | (2.601) | (1.008) | (2.320) | (1.915) |
| $\%LO_{t-5}$ | .483[b] | .233 | .360[b] | .355[b] |
| | (2.217) | (.903) | (1.815) | (1.443) |
| $\%LO_{t-6}$ | .453[b] | .207 | .360[b] | .354[b] |
| | (2.078) | (.803) | (1.871) | (1.533) |
| $\%LO_{t-7}$ | .221 | –4.67 | .174 | .163 |
| | (.996) | (.179) | (1.00) | (.837) |
| $\%LO_t \cdot$ NOTICE | –.488[b] | –1.323[b] | –.422[b] | –.728[b] |
| | (–1.658) | (–3.600) | (–1.867) | (–2.755) |
| $\%LO_{t-1} \cdot$ NOTICE | –.206 | –.934[b] | –.287 | –.840[b] |
| | (–.699) | (–2.510) | (–1.071) | (–2.375) |
| $\%LO_{t-2} \cdot$ NOTICE | –.155 | –.867[b] | –.236 | –.940[b] |
| | (–.389) | (–2.338) | (–.822) | (–2.266) |
| $\%LO_{t-3} \cdot$ NOTICE | –.246 | –.980[b] | –.313 | –1.190[b] |
| | (–.832) | (–2.619) | (–1.041) | (–2.513) |
| $\%LO_{t-4} \cdot$ NOTICE | –.231 | –.255 | .105 | –6.24 |
| | (–.507) | (.538) | (.276) | (1.292) |
| $\%LO_{t-5} \cdot$ NOTICE | –.263 | –.478 | .822 | –.434 |
| | (–.578) | (1.026) | (.210) | (–.922) |
| $\%LO_{t-6} \cdot$ NOTICE | –.405 | –.534 | –.272 | –.735 |
| | (–.891) | (–1.140) | (–.706) | (–1.655) |
| $\%LO_{t-7} \cdot$ NOTICE | –.081 | –.151 | –.145 | –.465 |
| | (–.176) | (–.318) | (.418) | (–1.220) |
| Constant | –.119 | –1.527 | 3.691 | 4.658 |
| N | 1140 | 130 | 1125 | 115 |
| $R^2$ | .795 | .861 | .871 | .849 |
| Standard Error | 1.298 | 1.245 | 1.015 | 1.035 |
| F | 153.425 | 43.421 | 266.323 | 37.69 |
| Durbin-Watson | .828 | 1.263 | 2.161 | 2.450 |

[a]Coefficients for LMAREA are not shown. For definitions of the other variables, see text.

[b]Statistically significant at the 0.01 level in a two-tailed test.

Estimates for the sample as a whole, shown in column 1 of the table, indicate that advance notice significantly diminished the unemployment impact in the month of the closing. When advance notice was given, the coefficient of $\%LO_{MONTH_t}$ decreased by more than 50 percent to 0.326. In other words, when advance notice was given, 0.3 rather than 0.8 workers were unemployed in the local area for every worker laid off, and advance notice lowered the overall unemployment resulting from a closing from five worker-months of unemployment to four worker-months for every worker laid off.

As one might expect, a larger percentage of the variation in monthly unemployment rates is explained when the model is applied to the smaller sample, which eliminates observations for months in which no plant closing occurred or in months having seen no closings during the preceding seven-month period (see column 2). The coefficients for the $\%LO$ variables are smaller than those in column 1, and their significance declines more rapidly over time. Interestingly, the effect of advance notice in this smaller sample appears much more pronounced and persists for four months.

Indeed, in this subset of observations, notice of a closing actually had a negative effect on unemployment rates, whereas no notice of a closing significantly increased unemployment rates. A possible explanation of this result is that lack of adequate data made it impossible to include any measure of the number of workers laid off in closings or layoffs that involved fewer than 100 employees. Omission of this kind of variable biases the coefficients of all the $\%LO$ variables downward, since the months in which a plant with fewer than 100 employees closed were treated as though the percentage of workers laid off were zero. A lower frequency of notification among closings affecting fewer than 100 employees would lead to greater downward bias in the estimated coefficients for no notification than in those for notified closings.

This omission also helps explain the high degree of serial correlation in both sets of estimates (with Durbin-Watson statistics of 0.828 and 1.263, respectively). The effect of a small closing, likely to be felt over a period of at least six months, is reflected in error terms that are highly correlated for at least six successive periods.

Despite this flaw in the specification of the equation, the results are quite robust when corrected for serial correlation. After pooling the cross-section and time-series data, we performed a separate estimate of first-order correlation of the residuals for each of the fifteen labor market areas. The results in columns 3 and 4 confirm the significance of advance notice. Although it is doubtful that the estimated coefficients provide a completely accurate guide to the unemployment impact of closings, they are consistent with the hypothesis that advance notice significantly lowers the unemployment resulting from a closing.

Advance notice may affect unemployment rates in a number of different ways: through an increase in employment; through an increase in migration from the local area; or through withdrawal from the labor force. A full

**Table 23–2**
**The Effect of Closings of Plants in Maine**
**with More Than 100 Employees on the Size of**
**the Local Labor Force, 1973–82**[a]
*(t-statistics in parentheses)*

| Variables | All Months (1) | Months in Which a Closing Occurred During the Period $MONTH_t$-$MONTH_{t-7}$ (2) |
|---|---|---|
| $SUNEMP_t$ | 18.950[b] | 6.435 |
| | (2.523) | (.594) |
| $T$ = August 1, 1974 | 28.501[b] | 19.779[b] |
| | (18.31) | (7.941) |
| $T^2$ | .847[b] | −.058[b] |
| | (2.523) | (1.777) |
| $\#LO_t$ | .102 | .075 |
| | (.741) | (1.015) |
| $\#LO_{t-1}$ | −.198 | −.096 |
| | (−.000) | (−1.056) |
| $\#LO_{t-2}$ | −.073 | −.206[b] |
| | (−.439) | (−2.103) |
| $\#LO_{t-3}$ | −.041 | −.120[b] |
| | (−.243) | (−1.998) |
| $\#LO_{t-4}$ | −.013 | −.142[b] |
| | (−.077) | (−1.402) |
| $\#LO_{t-5}$ | −.095 | −.060 |
| | (.571) | (−.625) |
| $\#LO_{t-6}$ | .128 | −.003 |
| | (.805) | (−.032) |
| $\#LO_{t-7}$ | −.071 | −.007 |
| | (−.517) | (−.105) |
| $\#LO_t$ · NOTICE | −.144 | −.176[b] |
| | (−.567) | (1.374) |
| $\#LO_{t-1}$ · NOTICE | −.455 | −.056 |
| | (−.148) | (−.356) |
| $\#LO_{t-2}$ · NOTICE | −.045 | −.088 |
| | (−.148) | (−.515) |
| $\#LO_{t-3}$ · NOTICE | .090 | .107 |
| | (.563) | (.604) |
| $\#LO_{t-4}$ · NOTICE | .116 | −.045 |
| | (.569) | (−.213) |
| $\#LO_{t-5}$ · NOTICE | −.062 | .031 |
| | (−.577) | (.167) |
| $\#LO_{t-6}$ · NOTICE | −.033 | −.066 |
| | (−.583) | (−.339) |
| $\#LO_{t-7}$ · NOTICE | −.041 | −.043 |
| | (−.588) | (−.207) |
| Constant | 906.007 | 1331.854 |
| $N$ | 1125 | 115 |
| $R^2$ | .919 | .984 |
| $S$ | 192.833 | 95.330 |

| | | |
|---|---|---|
| *F* | 420.152 | 391.871 |
| *D.W.* | 1.998 | 2.287 |

[a]The dependent variable is the labor force, corrected with estimates of rho. #*LO* represents the number of workers laid off. *NOTICE* is as defined in text. The coefficients for LMAREA are not shown here.
[b]Statistically significant at the 0.01 level in a two-tailed test.

explanation of these effects is complicated by the fact that unemployment rates themselves have a significant effect on cyclical changes in labor-force participation. Since Maine does not collect data on the population of labor market areas, changes in labor-force participation resulting from a plant closing cannot be calculated. Although this gap in the data makes it impossible to distinguish empirically between withdrawal from the labor force (a "discouraged worker" effect) and migration from the local area, it is possible to explore the impact of notified and unnotified closings on the absolute size of the local labor force.

The equation used to estimate this impact was identical to that described above, except for the addition of variables for time $(T)$ and time squared $(T^2)$ to capture secular trends. The dependent variable was the size of the local labor force, and independent variables were the number of workers laid off in different periods $(\#LO_{MONTH_t}-\#LO_{MONTH_{t-7}})$. (These equations were corrected for serial correlation in the same way as were the previous equations.)

As shown in table 23–2, the coefficients are uniformly insignificant for the total sample, but the restricted sample shows a decrease in the size of the local labor force, peaking during the second month after a closing. For every 100 workers laid off two months previously, 20 workers dropped out or migrated from the local labor force during those two months. In this sample, advance notice is associated with a significant reduction in the size of the local labor force in the month of the closing, rather than two to four months later. Advance notice apparently did not increase out-migration or withdrawal from the labor force, but it did speed up these forms of labor market adjustment.

## Conclusion

Econometric analysis of the effects of plant closings on local unemployment rates and on the size of the local labor force strongly suggests that Maine workers significantly benefited from advance notification of job loss. Ironically, poor enforcement of Maine's regulations and the confidentiality permitted for advance notice resulted in benefits being provided primarily by employers who voluntarily notified their employees of a closing. The 1981 change in the law to require direct notification to workers in the event of relocation out of state may help guarantee that more workers receive such benefits in the future. Unfortunately, many workers affected by closings of plants not relocating out of state may continue to be uninformed of impending closings.

This case study of plant closings and the relevant legislation in Maine clearly demonstrates the need for more systematic exploration of the broader social costs of plant closings, as well as more empirical research on the costs of regulation. The methodology developed here may prove useful in analyzing the economic impact of plant closings in other states. Our own conclusions lend support to the claim that regulations that impose relatively insignificant costs on firms may significantly lower the costs of major plant closings to society as a whole.

# 24
# Federal Legislation

*William D. Ford*

For many years, I have introduced legislation to enable the federal government to help prevent plant closings and avoid mass layoffs and to assist businesses, communities, and workers in the event that such dislocations could not be prevented. Despite the unqualified support of organized labor and a growing perception that business closings, and particularly manufacturing plant closings, have become epidemic, the plant closing bills I have introduced have never attracted more than eighty-five cosponsors and have never reached the House floor. Because they were comprehensive in approach and sought to regulate extensively how businesses conduct closures, relocations, and major work force reductions, those bills generated tremendous opposition in the business community and never received support from Republican Members of Congress.

In light of this history, it has become clear that a new, more politically viable approach must be taken if any assistance is to be provided to the millions of American workers who lose their jobs in mass layoffs and plant closings every year. That new approach is embodied in the proposed Labor-Management Notification and Consultation Act of 1985 (H.R. 1616), which I have introduced with Mr. Silvio Conte (R-Mass) and Mr. William Clay (D-Mo).

## Advance Notice

The proposed act does not attempt to solve all of the serious problems caused by economic dislocation. It does not deal with such important issues as severance pay, transfer rights, local government tax losses, or employee ownership. Rather, it is a modest attempt to address only two of the worst problems involved in plant closings: that workers and communities are often taken by surprise by sudden closings; and that their representatives usually have no meaningful input into decisions that can drastically change their lives.

---

William D. Ford, "Coping with Plant Closings," *Labor Law Journal,* 36, no. 6 (June 1985), pp. 323–326.

To prevent surprise and the harm that flows from it, the proposed act requires employers to notify their workers and the Federal Mediation and Conciliation Service ninety days in advance of any permanent layoff of fifty or more employees. This requirement applies to closures, relocations, subcontracting, the introduction of new technologies, reductions because of a falloff in orders, in short, to any case where a business intends to lay off fifty or more workers in a twelve-month period. Most businesses will have no trouble meeting this requirement, but if an employer cannot provide ninety days prenotification because of unavoidable business circumstances, the notice period may be reduced or eliminated.

Ninety days notice that a major employer in a community intends to close its doors and eliminate all of its jobs is not much. In many cases, it may not be enough time for the community to put into place an adjustment process to help the displaced employees, let alone to find a buyer for the closed facility. In every case, however, it will provide time for individual employees and their families to begin searching for new employment or training, to begin adjusting their budgets for the income loss they will sustain, and for government agencies to begin planning the delivery of adjustment services. It is not just humane; it is economically efficient for society to require some minimal amount of advance notice before employers terminate large groups of employees.

As the President's Commission on Industrial Competitiveness pointed out, the most effective time to provide assistance to displaced workers is before they are laid off. "Where possible, early identification of the worker to be displaced should be encouraged. Delay in identifying these individuals directly contributes to prolonging the adjustment process—a process already made difficult by the individual's denial of the problem, lack of job search skills, and absence of alternative job or occupation at a comparable wage. Employers should be urged to provide early notification of plant closings, and joint public-private efforts providing layoff assistance (such as those authorized by the Job Training Partnership Act) should be emphasized."

The usefulness of advance notice is more than just common sense. Researchers who have investigated the process by which laid off workers adjust and find new jobs learned 20 years ago that advance notice was crucial to success. George Shultz, now Secretary of State of the United States, was a consultant to the Armour Company and the Meatcutters Union during a period of massive consolidation and technological change in the meat processing industry. In *Strategies for Displaced Workers*, Shultz and his coauthor, Arnold Weber, concluded that six months, or preferably a year, of advance notice was

> a procedural prerequisite for constructive action. It gives the various organizations some time to organize their programs and permits individuals to adjust their own plans, as well as to consider the various options with care.

More recently, a group of researchers in Maine quantified the impact of prenotification. They showed that, when companies give advance notice of a closure, the resulting unemployment is significantly less severe. If as little as one month of advance notice were given, the overall unemployment resulting from the closure, both direct and indirect, would decline from five worker-months of unemployment to four, a 20 percent decline.

## Consultation with Employees

The second requirement the proposed act would impose on employers, that they consult with their employees during the notice period about alternatives to a closure or layoff, is also economically efficient. Nearly everyone today recognizes that labor-management cooperation and the solicitation of employee input into workplace decisions are essential to improvement of American productivity and international competitiveness.

Certainly, there is no subject of greater interest to employees than whether or not they will continue to be employed, whether they will have an income to support their families and health insurance to protect them in the event of illness. If management will not accept employee input into decisions concerning the very existence of the employees' jobs, why should management, or the American public, expect employees to provide input into decisions concerning productivity, product quality, or efficiency? Cooperation has to be a two-way street.

The potential benefits of the consultation process are enormous. When employees are given sufficient notice of a proposed closing, they may be able to find serious flaws in the assumptions and analysis of company management and to dissuade the company from its plans. They may be able to arrange a sale to a new owner or to themselves. At the very least, they may understand the need for and agree to make wage or benefit concessions that could maintain the profitability of the operation. The last few years provide several examples of the potential value of such consultation.

In the autumn of 1982, the General Motors Corporation, which should have enough resources to make decisions in its own best interest, decided that its carburetor plant in Tuscaloosa, Alabama, was no longer profitable. Despite GM's best cost-cutting efforts, the plant was $500,000 short of meeting GM's profitability goals and would have to be closed. Two hundred jobs would be lost if the plant closed, so GM agreed to give the University of Alabama and the United Auto Workers a chance to find ways to cut operating costs. The university paid GM a rent of $500,000 a year for access to the plant, with any operating savings to be deducted from the rent. Within seven months, the university's students had identified $470,000 in annual cost savings, $175,000 in potential savings, and had convinced GM to move new product lines into the plant.

At Chrysler's Detroit trim plant, local management and the UAW local union were told by corporate officials in 1982 that the plant was costing the company $20 million a year that could be saved by contracting out. The plant would close and 709 employees, mostly women, would lose their jobs and be thrown into one of the nation's bleakest labor markets.

The plant manager did his own analysis and found that Chrysler had overstated the plant's costs by nearly 100 percent. Even so, the plant would have to be closed unless another $6.4 million in savings could be found within six months. The company and the union jointly hired an outside consultant who developed a successful plan to increase productivity and cut costs. Without new machines, robots, quality circles, or a plant closing, Chrysler saved $6.4 million a year, 700 workers saved their jobs, and we taxpayers were spared paying for unemployment benefits, food stamps, and Chrysler's tax write-offs.

A third case from 1982 is somewhat less dramatic, but it shows, as the other cases do, what can be done with a little time. The Dan River Corporation announced in December of 1982 that it planned to close its unprofitable print works in Chicamauga, Georgia, a textile mill which was the town's major employer.

City officials worked feverishly for two months to find a buyer for the plant, for they knew that shutdown would wreck the local economy. The property tax rate would have to be raised 20 percent, the local power system would have to be closed, and the school system would have to lay off teachers. The city's mayor raised $1.5 million, putting up his own house as collateral the day before the scheduled closing in order to raise the last $100,000. A year later, 250 of the original 300 employees were at work, and the company was profitable.

## Employee Stress

It is time for everyone to realize that the managers of American business do not always know what is best for their own firms, much less what is best for their employees and their communities. There is far too much at stake to let disinvestment decisions be made in distant corporate boardrooms without any input from those whose lives will be most directly affected. Like second opinions in surgery, consultation can be a means to prevent unnecessary trauma, misguided solutions, and otherwise avoidable deaths of ailing corporations.

The stakes in plant closings are much greater for employees than many people realize. Not just the employees' income and retirement security are at risk but also their mental and physical health. Victims of plant closings typically suffer from hypertension, abnormally high cholesterol and blood sugar levels, a higher incidence of ulcers, respiratory diseases, unduly high propensities to gout and diabetes, and hyperallergic reactions. The mental health effects can be even more critical: depression, anxiety, substance abuse, and aggressive feelings frequently translate into spouse abuse, child abuse, crime, or suicide.

Research by Professor Charles Craypo of Cornell University provides dramatic evidence of the impact of plant closings on the affected workers.

Professor Craypo studied the effects of a brewery closing in South Bend, Indiana, on the workers who lost their jobs. His most significant finding: the displaced workers' mortality rate following the closing was sixteen times higher than normal for people of their ages.

The proposed act has one other major feature: it would establish a National Commission on Plant Closings and Worker Dislocation, a fifteen-member bipartisan panel appointed by the Speaker of the House and the Senate Majority Leader. The commission's purpose would be to study the many issues we have chosen not to address in the proposed legislation. These issues include health insurance for displaced workers, retraining, severance pay, transfer rights, and pension protections, among other things. In the tradition of the President's Bipartisan Commission on Social Security, this commission would recommend administrative or legislative action to address these issues.

Mr. Conte, Mr. Clay, and I see the notice and consultation requirements of this proposed act as the least common denominator for good business practices. It is my hope that the commission will forge a consensus for greater responsibility by business and government. Ultimately, I intend to support any recommendation of the commission that receives bipartisan, consensus support.

# 25
# Federal and State Legislation

*Bennett Harrison*

A s far back as 1974, then-Senator Walter Mondale from Minnesota and Congressman William Ford from Michigan introduced the National Employment Priorities Act into Congress. This was the first plant closing bill, calling for mandatory prenotification of intended closures. When Ford-Mondale failed to gain support at the congressional level, the labor movement's effort to gain mandatory and universal prenotification and severance arrangements in the event of a plant shutdown or relocation shifted to the states.

The lead was taken by the Ohio Public Interest Campaign (OPIC). In July 1977, a new plant closing law drafted by OPIC, with the backing of the UAW and other unions, was introduced into the Ohio legislature. The bill set the pattern for virtually all of the state and federal legislative attempts that were to follow over the next seven years, by proposing specific legal language around three basic principles: advance notification, income maintenance, and job replacement.

The OPIC bill was taken up again, along with proposals drafted by the United Auto Workers, and reworked into the second version of the National Employment Priorities Act, which has in some respects become the American counterpart of Vredeling. H.R. 5040, the N.E.P.A. of 1979, gained a good deal of initial support and prompted substantial and predictably negative response from private industry. It ran afoul of the conservative political tide of 1980 and again failed to reach the floor of Congress for a vote.

In May of 1983, 59 members of Congress co-introduced "NEPA-III" (H.R. 2847). This time, the bill was drafted by the Industrial Union Department of the AFL-CIO. It was successfully reported out of the House Subcommittee on Labor Management Relations on October 5, 1983. Even its strongest supporters are not optimistic about early passage on the floor of Congress, although it seems likely that plant closing regulations will be widely

Bennett Harrison, "The International Movement for Prenotification of Plant Closures," *Industrial Relations* 23, no. 3 (fall 1985), pp. 400–407.

discussed over the next several years as part of the general debate on industrial policy.

## The Bill's Main Provisions

### Notification

Prenotification is required for establishments with 50 to 100 employees, at least 15 per cent of whom will lose their jobs; the required advance notice is six months. If there are more than 100 employees, at least 15 per cent of whom will be laid off permanently, then a prenotification of one year is mandated. Notice must be provided simultaneously to the employees, to local government, and to the U.S. Secretary of Labor.

### Consultation

Within sixty days after receipt from the managers of notice of intent to close the plant (or store or office), the union (or at least 10 per cent of the workers in an establishment without a union) may request an investigation by the U.S. Secretary of Labor. Or the secretary may initiate such an investigation her/himself. In addition, if more than half of the workers in a plant suspect that a closing is imminent but has not been announced, they may petition the Secretary of Labor to initiate an inquiry.

### Federal Adjustment Assistance

As much as possible prior to any actual closure, the government is authorized to target retraining, job placement, education, and other services to the employees at risk, and to provide funds for the purpose of financing individual job search, relocation, or schooling. Present or prospective employers or workers' organizations which are considering a buy-out (including cooperatives) may apply for federal loans, loan guarantees, interest subsidies, and even the assumption by the U.S. government of the firm's outstanding debt. Research and development grants and contracts are available, as is a targeted federal procurement subsidy. The local governments affected by a private closure are also eligible for federal financial assistance to support local social service and public works projects.

### Individual and Community Income Maintenance

The proposed legislation provides for extensive severance pay to individual workers: 85 per cent of their former average wage (reduced by any earned income, unemployment compensation, and trade adjustment assistance) for a period of up to one year, or to a maximum of $25,000. To qualify, workers need not accept another job if it pays lower wages, offers substantially fewer benefits, and does not utilize "substantially similar skills." Displaced workers have transfer rights, health benefits must be continued for a stipulated period, and laid-off employees with at least five years' worth of company pension experience retain full vesting of their claims on that pension. The company is required to pay to each unit of local government in which it shuts down (or from which it relocates) a sum equal to 85 per cent of one year's normal tax revenues. If operations are transferred to a location outside the United States "when an economically viable alternative exists," the company must pay the U.S. Treasury 300 per cent of that portion of the firm's next year's expected federal tax liability foregone as a result of the relocation or transfer of work.

*Penalties*

The penalties for noncompliance include five-year prison terms for business executives falsifying (or failing to make) statements concerning prenotification of closure decisions.

*Administration*

The act would create a National Employment Priorities Administration within the U.S. Department of Labor, having offices in the capitol and in the regions. This office would administer the law, conduct research into the causes and consequences of closings and relocations, and design new packages of government assistance for those firms that did comply with the law.

## State and Local Legislation

By the end of 1983, plant closing legislation had been passed or was being considered in seventeen states and two cities. Maine has had a modest law on its books since 1971. Wisconsin first legislated in this area in 1975. In 1983, Connecticut passed a law requiring a modest continuation of the health benefits of certain eligible workers displaced by plant closures. Philadelphia's 1982 law mandated a sixty-day prenotification period. In July 1983, the city council of Pittsburgh passed a three to nine month advance notice bill over the veto of the mayor, only to have it disallowed in August by two local judges. Fourteen other states and Connecticut are currently debating advance notification and positive adjustment legislation, including California, Illinois, Indiana, Iowa, Kentucky, Maryland, Massachusetts, Michigan, Minnesota, Missouri, New Jersey, New York, Pennsylvania, and Rhode Island. These bills call for an average prenotification period of from two to six months, depending on the size of the facility.

Employers' statehouse lobbying associations usually oppose these state and local initiatives, although the business community may be less monolithic on the question of prenotification than advocates seem to think. For example, in a May 1980 survey of more than 100 Fortune 500 companies, the editors of *Forbes* magazine discovered that three out of five executives thought a prenotification period of at least three months was quite feasible, while "over a third of the respondents considered six months to a year to be the ideal period." Similarly, one of the more common fears associated with prenotification requirements is the concern that advance notice will wreak havoc with plant productivity. While more research is needed, preliminary inquiries into the Canadian experience with prenotification indicate that executives of individual private corporations and officials of the Canadian Manufacturing Association have, from personal experience, found such fears to be unjustified.

In an era of slow secular economic growth, the concern of state and local public officials about policies that might pollute their "business climates" may be undermining the momentum for plant closing laws. Thus, Wisconsin recently repealed its mandatory prenotification law. And California, Rhode Island, Massachusetts, Wisconsin, Ohio, and Michigan have all created (or are actively considering the creation of) "economic adjustment teams" or "industrial extension services." The policy instruments being discussed consist of subsidies, incentives, targeted government procurement, and moral suasion by governors—but minimal coercion on employers. The goals of these experimental programs are to use state power to help firms, employees, and local governments to work out plans for restructuring businesses in trouble, for finding new buyers (including assisting workers in considering whether to buy the plant themselves), or for effecting an orderly redeployment of the displaced labor if a shutdown cannot be avoided. Some governments have contracted with private consultants to produce the sort of business reorganization plans for the state that are already a standard aspect of management practice in the private sector. It remains to be seen whether such private sector strategic planning can be successfully and equitably employed in public settings.

## American and European Experiences: Similarities and Differences

In the United States, the public policy debate has become extremely confused, if not actually polarized, in terms of objectives. The unions have tended to be oblivious to efficiency questions, stressing instead the workers' right to know. Business and government officials talk almost exclusively about efficient adjustment—whether of labor, capital, or product markets. When unions raise the question of fairness, they are invariably criticized for advocating protectionism which will impede efficient adjustment. In Europe, the language of the debate is much more likely to intertwine both the efficiency and the equity objectives.

U.S. legislation has never called for European-style periodic disclosure of company operating information (although it is beginning to emerge from concession bargaining, as in the 1983 agreement between Eastern Airlines and its three unions). Instead, in this country, plant closing laws are usually seen as a mechanism for dealing with emergencies. Their provisions take effect only when management has already made a decision to shut down. I find this contrast striking.

In the U.S., prenotification requirements and positive adjustment assistance are coming to be treated by many as though they were very nearly mutually exclusive. In Europe, by contrast, subsidies and other aids to business are often tied to (or triggered by) prenotification.

The very demand for notification and consultation puts an enormous responsibility on the unions to learn how to combine advocacy and direct pressure with the acquisition of technical expertise in the evaluation and use of the information which they seek. This, in my judgment, is the issue that is most common to the situations in Europe and North America. Whether for the purpose of facilitating knowledgeable participation in formal, high-level, neo-corporatist codetermination policies, to enable a union local or works council to assess the chances that a firm is bluffing when it threatens a closure if wage concessions are not forthcoming, or to plan a strategy of positive resistance to closure, workers and their representatives are going to have to learn a *lot* more about the financial, legal, and engineering aspects of investment and production decision making than they know at present.

Some will argue that the labor movement is incapable of such self-education. Others will say that the result would only be the further bureaucratization of the labor movement and the consequent erosion of its will to pursue direct action. As far as I can see, workers and their organizations really have no choice. If progress is to be made in the extension of workers' rights to include protection from the consequences of unannounced corporate investment decisions, then the labor movement is going to have to learn how to combine direct political action, increasing sophistication in legal bargaining, and greater technical competence in making productive use of the information whose disclosure they are seeking.

Finally, there is a powerful contradiction developing in the United States between the decentralized federalism that drives our political processes and the increasing mobility and multilocationality of capital. In Europe, the Common Market-wide legislative drive to pass the Vredeling Directive is an attempt to rationalize *existing* local (that is, individual country) legal practices. In this respect, at least, the movement for continent-wide federal legislation in America might be considered premature. And it is true that if one looks back at the history of social security, unemployment insurance, minimum wages, restrictions against child labor, and all the other elements of the so-called American social safety net, the pattern was typically one of local agitation resulting in the passage of state laws whose differences (and even inconsistencies) eventually created pressure for the passage of uniform national laws and regulations. Even the private companies that would be regulated by the new laws sometimes came to prefer rational, centralized regulation to chaotic decentralized regulation.

But the historic pattern of decentralized experimentation is undermined in the current era by a qualitative increase in the concentration and centralization of capital. Large corporations have the resources with which to finance the rapid relocation of facilities. Many already operate in multiple jurisdictions, and are capable of shifting work among their different branch plants and subsidiaries. This allows them to threaten individual states and localities with a capital strike, or boycott, should those jurisdictions attempt on their own to augment the local social wage. The current competition among state and local governments in the U.S. to legislate good business climates is evidence of this striking imbalance of power.

In short, what Barry Bluestone and I call the "hypermobility" of capital may have effectively undermined the traditional political mechanisms by which social legislation gets passed in the United States. If so, then this places an unprecedented pressure on Congress, and greatly increases business' political advantage vis-a-vis labor in the struggle over preservation (let alone extension) of the social wage.

## Conclusion

The particular social problem to which plant closing regulatory policy is addressed is unlikely to disappear, certainly not in Europe and not in North America, either, even with its huge internal market. To be sure, there is a cyclical component to the rate of job loss associated with establishment closures. But beneath those cyclical variations lie profound structural transformations in all of the capitalist economies.

The product cycle refers to the pattern by which new products and production processes develop from being characterized by innovation, high costs, low volume, and frequent modification for customers into becoming standardized, low-cost, mass-produced commodities (or services), capable of being spatially decentralized in keeping with the geographic equivalent of the old Babbage principle that the profit-maximizing firm should reorganize the work (or change the technology) so as to make maximum use of the cheapest and most plentiful sources of the least skilled labor power. In an increasingly interdependent global economy, capital markets have been virtually perfected in the sense of economic efficiency through the combination of permissive telecommunications and transportation technologies and extraordinary increases in the organizational capacities of the modern corporation (decentralized in terms of the location of individual workplaces, yet increasingly centralized in terms of strategic control).

At the same time, it is becoming increasingly difficult for government planners to manage Keynesian economic stabilization policies within any one country, given heightened international interdependence. Commodity trade and financial flows are highly sensitive to interest rate differentials, directly and via the effect of interest rates on currency exchange rates. A likely consequence is chronic uncertainty about the movement of domestic interest rates. And that, in turn, is likely to give rise to further periodic bouts of merger mania—or what Robert Reich calls "paper entrepreneurialism"—as a rational short-term corporate alternative to real investment in the updating of older plants or the building of new facilities.

Taken together, these market forces point to a secular acceleration of the product cycle—a further manifestation of the hypermobility of capital. Nor is this a forecast based solely on analyzing market phenomena. Most of the American advocates of industrial policy have as an explicit objective the speeding up of the shift of capital from "sunset" to "sunrise" industries.

Thus, in my judgment, by any stretch of the imagination, structural displacement that is unreachable by purely aggregate demand stimulation is going to be a problem of increasing severity in the years ahead in all countries. To the degree that this forecast is at all accurate, we may expect continued concern throughout the industrial world with the problem of plant closings, and continued efforts to institutionalize procedures for regulating the process of shutdown.

# 26
# The Case against Legislation

*James J. Chrisman*
*Archie B. Carroll*
*Elizabeth J. Gatewood*

T he current wave of plant closings, which resulted in the loss of between 32 and 38 million jobs in the U.S. during the 1970s, has prompted considerable public debate. Some real social costs are involved when a plant shuts down, and many of these costs are borne by employees and the local community.

In response to this problem, at least 24 states currently are considering plant-closing legislation. Wisconsin, Maine, and the city of Philadelphia already have passed bills or ordinances requiring companies to give affected parties advance notice of any impending plant closing. In addition, because the recent wave of plant closings reflects, in part, a fundamental decline in several basic industries (steel, automobiles) and an overall decline in world market share, competitive position, and relative productivity, many have suggested a need for national economic planning—in other words, an industrial policy for the United States. Thus, besides being an important concern in its own right, the plant-closing issue is part of the larger industrial policy debate now taking place.

Given the importance of the plant-closing issue, more research is needed to clarify the specific factors involved and the perceptions of the affected constituents. Furthermore, because the plant-closing issue is directly related to the broader debate over U.S. industrial policy, it is useful to analyze the problem from a long-term economic and social perspective rather than a quick-fix, short-term approach.

This chapter has two purposes. First, it will show how managers answer the question, What is the proper role of states, communities, and the federal government in plant closings? It presents the results of a survey, conducted among corporate public affairs officers and other business managers, to gain insights into the plant-closing issue. Second, it discusses the problems and

James J. Chrisman, Archie B. Carroll, and Elizabeth J. Gatewood, "What's Wrong with Plant-Closing Legislation and Industrial Policy?" *Business Horizons* 28, no. 5 (Sept.–Oct. 1985), pp. 28–35.

implications of increased government intervention in the private economic sector, whether this intervention be through plant closing legislation or, on a broader front, in the form of a national industrial policy.

On this broader front, if government control over industry increases, what will be the future scenario for American business? To explore this question, we will examine an actual case from Great Britain during the early 1970s. We will conclude with a set of recommendations that may offer better long-term solutions than ones that mirror the policies of our Western or Japanese allies.

## How Should State and Local Governments React to Plant Closings? What Should Be the Role of Federal Government?

The topic of plant closings can be viewed from at least two perspectives: their impact in local communities and the broader issue of industrial policy.

Taking a regional perspective, one might build a reasonable argument for the need for plant-closing legislation or, at the very least, some proactive responses from management. The shutdown of a plant has many negative effects on its employees and the surrounding community. Communities lose tax revenues, unemployment rises, and the need for public aid increases. Other businesses directly and indirectly related to the prosperity of the closed plant are adversely affected. We asked managers what they think state and local governments should do to address this issue.

However, because plant closings are a symptom of the malaise of American business, the problem also should be addressed from a broader viewpoint of industrial policy. Thus, managers also were asked to give their opinions on the various programs and plans that have been proposed for the federal government to correct the plant-closing problem.

To gain insights into what managers think are the responsibilites of community, state, and federal governments in plant closings, we conducted a survey using a random sample of 500 corporate public affairs officers drawn from the 1983 *National Directory of Corporate Public Affairs* and the 1982 *O'Dyer's Directory of Corporate Communications*. About half our respondents were public affairs officers, and about half were other managers. Because the plant-closing issue has evolved from a purely economic concern to a subject of much public debate, we thought it appropriate to concentrate on the views of public affairs officers. Part of their job is to communicate their company's point of view to outsiders. Thus, public affairs officers not only would be knowledgeable about their companies' positions, but they also would be more likely to complete the questionnaire than other corporate officials. A MANOVA (multivariate analysis of variance) test showed no significant differences between responses of public affairs officers and other managers.

Questionnaires were distributed to managers through two mailings, resulting in 210 usable responses (42 percent). Of the respondents, 137 (65 percent) indicated that their companies had experienced a shutdown in the past five years. Approximately 68 percent of these firms lost less than 5 percent of their total work force due to plant closings.

Managers were asked to evaluate the desirability of a number of possible actions which federal, state, and community governments could take, either to soften the blow of plant closings on the affected employees and region or to deter plant closings in the first place.

## Survey Results

Table 26–1 provides managerial perceptions on actions that communities or state governments could take to prevent plant closings. Table 26–2 provides the opinions of managers concerning possible actions that the federal government could take to deter plant closings, or at least lessen their effects. Both tables provide aggregate breakdowns of responses. A MANOVA test indicated that no significant differences existed between the managers whose companies experienced shutdowns and managers whose companies did not.

Managers seem willing to accept action on the part of local governments when that action comes in the form of property tax reductions, changes in

**Table 26–1**
**Managers' Perceptions of Actions Communities or State Governments Could Take to Deter Plant Closings**
*(percentages)*

|  | Strongly Agree | Agree | Neither Agree Nor Disagree | Disagree | Strongly Disagree |
|---|---|---|---|---|---|
| Property tax reductions | 14.1 | 52.4 | 21.4 | 11.2 | 1.0 |
| Worker compensation changes | 14.6 | 45.4 | 29.3 | 9.3 | 1.5 |
| State income tax reductions | 13.7 | 48.8 | 24.9 | 10.7 | 2.0 |
| Worker retraining programs | 6.3 | 48.1 | 28.6 | 15.5 | 1.5 |
| Creation of enterprise or high-tech zones | 7.7 | 42.5 | 33.8 | 13.5 | 2.4 |
| Joint effort industrial boards | 1.0 | 41.6 | 45.0 | 10.4 | 2.0 |
| Downtown revitalization/ community development programs | 3.4 | 33.8 | 44.0 | 16.9 | 1.9 |
| Easing of pollution control standards | 7.8 | 26.8 | 42.4 | 20.5 | 2.4 |
| Zoning law changes | 4.9 | 27.6 | 46.8 | 19.2 | 1.5 |
| Plant-closing legislation | 1.5 | 1.9 | 18.0 | 37.4 | 41.3 |

*Notes:* Listed in descending order from most to least agreement with the alternative. Percentages based on sample size of 210. Percentages may not add to 100 because of rounding.

**Table 26-2**
**Managers' Perceptions of Actions the Federal Government Could Take to Deter Plant Closings**
(*percentages*)

| | Strongly Agree | Agree | Neither Agree Nor Disagree | Disagree | Strongly Disagree |
|---|---|---|---|---|---|
| Joint R&D efforts | 7.4 | 42.9 | 32.5 | 12.8 | 4.4 |
| Worker retraining programs | 6.0 | 40.8 | 33.8 | 17.4 | 2.0 |
| Easing of pollution control standards | 10.8 | 30.0 | 35.0 | 20.7 | 3.4 |
| Creation of enterprise or high-tech zones | 4.9 | 37.6 | 36.6 | 16.1 | 4.9 |
| Tax law changes on foreign income investment | 6.4 | 31.2 | 41.1 | 18.3 | 3.0 |
| Joint effort industrial boards | 3.9 | 29.1 | 45.8 | 17.7 | 3.4 |
| Tax credits to older industries | 5.4 | 35.0 | 22.2 | 29.1 | 8.4 |
| Capital mobility laws | 5.1 | 13.7 | 52.8 | 18.3 | 10.2 |
| Financial assistance to unemployment-prone urban centers | 2.5 | 24.3 | 32.2 | 33.2 | 7.9 |
| Changes in federal grant formulas to offset regional shifts in national wealth | 2.9 | 15.7 | 39.7 | 29.4 | 12.3 |
| Import quotas | 5.9 | 22.1 | 24.5 | 28.4 | 19.1 |
| Bailouts for ailing companies | 1.0 | 3.9 | 22.0 | 43.9 | 29.3 |
| Plant-closing legislation | 1.0 | 1.5 | 14.1 | 37.9 | 45.6 |

*Notes:* Listed in descending order from most to least agreement with the alternative. Percentages based on sample size of 210. Percentages may not add to 100 because of rounding.

worker compensation laws, or state income tax reductions. Businesses also appear willing to adopt proactive responses to improve the lot of workers and communities, such as worker retraining, downtown revitalization, or community development programs. In fact, managers appear to be at least slightly in favor of any proposed remedies or corrective actions that states or communities might take, except plant-closing legislation.

These responses suggest at least two conclusions:

1. Managers recognize that the plant-closing issue is important and are willing, and perhaps anxious, for communities and states to provide an atmosphere conducive to business and to the needs of the region; and

2. Managers are adamantly against any proposals that would reduce or constrict their discretion in business decision making.

This first conclusion is consistent with the findings of another survey. Managers of 204 new plants opened in Virginia, North Carolina, and South

Carolina cited state and local industrial climate as the most important factor, overall, in the location decision.

The second conclusion is supported by the following comments from managers who completed the questionnaire:

> Legislation against closings is illogical and impractical—this would discourage construction of new plants in the first place. Management decides on plant locations based on their best view of the future at that time. If conditions change, due to outside forces, a company must be allowed the flexibility of taking steps necessary, including shutdown.
>
> Government involvement in plant-closing decisions is like involving a policeman in a divorce—ineffective, inappropriate, and fruitless . . . . No businessman closes a plant unless he has exhausted all reasonable alternatives. If government can change the bottom line by tax relief or low-interest loans, it may provide additional options, but short of that, can't really be of any help.

Tables 26–1 and 26–2 present a dichotomy. Table 26–1 shows managers giving moderate support to all state and community responses (short of legislation) to deter plant closings. But Table 26–2 shows a far different reaction to intervention by the federal government. When asked about possible federal government action, managers approved of only six of thirteen measures proposed.

The six measures managers approved of are the more proactive responses to the problem, such as joint R&D efforts, worker retraining programs, easing of pollution standards, and the creation of enterprise zones. Business managers seemed slightly opposed to such proposals as bailouts for ailing companies, import quotas, changes in federal grants or assistance to unemployment-prone urban centers, and tax credits to older industries. Opponents of business usually argue just the opposite, but companies seem to want less government involvement in business affairs, even when the involvement is meant to assist rather than restrict or regulate. This response to government interference on the part of managers of large businesses is quite similar to the responses of small business managers to a 1982 survey on the roles and responsibilities of the government. One public affairs executive put it this way:

> The primary issue is economic freedom to respond to changing conditions without cumbersome state or federal intervention. Capitalism must recognize the possibility of failure which brings pain to employees and investors. A business has a responsibility to cushion the shock through severance pay, etc., but it cannot be held accountable for every negative consequence of a business decision driven by external economic developments. A plant closing also brings adverse financial results to the company in terms of changes to current earnings.

Nowhere is the desire of business for unencumbered activity better illustrated than by the near unanimous disapproval of plant-closing legislation at the federal level. Almost 85 percent of the respondents (172 managers) either disagreed or strongly disagreed with plant-closing legislation. Only five managers, 2.5 percent of the respondents, indicated that they agreed or stongly agreed with this type of action. Perhaps the biggest surprise is not the lopsided opposition but finding five managers who support such legislation. The majority view is expressed as follows:

> Plant closures for the most part are avoided by use of timely, present-day management techniques which recognize change in the marketplace. Change *is* real and inevitable, and those who fail are those who do not respond effectively. Firms have been put out of business simply because there no longer exists a need for a product. Management's responsibility to owners, employees, and consumers is to 'anticipate' changing attitudes and life-styles. The less government intervention—local, state, or federal, the better. Closures are sad lessons to learn, but our task is to prevent them, i.e., planning rather than reacting.

The research findings presented above suggest several important conclusions. Managers are not opposed to state and local programs that would improve the business climate for industries. Businesses also support the idea of states and communities taking steps to help workers and revitalize declining regions. And managers welcome programs that facilitate communication between business and government at local, state, and federal levels.

What business does not want, in any way, shape, or form, is for the government to become involved in its affairs. And this is especially true at the macro level of federal policy making. This leads us to the following proposition, one deserving of further investigation: Since businesses oppose the intervention of federal government in business matters, they will oppose an explicit industrial policy for the United States.

Though we have no concrete evidence to support this proposition at this time, we can discuss why businesses should oppose plant-closing legislation or an explicit industrial policy at the national level. We can also present one rather shocking example of what could happen if such legislation or policies were to be enacted. The basic question is, Do we want and do we need plant-closing legislation or, more importantly, a formal industrial policy?

## Plant-Closing Legislation and the Industrial Policy Debate

Despite the fact that plant closings have many harmful effects at the state and local levels, the basic causes of the current wave of plant closings, and hence the fundamental solutions to the problem, lie at the national level. At the national level the problem of plant closings is subsumed by the more important debate over whether the U.S. should have an industrial policy. Some have argued we should have—in fact already *do* have—an industrial policy, especially in regard to defense.

According to economists Barry Bluestone and Bennett Harrison, poor long-term decision making on the part of American business managers is to blame for many of the recent declines in U.S. economic prosperity. Therefore, argue Bluestone and Harrison, it is right and just for business to absorb some of the social costs involved in plant closings. They also suggest that one cannot expect businesses to take the required steps to correct the problem of their own accord.

The remedy Bluestone and Harrison present to correct the problem is based on a national program of "democratic socialist reindustrialization" involving:

1. Plant-closing legislation;
2. Corporate and personal tax reforms;
3. An industrial policy to channel investments, adjust the transition from declining to growing industries, and deal with corporate bail-outs; and
4. A strengthened social wage.

As Bluestone stated in a 1983 *Forbes* interview:

> The price system gives the private market some fairly decent signals about where new investment or disinvestment should be occurring. But it doesn't tell us at what speed that should occur. That's where you need to merge markets with planning in our society . . . it is to try to help in determining the optimal velocity of capital movement.
>
> We need a new set of rules. Some of them will tie the hands of managers, labor unions, even economists at Jesuit universities.[1]

## Arguments against Government Industrial Policy

Given our decline in competitiveness in world and domestic markets, Bluestone and Harrison's arguments concerning the mistakes of American managers are well founded. However, if we consider the problem of plant closings from the broader perspective of industrial policy, we find some interesting and important considerations that Bluestone and Harrison's arguments do not adequately address. These considerations lead to four familiar but important reasons why the United States must avoid an explicit industrial policy.

*Reason 1: Government Cannot "Pick and Choose"*

An industrial policy should:

1. Speed up and ease transition from one set of industries to another, providing finances for modernization through investment in physical and human capital;
2. Provide incentives and stimuli necessary for industrial development;
3. Assure that industry structure gives priority to fields that have the greatest future potential: high value added, growing productivity, growing markets; and
4. Assure efficient production and competitive products in global markets.

The problem is that plant-closing legislation and more comprehensive programs for industrial policy, such as those proposed by Bluestone and Harrison, could slow the transition from declining to growing industries, tie up capital in fields with limited potential, and inhibit improvements in our relative efficiency and competiveness in world markets. As Charles Schultze, former chairman of the Council of Economic Advisors, aptly pointed out:

> The U.S. Government, in particular, is inherently incapable of doing a decent job of picking and choosing among individuals—be they individual regions, firms, or people . . . [For example,] we have an Economic Development Administration to assist depressed areas. By the time it got through Congress, 85 percent of the counties in the nation were counted as 'depressed.'
>
> For a very fundamental reason, therefore, there is not only a substantive problem but also a very strong governmental problem in trying to execute an industrial policy. No matter that for every one winner in the race, there are nine losers, you can be sure the U.S. Government's portfolio, 20 years later, would still have all ten.[2]

This very basic problem of implementing an effective industrial policy (or any part thereof, such as plant-closing legislation) is one that has received scant attention. Industrial policies in other Western countries, such as Great Britain, have been largely unsuccessful, even for such popularly acclaimed programs as "enterprise zones." Japan's economic success is *not* largely attributable to its industrial policies. Basic impediments could prevent the U.S. government from making rational choices regarding industries to invest and disinvest in, and hence, what types of plants should open and which ones should be allowed to close. These impediments are inherent in our pluralistic society and our political mechanisms.

As Tyson and Zysman point out, "No company or industry will calmly accept its own demise as a sacrifice to be made in the national interest."[3] The same conclusion unquestionably holds true for the other parties which would be affected by a decision to exit an industry or close a plant—namely the employees, communities, and even the state which depend upon that particular industry or factory for their livelihood or well-being.

Thus, every time the possiblity of a plant closing or disinvestment in an industry arises, we can expect the interest groups which will suffer most to hotly contest a decision that would work against them, whether the decision is justified or not. And we can be sure that these interest groups will make government aware of their feelings. Our experiences should have taught us that, once special interest groups appear, politicians find it very difficult to resist the pressure since those with opposite interests often provide no counter-vailing support. (One might recall Machiavelli's argument that those who oppose a particular leader or policy are usually strongly committed in their opposition, while those who do not represent opposing viewpoints usually can be counted on for only lukewarm support at best.) After all, the foremost objective of politicians is to get re-elected, and therefore they are quite con-scious of the voting power and opinions of their more militant and vocal con-stituents. As a result, unless the United States government takes steps to alter its political mechanisms, its direct involvement in the affairs of business is likely to hasten the economic and competitive decline the nation has suffered through in recent years.

*Reason 2: The United States Is Not Deindustrializing*

The second argument against government's setting industrial policies or mak-ing plant-closing decisions relates to what Bluestone and Harrison term the "deindustrialization" of the United States. However, though the U.S. is obvi-ously suffering from inappropriate decisions that businesses, government, and labor made in the past and though the structure of American industry has weakened, one problem we do not face is the deindustrialization of the U.S. economy. For each job lost in the 1970s due to plant closings or business failures, 1.1 jobs were created by new business starts or new plant openings. Though the relative proportion of manufacturing jobs to total jobs has de-creased, the number of manufacturing jobs in the U.S. increased by about 8 percent between 1970 and 1979.

There are several reasons why the problem seems greater on a macro level than it really is. One simple reason is that a plant closing gets much greater attention in the media than a plant opening. A more fundamental reason has to do with the basic shift of investments from basic industries to high technology industries and from the northern U.S. to the South. Futhermore, U.S. industry may well be in the midst of a fundamental shift in corporate strategy, moving away from conglomerate or agglomeration strategies of unrelated diversifica-tion toward a strategy of rationalized diversity. After firms have gone through the natural progression of geographic and product line expansion, vertical inte-gration, and related diversification, often the only avenue for growth remaining for the firm is unrelated diversification. And just as unrelated diversification may be a logical solution to the slowing growth, intense competition, and falling profits which accompany the maturation of a company's core business, planned divestitures and a realignment of resources may be natural solutions for directing the company toward a revised set of missions, markets, and products.

In fact, it is precisely this phenomenon that reflects the market mechanisms that move capital from one sector of the economy to another. That economists and politicians fail to understand or refuse to acknowledge this fact is beside the point, though this lack of acceptance undoubtedly could adversely affect the transition.

What all this implies is that the marketplace already is in the process of accomplishing many of the objectives cited as appropriate for industrial policy. It is doing so without the messy governmental apparatus, red tape, and politically motivated decision making that characterize government-instituted programs.

## Reason 3: Timing

A third problem related to the industrial policy debate and more specifically to the plant-closing issue is the question of the timing of responses and actions. Derek Abell stresses the importance of timing:

> There are only limited periods during which the "fit" between the key requirements of a market and the particular competencies of a firm competing in that market is at an optimum. Investment in a product line or market area should be timed to coincide with periods in which such a strategic window is open. Conversely, disinvestment should be contemplated if what was once a good fit has eroded—i.e., if changes in market requirements outstrip the firm's capability to adapt itself to them.[4]

What Abell means, of course, is that often there comes a time when management must decide to "fish or cut bait" in its product/market areas. Postponing decisions of this nature can be a serious strategic error. It is hard to imagine the typically sluggish U.S. government quickly making appropriate decisions for industry as a whole, let alone determining in a timely fashion the fate of an individual company or facility.

A real-world example helps emphasize this point. In 1971, Tracor, Inc., a high-technology firm which originated in 1955 as a research and development operation, was on the verge of bankruptcy due to an unsuccessful venture into the computer business. Around this time Frank McBee, one of the organization's founders, assumed control and instituted a number of management and operating changes that helped save the company. However, McBee also recognized the immediacy of Tracor's business problems. He took the necessary steps to get out of the computer business and sell off those assets. In all, Tracor lost nearly $28 million from discontinued operations, and its debt-equity ratio stood at an unhealthy 12.4 at the end of 1971.

Despite its problems in the early 1970s, Tracor recovered and in fact has prospered. In 1983 the firm enjoyed profits of $24.6 million (up 21 percent from 1982) on sales of $424.6 million (up 8 percent from 1982), compared to $71.1 million sales and $1.3 million profits before discontinued operations in 1971.

Since moving out of computers in 1971, Tracor has steadily expanded its base businesses and entered new fields, such as aerospace. Many jobs were undoubtedly lost when Tracor divested its computer operations. But as the company recovered, these jobs eventually were replaced, and more created, in new business areas. The question is, Could Tracor have accomplished its impressive turnaround had government determined where investments should be directed and when a plant closing or business divestiture was warranted? Would Tracor have been allowed to rid itself of its unsuccessful computer business in time to make a strategic exit from that business and timely entry into new product/market areas? And would an environment where capital was targeted for select industries, such as the computer industry, have allowed the firm to secure the funds needed to achieve its remarkable comeback?

The answer to these questions is, Probably not. Government is surely no better equipped to read market signals than managers trained in that art, and it most assuredly would be slower in reacting to them. Further, "regulatory" or "market rational" countries such as the U.S. tend to be concerned more with the "rules" of economic competition (forms, procedures, and so forth) than with substantive matters. Had industrial policy existed in the United States in 1971 and had computers been targeted for investment, it is likely that the government would have offered Tracor loans to *remain* in the computer industry instead of facilitating its exit.

*Reason 4: Business Can Do It Better*

The Tracor case is an appropriate transition to the fourth argument against government intervention through a formal industrial policy or a "democratic socialist reindustrialization" program. This final argument rests on the most basic and sacred premise of the capitalistic society that has served us well for more than 200 years: the unfettered workings of a "free enterprise" system will yield results superior to the manipulations of government or any other group that makes decisions based on motives unrelated to the marketplace. Given the time, motivation, and freedom to do so, businesses can deal with the problems of plant closings, and can deal with them more effectively than can government.

Though many companies have treated employees and nearby communities less than admirably, many companies have taken quite proactive stances in dealing with plant and business shutdowns. The Brown & Williamson Tobacco

Company, American Hospital Supply, Olin Corporation, General Motors, National Steel, Sperry Rand, Rath Packing Company, and Kellogg's are just a few examples of companies that have been praised for their innovative responses to plant-closing decisions.

One of the best company responses to an impending plant closing was Chrysler's solution in regard to its highly inefficient Detroit Trim plant, which makes seat covers for Chrysler's automobiles. A company survey in 1981 revealed that Detroit Trim's output, which would have cost Chrysler slightly under $31 million if purchased from outside vendors, would cost Chrysler $51.5 million. Factoring in the $9 million cost of closing Detroit Trim, Chrysler could have saved approximately $11.5 million by shutting down the facility. That was no small amount for a company which lost $476 million in 1981 on $10 billion sales and made only $170 million in 1982 with essentially the same sales volume.

However, due to the combined efforts of company management, UAW union officials, and outside consultants, the plant was saved. Today Detroit Trim operates with a labor force of 528, as opposed to 709 employees previously. Its productivity is substantially higher—piece rates were increased from 15 to 28 percent. In addition, Detroit Trim has reduced its operating costs so that they are $6.4 million less than the projected $51.5 million. Since there are certain advantages to having in-house facilities, all this convinced Chrysler to keep the plant. According to Charles Howell, the Arthur D. Little consultant who helped labor and management work out an agreement, "The typical U.S. factory probably has just as much room for improvement."[5]

Chrysler offers an example of private business accomplishing what an industrial policy or plant-closing legislation is supposedly going to do. Though some may feel that the labor-managment cooperation exhibited in the Chrysler example is really an exception, a survey of corporate human resources management vice presidents suggests that, on the average, when compared to the 1970s the relationship between management and labor has improved in the 1980s. This improvement has not been accomplished by managers turning over their responsibilities and prerogatives to any other group, whether it be labor, government, or someone else. The comments of one vice president who responded to the survey emphasize this point:

> The basic problem in both the U.S. and other countries is managements who have given up the will and determination to manage. They have given away too much and now have to negotiate and make concessions to regain the right to manage.[6]

## Notes

1. James Cook, "The Argument for Plant Closing Legislation: An Interview with Economist Barry Bluestone," *Forbes* (June 20, 1983), p. 84.

2. Charles L. Schultze, "Industrial Policy: A Solution in Search of a Problem," *California Management Review* 25, no. 4 (summer 1983), pp. 11–12.

3. Laura Tyson and John Zysman, "American Industry in International Competition: Government Policies and Corporate Strategies," *California Management Review* 25, no. 3 (spring 1983) p. 51.

4. Derek F. Abell, "Strategic Windows," *Journal of Marketing* (July 1978), p. 21.

5. Jeremy Main, "Anatomy of an Auto-Plant Rescue," *Fortune* (April 4, 1983), p. 113.

6. William E. Fulmer, "Labor-Management Relations in the 1980s: Revolution or Evolution?" *Business Horizons* (Jan.–Feb. 1984), p. 31.

# References

Aaron, Benjamin. *Plant Closings: American and Comparative Perspectives.* Los Angeles: University of California Press, 1984.

Abernathy, William J., Kim B. Clark, and Alan M. Kantrow. *Industrial Renaissance: Producing a Competitive Future for America.* New York: Basic Books, 1983.

Aboud, Antone, ed. *Plant Closing Legislation.* Ithaca: ILR Press, New York State School of Industrial and Labor Relations, Cornell University, 1984.

Aronson, Robert L., and Robert B. McKersie. *Economic Consequences of Plant Shutdowns in New York State.* Ithaca: New York State School of Industrial and Labor Relations, Cornell University, 1980.

Bagshaw, Michael L., and Robert H. Schnorbus. "The Local Labor-Market Response to a Plant Shutdown." *Economic Review:* 16–24. Cleveland, Ohio: Federal Reserve Bank of Cleveland, January 1980.

Barker, Lawrence. "There Is a Better Way: Plant Closing and Legislation." *Labor Law Journal* 32, no. 8 (August 1981): 453–459.

Bendick, Marc Jr., and Mary Lou Egan. *Recycling America's Workers: Public and Private Approaches to Mid-Career Retraining.* Washington, D.C.: The Urban Institute, April 1982.

Berenbein, Ronald. *Company Programs to Ease the Impact of Shutdowns.* Report no. 878. New York: Conference Board, March 1986.

Birch, David. *The Job Generation Process.* Cambridge: MIT Program on Neighborhood and Regional Change, 1979.

Blasi, Joseph R. "The Sociology of Worker Ownership and Participation." In *Proceedings of the 37th Annual Meeting of the Industrial Relations Research Association.* Madison, Wisconsin, 1985.

Block, Richard N., and Kenneth McLennan. "Structural Economic Change and Industrial Relations in the United States' Manufacturing and Transportation Sectors since 1973." In *Industrial Relations in a Decade of Economic Change,* edited by Hervey Juris, Mark Thompson, and Wilbur Daniels, pp. 337–382. Madison, Wisconsin: Industrial Relations Research Association, 1985.

Bluestone, Barry, and Bennett Harrison. *Capital and Communities: The Causes of Private Disinvestment.* Washington, D.C.: The Progressive Alliance, 1980.

―――. "Why Corporations Close Profitable Plants." *Working Papers for a New Society* 7 (May–June 1980): 15–23.

────── . *The Deindustrialization of America: Plant Closings, Community Abandonment, and the Dismantling of Basic Industry.* New York: Basic Books, 1982.

Bluestone, Barry, Bennett Harrison, and Lawrence Baker. *Capital Flight.* Washington, D.C.: The Progressive Alliance, 1981.

────── . *Corporate Flight: The Causes and Consequences of Economic Dislocation.* Washington, D.C.: The Progressive Alliance, 1981.

Brenner, M. Harvey. *Mental Illness and the Economy.* Cambridge: Harvard University Press, 1973.

────── . *Estimating the Social Costs of National Economic Policy.* Report prepared for the Joint Economic Committee, U.S. Congress. Washington, D.C.: U.S. Government Printing Office, 1976.

Brockner, Joel, Jeanette Davy, and Carolyn Carter. "Layoffs, Self-Esteem, and Survivor Guilt: Motivational, Affective, and Attitudinal Consequences." *Organizational Behavior and Human Decision Processes* 36, no. 2 (October 1985): 229–244.

Bureau of National Affairs, Inc. *Labor Relations in an Economic Recession: Job Losses and Concession Bargaining.* Washington, D.C.: 1982.

────── . *Basic Patterns in Union Contracts.* Washington, D.C.: 1983.

────── . *Layoffs, Plant Closings and Concession Bargaining, Summary Report for 1982.* Washington, D.C.: 1983.

────── . *Directory of U.S. Labor Organizations 1984–85 Edition.* Washington, D.C.: 1984.

────── . *Unions Today: New Tactics to Tackle Tough Times.* Washington, D.C.: 1985.

*Business Week.* "The Reindustrialization of America." Special Issue. 30 June 1980, 56–146.

Buss, Terry F., and F. Stevens Redburn. *Shutdown at Youngstown: Public Policy for Mass Unemployment.* Albany: State University of New York Press, 1983.

Buss, Terry F., F. Stevens Redburn, and Joseph Waldron. *Mass Unemployment: Plant Closings and Community Mental Health.* Beverly Hills, Calif.: Sage Publications, 1983.

California. Employment Development Department. *Displaced Worker Evaluation Report.* Sacramento, Calif., November 1983.

────── . Office of Economic Policy, Planning, and Research. *The Importance of Plant Closings in the California Economy and Policy Directions for State Government.* San Francisco, 1980.

────── . Office of Planning and Policy Development. *Planning Guidebook for Communities Facing a Plant Closure or Mass Layoff.* Sacramento, Calif.: Plant Closures Response Unit, Spring 1982.

────── . Senate. Committee on Industrial Relations. *SB 1494, Plant Closures.* 3 vols. Sacramento: Committee on Industrial Relations, 1980.

California Chamber of Commerce. "Business Must Act to Halt Threat from Punitive Plant Closures Bills." *Small Business Advocate* (June 1982).

California Manufacturers Association. *Difficult Times . . . Difficult Decisions: An Employer Guide for Work Force Reductions.* Sacramento, Calif., 1983.

Campbell, Lewis B., J. Barry Mason, Joseph M. Mellichamp, and David M. Miller. "Unlikely Partners: Company, Town, and Gown." *Harvard Business Review* 63, no. 6 (December 1985): 20–28.

Carlisle and Redmond Associates. *Plant Closing Legislation and Regulation in the United States and Western Europe: A Survey.* Washington, D.C.: Federal Trade Commission, 1979.

Carnoy, Martin, and Derek Shearer. *Economic Democracy: The Challenge of the 1980's.* Armonk, N.Y.: M.E. Sharpe, Inc., 1980.

Chavez, Cynthia. *A Summary of Issues Relating to Plant Closures, Job Dislocation and Mass Layoffs.* Sacramento, Calif.: Assembly Publications Office, 1982.

Cipparone, Joseph A. "Advance Notice of Plant Closings: Toward National Legislation." *University of Michigan Journal of Law Reform* 14 (Winter 1981): 283–319.

Cobb, Sidney, and Stanislaw Kasl. *Termination: The Consequences of Job Loss.* Washington, D.C.: National Institute for Occupational Safety and Health, U.S. Department of Health, Education, and Welfare, June 1977.

Conference on Alternative State and Local Policies. *Plant Closings: Resources for Public Officials, Trade Unionists and Community Leaders.* Washington, D.C.: May 1979.

Craft, James A. "Controlling Plant Closings: A Conceptual Framework and Assessment." In Antone Aboud, ed., *Plant Closing Legislation.* Ithaca: New York State School of Industrial and Labor Relations, 1984.

Dickens, William T., and Jonathan S. Leonard. "Accounting for the Decline in Union Membership, 1950–1980." *Industrial and Labor Relations Review* 38, no. 3 (April 1985): 323–34.

Duensing, Edward. *Plant Closing Legislation in the United States: A Bibliography.* Number P-1691. Monticello, Ill.: Vance Bibliographies, 1985.

Ekstrom, Brenda, and F. Larry Leistritz. *Plant Closure and Community Economic Decline: An Annotated Bibliography.* Number P-1887. Monticello, Ill.: Vance Bibliographies.

Farber, Henry S. "The Extent of Unionization in the United States." In Thomas A. Kochan, ed., *Challenges and Choices Facing American Labor,* 15–43. Cambridge: MIT Press, 1985.

Feigen, Edward M., and Mona R. Hochberg. "Union Responses to Plant Disinvestment and Shutdowns." Master's Thesis, Department of Urban and Environmental Policy, Tufts University, 1984.

Forsyth, George. "Advance Notice Legislation: A New Challenge for Management's Right to Manage." *The Business Quarterly* 35, no. 3 (Autumn 1970): 75–79.

Freedman, Audrey. "A Fundamental Change in Wage Bargaining." *Challenge* 25, no. 3 (July-August 1982): 14–17.

Freeman, Richard B., and James L. Medoff. *What Do Unions Do?* New York: Basic Books, 1984.

Friedman, Sheldon. "Why Plant Closing Legislation Is Necessary." *Industrial and Labor Relations Report* 18 (Spring 1981): 15–19.

Fulmer, William E. "A Resurrection Plan for Dying Factories." *Business and Society Review* 54 (Summer 1985): 50–55.

Gerhart, Paul F. "Finding Alternatives to Plant Closings (Preliminary Report)." *Labor Law Journal* 35, no. 8 (August 1984): 469–473.

Givens, Alison. "Fighting Shutdowns in Sunny California." *Labor Research Review* 5 (Summer 1984): 15–27.

Gordus, Jeanne Prial, Paul Jarley, and Louis A. Ferman. *Plant Closings and Economic Dislocation.* Kalamazoo, Mich.: W.E. Upjohn Institute for Employment Research, 1981.

Green, Hardy. "When the Factory Shuts Down, Impacts on Workers—Impact on the Community—Proposed Federal Laws." *Labor Unity* (October 1979).

Greenfield, Patricia A. "Plant Closing Obligations under the National Labor Relations Act." In Antone Aboud, ed., *Plant Closing Legislation,* pp. 13–32. Ithaca: New York State School of Industrial and Labor Relations, 1984.

Haas, Gilda. *Plant Closures: Myths, Realities, and Responses.* Boston, Mass.: South End Press, 1985.

Hall, Robert E. "The Importance of Lifetime Jobs in the U.S. Economy." *The American Economic Review* 72 (September 1982): 716–724.

Hansen, Gary B. *Organizing the Delivery of Services to Workers Facing Plant Shutdowns: Lessons from California and Canada.* Logan: Center for Productivity and Quality of Working Life, Utah State University, July 1984.

———. *Executive Summary—Cooperative Approaches for Dealing with Plant Closings: A Resource Guide for Employers and Communities.* Logan: Utah State University, August 1984.

———. "Preventing Layoffs: Developing an Effective Job Security and Economic Adjustment Program." *Employee Relations Law Journal* 2, no. 2 (Autumn 1985): 239–268.

Hansen, Gary B., and Marion T. Bentley. *Mobilizing Community Resources to Cope with Plant Shutdowns: A Demonstration Project Final Report.* Logan: Business and Economic Development Services, Utah State University, August 1981.

Hansen, Gary B., Marion T. Bentley, Jeanni Hepworth Gould, and Mark H. Skidmore. *Life After Layoff: A Handbook for Workers in a Plant Shutdown.* Logan: Center for Productivity and Quality of Working Life, Utah State University, February 1981.

Hansen, Gary B., Marion T. Bentley, and Mark H. Skidmore. *Plant Shutdowns, People and Communities: A Selected Bibliography.* 2d ed. Logan: Utah State University, October 1982.

Hardy, Cynthia. *Managing Organisational Closure.* Aldershot, England: Gower Publishing Co., 1985.

Harris, Candee S. *Plant Closings: The Magnitude of the Problem.* Working Paper No. 13. Washington, D.C.: Business Microdata Project, The Brookings Institution, June 1983.

Harrison, Bennett. "Plant Closures: Efforts to Cushion the Blow." *Monthly Labor Review* (June 1984): 41–43.

Harrison, Bennett, and Barry Bluestone. "The Incidence and Regulation of Plant Shutdowns." *Policy Studies Journal* 10 (December 1981): 297–320.

Hekman, John, and John Strong. "Is There a Case for Plant Closing Laws?" *New England Economic Review:* 34–51. Boston: Federal Reserve Bank of Boston, July-August 1980.

Hickey, John V. "Plant Closing: Electrolux Program Is a Model." *World of Work Report* 10, no. 10 (October 1985): 1–3.

Hochner, Arthur. "Shutdowns and the New Jobs Coalitions: The Philadelphia Experience." *Labor Research Review* 5 (Summer 1984): 15–27.

———. "Worker Ownership and Reindustrialization: A Guide for Workers." In Donald Kennedy, ed., *Labor and Reindustrialization: Workers and Corporate Change,* pp. 95–120. State College: Department of Labor Studies, Pennsylvania State University, 1984.

Hochner, Arthur, and Daniel Zibman. *Plant Closures and Job Loss in Philadelphia: The Role of Multinationals and Absentee Control.* Philadelphia, Pa.: Department of Industrial Relations and Organizational Behavior, Temple University, 1981.

Irving, John S. Jr. "Closings and Sales of Businesses: A Settled Area?" *Labor Law Journal* 33, no. 4 (April 1982): 218–229.

Jacobson, Louis. *Earnings Loss Due to Displacement.* Working Paper No. CRC-385. Alexandria, Va.: Public Research Institute, Center for Naval Analyses, April 1979.

Katz, Harry C., Thomas A. Kochan, and Kenneth R. Gobeille. "Industrial Relations Performance, Economic Performance, and QWL Programs: An Interplant Analysis." *Industrial and Labor Relations Review* 37, no. 1 (October 1983): 3–17.

Kelly, Ed. *Industrial Exodus.* Washington, D.C.: Conference on Alternative State and Local Policies, 1977.

Kelly, Edward. *Plant Closings: Resources for Public Officials, Trade Unionists and Community Leaders.* Washington, D.C.: Conference on Alternative State and Local Policies, 1979.

Kennedy, Donald, ed. *Labor and Reindustrialization: Workers and Corporate Change.* University Park: Department of Labor Studies, Pennsylvania State University, 1984.

Kochan, Thomas A. *The Federal Role in Economic Dislocations: Toward a Better Mix of Public and Private Efforts.* Washington, D.C.: U.S. Department of Labor, October 1979.

Kochan, Thomas A., Harry C. Katz, and Nancy R. Mower. "Worker Participation and American Unions." In Thomas A. Kochan, ed., *Challenges and Choices Facing American Labor,* pp. 271–306. Cambridge: MIT Press, 1985.

Kochan, Thomas A., Robert B. McKersie, and Harry C. Katz. "U.S. Industrial Relations in Transition: A Summary Report." In *Proceedings of the 37th Annual Meeting of the Industrial Relations Research Association,* 261–90. Madison, Wisconsin, December 1984.

Kovach, Kenneth A., and Peter E. Millspaugh. "The Plant Closing Issue Arrives at the Bargaining Table." *Journal of Labor Research* 4, no. 4 (Fall 1983): 367–374.

Krotseng, Richard Van M. "Judicial and Arbitral Resolution of Contractual Plant Closing Issues." *Labor Law Journal* 35, no. 7 (July 1984): 393–406.

Lauria, Mickey, and Peter Fisher. *Plant Closings in Iowa: Causes, Consequences, and Legislative Options.* Iowa City: Institute of Urban and Regional Research, University of Iowa, 1983.

Lawrence, Anne T. *Plant Closing and Technological Change Provisions in California Collective Bargaining Agreements.* San Francisco: Division of Labor Statistics and Research, California Department of Industrial Relations, 1985.

———. "Organizations in Crisis: Labor Union Responses to Plant Closures in California Manufacturing 1979–1983." Ph.D. Dissertation, Department of Sociology, University of California, Berkeley, 1985.

Lawrence, Anne, and Paul Chown. *Plant Closings and Technological Change: A Guide for Union Negotiators.* Berkeley: Center for Labor Research and Education, University of California, 1983.

Lawrence, Paul R., and Davis Dyer. *Renewing American Industry.* New York: The Free Press, 1983.

Lawrence, Robert Z. *Can America Compete?* Washington, D.C.: The Brookings Institution, 1984.

Leary, Thomas J. "Deinstitutionalization, Plant Closing Laws, and the States." *State Government* 58, no. 3 (1985): 113–118.

Levie, Hugo, Denis Gregory, and Nick Lorentgen, eds. *Fighting Closures: Deindustrialization and the Trade Unions 1979–1983.* Nottingham, England: Spokesman, 1984.

Liem, Ramsay, and Paula Rayman. "Health and the Social Costs of Unemployment." *American Psychologist* 37, no. 10 (1982): 1116–23.

Littman, Daniel A., and Myung-Hoon Lee. "Plant Closings and Worker Dislocation." *Economic Review* (fall 1983), 2–18.

Lynd, Staughton. *The Fight against Shutdowns: Youngstown's Steel Mill Closings.* San Pedro, Calif.: Singlejack Books, 1982.

Mann, Eric. "What GM Owes the Freeway Capital." *Los Angeles Times,* 17 April 1983.

———. "The Van Nuys Campaign: Workers and Community Take on G.M." *The Nation,* 11 February 1984.

Mazza, Jacqueline, ed. *Shutdown: A Guide for Communities Facing Plant Closings.* Washington, D.C.: Northeast-Midwest Institute, 1982.

McKenzie, Richard B. *Fugitive Industry: The Economics and Politics of Deindustrialization.* San Francisco: Pacific Institute for Public Policy Research, 1984.

———. *Competing Visions: The Political Conflict over America's Economic Future.* Washington, D.C.: Cato Institute, 1985.

———. "Hostage Factories: The Hidden Costs of Plant Closing Laws." *Reason* 14 (April 1985): 35–39.

McKenzie, Richard B., ed. *Plant Closings: Public or Private Choices?* Washington, D.C.: Cato Institute, 1982.

———. *Restrictions on Business Mobility: A Study in Political Rhetoric and Economic Reality.* Washington, D.C.: American Enterprise Institute, 1979.

McKenzie, Richard B., and Bruce Yandle. "State Plant Closing Laws: Their Union Support." *Journal of Labor Research* 3 (Winter 1982): 101–110.

McKersie, Robert B., and William S. McKersie. *Plant Closings: What Can Be Learned from Best Practice.* Washington, D.C.: U.S. Department of Labor, 1982.

McKersie, Robert B., and Werner Sengenberger. *Job Losses in Major Industries: Manpower Strategy Responses.* Paris, France: Organization for Economic Cooperation and Development, 1983.

Meacham, Melva. "Coping with Long-Term Unemployment: The Role of the Trade Union Movement." In *Proceedings of the 36th Annual Meeting of the Industrial Relations Research Association,* pp. 138–44. Madison, Wisconsin, 1984.

Metzgar, Jack. "Plant Shutdowns and Worker Response: The Case of Johnstown, Pa." *Socialist Review* 10, no. 5 (September-October 1980): 9–49.

Millspaugh, Peter E. "The Campaign for Plant Closing Laws in the United States: An Assessment." *Corporation Law Review* 5, no. 4 (Fall 1982): 291–307.

Miscimarra, Philip A. *The NLRB and Managerial Discretion: Plant Closings, Relocations, Subcontracting and Automation.* Philadelphia: Industrial Research Unit, University of Pennsylvania, 1983.

Naffziger, Fred J. "Partial Business Close-downs by an Employer and the Duty to Bargain under the National Labor Relations Act." *American Business Law Journal* 20 (Summer 1982): 223–243.

National Association of Manufacturers. *When a Plant Closes: A Guide for Employers.* Washington, D.C.: Industrial Relations Department, 1983.

National Institute of Mental Health. "Unemployment and Mental Health: A Report on Research Resources for Technical Assistance." Washington, D.C., March 31, 1985.

National Lawyers Guild. *Plant Closings and Runaway Industries: Strategies for Labor.* Washington, D.C.: National Labor Law Center, 1981.

O'Connell, Francis A. Jr., and Richard B. McKenzie. *The Politics and Economics of Barriers to Plant Closures: Unions, NLRB, and the Courts.* Clemson, S.C.: Department of Economics, Clemson University, 1983.

Ognibene, Charles A. "Plant Closings and the Duty to Consult under Britain's Employment Protection Act of 1975: Lessons for the United States." *Boston College International and Comparative Law Review* 5 (Winter 1982): 195–223.

Parnes, Herbert S., and Randy King. "Middle Aged Job Losers." *Industrial Gerontology* 4, no. 2 (Spring 1977): 77–95.

Parzen, Julia, Catherine Squire, and Michael Kieschnick. *Buyout: A Guide for Workers Facing Plant Closings.* Sacramento, Calif.: Office of Economic Policy, Planning and Research, December 1982.

Raines, John C., Lenora E. Berson, and David McI. Gracie, eds. *Community and Capital in Conflict: Plant Closings and Job Loss.* Philadelphia, Pa.: Temple University Press, 1982.

Reich, Robert B. "Regulation by Confrontation or Negotiation?" *Harvard Business Review* 59 (May–June 1981): 82–92.

———. *The Next American Frontier.* New York: Times Books, 1983.

Rhine, Barbara. "Business Closings and Their Effects on Employees: The Need for New Priorities." *Labor Law Journal* 35, no. 5 (May 1984): 268–280.

Rosen, Corey M., Katherine J. Klein, and Karen M. Young. *Employee Ownership in America: The Equity Solution.* Lexington, Mass.: D.C. Heath/Lexington Books, 1985.

Rothschild-Whitt, Joyce. "Who Will Benefit from ESOPs?" *Labor Research Review* 1, no. 6 (Spring 1985): 71–80.

Rothstein, Lawrence E. *Plant Closings: Power, Politics, and Workers.* Dover, Mass.: Auburn House Publishing, 1986.

Schervish, Paul G. *The Structural Determinants of Unemployment: Vulnerability and Power in Market Relations.* New York: Academic Press, 1983.

Schmenner, Roger W. *Making Business Location Decisions.* Englewood Cliffs, N.J.: Prentice-Hall, 1982.

———. "Every Factory Has a Life Cycle." *Harvard Business Review* 61 (March–April 1983): 121–129.

Schweickart, David. "Plant Relocations: A Philosophical Reflection." *Review of Radical Political Economics* 16, no. 4 (Winter 1984): 32–51.

Schweke, William. *Plant Closings: Issues, Politics, and Legislation.* Washington, D.C.: Conference on Alternative State and Local Policies, 1980.

Shapira, Philip. "Plant Shutdowns and Job Loss in California." *California Data Brief* 7, no. 2. Berkeley: Institute of Governmental Studies, University of California, September 1983.

———. *The Crumbling of Smokestack California: A Case Study in Industrial Restructuring and the Reorganization of Work.* Berkeley: Institute of Urban and Regional Development, University of California, 1984.

Sheehan, Michael F. "Plant Closings and the Community: The Instrumental Value of Public Enterprise in Countering Corporate Flight." *American Journal of Economics and Sociology* 44, no. 4 (October 1985): 423–433.

Sheets, Robert, Russell Smith, and Kenneth Voyteck. *Corporate Disinvestment and Metropolitan Manufacturing Job Loss.* DeKalb: Center for Governmental Studies, Northern Illinois University, 1983.

———. *Corporate Disinvestment: An Empirical Examination of Capital Shift.* DeKalb: Center for Governmental Studies, Northern Illinois University, 1983.

Shostak, Arthur. "The Human Cost of Plant Closings." *The Federationists* (August 1980).

Shultz, George P., and Arnold R. Weber. *Strategies for the Displaced Worker.* New York: Harper & Row, 1966.

Simon, Sharon. "Plant Closings and the Law of Collective Bargaining." In Donald Kennedy, ed., *Labor and Reindustrialization: Workers and Corporate Change,* 69–94. State College: Department of Labor Studies, Pennsylvania State University, 1984.

Stern, Robert N., K. Haydn Wood, and Tove Helland Hammer. *Employee Ownership in Plant Shutdowns: Prospects for Employment Stability*. Kalamazoo, Mich.: W.E. Upjohn Institute for Employment Research, 1979.

Sutton, Robert I. *Organizational Closings: A Bibliography*. Number P-1103. Monticello, Ill.: Vance Bibliographies, 1982.

Teague, Carroll H. "Easing the Pain of Plant Closure: The Brown and Williamson Experience." *Management Review* 70 (April 1981): 23–27.

Thompson, William Irwin. *Pacific Shift*. San Francisco: Sierra Club Books, 1986.

Thurow, Lester C. *The Zero-Sum Society: Distribution and the Possibilities for Economic Change*. New York: Basic Books, 1980.

——— . *The Zero-Sum Solution: Building a World-Class American Economy*. New York: Simon & Schuster, 1985.

Tykulsker, D.A. "For a Reformed Labor-Law to Limit Plant Closings." *Columbia Human Rights Law Review* 12, no. 2 (1981): 205–246.

United States. Bureau of Labor Statistics. *Major Collective Bargaining Agreements: Plant Movement, Interplant Transfer, and Relocation Allowances*. Bulletin 1425-20. Washington, D.C.: July 1981.

——— . Commission on Civil Rights. Illinois Advisory Committee. *Shutdown: Economic Dislocation and Equal Opportunity*. Washington, D.C.: 1981.

——— . Congress. Office of Technology Assessment. *Technology and Structural Unemployment: Reemploying Displaced Adults*. OTA-ITE-250. Washington, D.C.: U.S. Government Printing Office, February 1986.

Wachter, Michael, and William L. Wascher. *Labor Market Policies in Response to Structural Changes in Labor Demand*. Paper presented at Symposium on Industrial Change and Public Policy. Kansas City, Kans.: Federal Reserve Bank of Kansas City, August 1983.

Wallace, W.E. "Industrial Relations in a Job-Loss Environment: The Telephone Industry in Pennsylvania." *Labor Law Journal* 31 (August 1980): 473–477.

Weber, Arnold R., and David P. Taylor. "Procedures for Employee Displacement: Advance Notice of Plant Shutdown." *Journal of Business* 36 (July 1963): 302–315.

Wendling, Wayne R. *The Plant Closure Policy Dilemma: Labor, Law, and Bargaining*. Kalamazoo, Mich.: W.E. Upjohn Institute for Employment Research, 1984.

Whyte, Terrence. "Shutdown." *Canadian Business* 55 (September 1982): 90–96.

Whyte, William Foote, and Joseph R. Blasi. "Employee Ownership and the Future of Unions." *Annals of the American Academy of Political and Social Science* 473 (May 1984): 128–140.

Wintner, Linda. *Employee Buyouts: An Alternative to Plant Closings*. Research Bulletin No. 140. New York: The Conference Board, 1983.

Young, John, and Jan Newton. *Capitalism and Human Obsolescence: Corporate Control vs. Individual Survival in Rural America*. Montclair, N.J.: Landmark Studies, 1980.

# Index

# About the Editors

**Paul D. Staudohar** is Professor of Business Administration at California State University, Hayward. He received his B.A. degree (1962) from the University of Minnesota, and M.B.A. (1966), M.A. (1968), and Ph.D. (1969) degrees from the University of Southern California. Dr. Staudohar is the author of seven books and fifty journal articles. Among his books are *Personnel Management and Industrial Relations*, 7th edition (with Dale Yoder); *Labor Relations in Professional Sports* (with Robert C. Berry and William B. Gould); *The Sports Industry and Collective Bargaining: Economics of Labor in Industrial Society* (with Clark Kerr); and *Industrial Relations in a New Age: Economic Social, and Managerial Perspectives* (with Clark Kerr). Dr. Staudohar serves as a labor arbitrator in private industry and government, and is a member of the board of editors of the *Journal of Collective Negotiations in the Public Sector* and the *Personnel Journal*.

**Holly E. Brown** is Associate Director of the California Policy Seminar. The California Policy Seminar is a joint University of California/state government program that sponsors research, conferences, seminars, and publications pertaining to current and emerging public policy issues in California. The Seminar is based in the Institute of Governmental Studies at the University of California, Berkeley, and represents all nine campuses in the University of California system. Ms. Brown received her B.S. in Business Administration (magna cum laude) from California State University, Hayward, where she specialized in industrial relations.